I0033414

Models That Matter

Selected Writings on
System Dynamics 1985-2010

Models That Matter
Selected Writings on
System Dynamics 1985-2010

Jack B. Homer

Models That Matter
Selected Writings on System Dynamics 1985-2010

Cover image: Untitled by Jannis Kounellis (1987) © 2012 Artists Rights Society (ARS), New York / SIAE, Rome. Lead and wax on steel, 200.7 x 180.5 x 19 cm. Solomon R. Guggenheim Museum, New York; image number 87.3515.

Grapeseed Press graphic is by Diana Kleine, from IAN Image Library, available at ian.umces.edu/imagelibrary/.

ISBN-13: 978-0615679280 (Grapeseed Press)
ISBN-10: 0615679285
Library of Congress Control Number: 2012946100
Grapeseed Press, Barrytown, New York

Copyright © 2012 by Jack B. Homer.

All rights reserved. Except as permitted under the United States Copyright Act of 1976, no part of this publication may be reproduced or transmitted in any form or by any means, or stored in a database or retrieval system, without permission of the author. Printed in the United States of America.

For My Parents

About the Author

Jack Homer is an internationally recognized expert in system dynamics policy modeling, a former faculty member at the University of Southern California (1984-1989), and a full-time consultant to private and public organizations since 1990. In the public sphere, his clients on modeling projects have included the Centers for Disease Control and Prevention (CDC), the Centers for Medicare and Medicaid Services, the Food and Drug Administration, the Veterans Health Administration, state and local health departments, the New Zealand Ministry of Health, and several foundations and other non-governmental organizations. In 1997, he received the International System Dynamics Society's Jay W. Forrester Award for best recent contribution to the field. (The cited work is Chapter 4 in this book, and the acceptance speech is Chapter 6.) He has been lead modeler with the CDC on award-winning projects addressing diabetes policy (see Chapter 11), cardiovascular disease policy (Chapter 15), and national health reform (Chapter 16). Dr. Homer has degrees in applied mathematics and statistics from Stanford University and a PhD in management from MIT, where for his dissertation he modeled the adoption and changing use of new medical technologies (Chapter 2).

Contents

It's not what you look at that matters, it's what you see.

- Henry David Thoreau from the Journals

For here we are not afraid to follow truth wherever it may lead, nor to tolerate any error so long as reason is left free to combat it.

- Thomas Jefferson to William Roscoe, 27 December 1820

Truth, like surgery, may hurt, but it cures.

- Han Suyin (Elizabeth Comber, 1917-)

My opinion is that in the world of knowledge the idea of good appears last of all, and is seen only with an effort…and that this is the power upon which he who would act rationally, either in public or private life, must have his eye fixed.

- Plato, The Republic, Book VII

Introduction and Acknowledgments

The 16 articles reprinted in this book reflect my work in system dynamics modeling over a 25-year period, from my time in the 1980s as a graduate student and young professor, followed by the 1990s when I consulted primarily to large companies, and then the past dozen years when I have consulted primarily to public organizations on questions of health policy. Fourteen of the papers in this collection describe original models on which I was sole or lead modeler, while the other two (Chapters 9 and 14) are thought pieces in response to the writings of others.

System dynamics is a technique for business and policy simulation modeling based on feedback systems theory. It was invented in the late 1950s by Jay Forrester, a pioneer in engineering and computer design. Since then, SD has developed as its own field distinct from the larger fields of operations research and management science to which it is related. Millions of people around the world have been introduced to SD through courses and books or through involvement in modeling applications at their workplaces. But the field is small at its core, with a society membership of not much more than a thousand and only a small number of full-fledged academic programs worldwide.

SD is a challenging discipline, uniting social and behavioral science with the nitty-gritty details of planning and accounting, and requiring the careful design and construction of original models with scores or hundreds of equations. Modeling requires patience and craft, as it is a fundamentally iterative process (Chapter 5). An initial model is developed in the first weeks of a project, setting the stage for repeated rounds of information gathering and model refinement in the ensuing months. By stages, a model emerges that is both flexible and realistic in its portrayal of potential futures and that can be used reliably by an organization looking to chart a new course.

Although SD modeling is technically demanding, the logic and results of a good SD model are neither esoteric nor hard for decision makers to understand. SD is used by organizations seeking an integrated view of the major forces that can affect key outcomes years or decades into the future. It helps these organizations to better weigh the pros and cons of various options they have been considering or might consider. Modeling may show that an option that looks good at first glance may in fact be likely to hit roadblocks or resistance that negate its impact or even make things worse. It may also show that an option that initially looks weak or too costly is likely to become a winner as its good points build upon themselves and its limitations diminish. By looking at not only primary effects but also secondary and feedback effects, SD models can improve our ability to make smart choices that will stand the test of time.

The papers selected for this collection address the how, why, and what of SD applications in many contexts. They span topics in management and operations, psychology, medical sociology, microbiology, disease and drug abuse epidemiology, and health policy. Some of the papers offer ideas on methodology and how the field of SD may best move forward. These ideas reflect my interests as a researcher and my experiences working with clients, often asking how we can be rigorous, how we can be useful, and how we can do both within the real-world pressures of time and budget. For example, I have written that SD should be more thorough and systematic in its use of data and exploration of alternative hypotheses (Chapter 5), should find insights not only in feedback loops but also in the dynamics of stocks and flows as tempered by data (Chapter 6), should seek to quantify and simulate even when some data are lacking (Chapter 9), and should when appropriate work in tandem with practitioners of other quantitative methods (Chapters 4 and 14).

Many thanks to all of my co-authors, clients, and colleagues for their inspired ideas, hard work, and friendship all these years. These people provided the ideas from which I crafted models, but they should not be held responsible for any errors or omissions—the buck stops with me. Thanks also to the journals for allowing authors like myself to easily reprint copyright materials; to the CreateSpace publishing team for guiding me with attention to detail through my first venture in book publishing; to Jay Forrester for his forward thinking and for continuing even in his nineties to encourage people young and old; and to Emily Hartzog for her unwavering support and encouragement.

September 2012

1

Worker Burnout: A Dynamic Model with Implications for Prevention and Control (1985)

Homer J. *System Dynamics Review*, 1(1): 42-62

Worker burnout: a dynamic model with implications for prevention and control

Jack B. Homer

This paper explores the dynamics of worker burnout, a process in which a hard-working individual becomes increasingly exhausted, frustrated, and unproductive. The author's own two-year experience with repeated cycles of burnout is qualitatively reproduced by a small system dynamics model that portrays the underlying psychology of workaholism. Model tests demonstrate that the limit cycle seen in the base run can be stabilized through techniques that diminish work-related stress or enhance relaxation. These stabilizing techniques also serve to raise overall productivity, since they support a higher level of energy and more working hours on the average. One important policy lever is the maximum workweek or work limit; an optimal work limit at which overall productivity is at its peak is shown to exist within a region of stability where burnout is avoided. The paper concludes with a strategy for preventing burnout, which emphasizes the individual's responsibility for understanding the self-inflicted nature of this problem and pursuing an effective course of stability.

In this world there are two tragedies. One is not getting what one wants, the other is getting it. (Oscar Wilde)

The survival of mankind as a species may well depend on the successful management of stress. (Greenwood 1979, xii)

The negative consequences of unrelieved stress have become increasingly evident over the last few decades. In a fast-moving, achievement-oriented society, individuals may easily become worn down and unable to function effectively if they are not careful. Excessive stress can lead not only to chronic fatigue but also to a wide variety of psychological, medical, and behavioral problems ranging from irritation, depression, and loss of appetite to violence, alcoholism, mental illness, and heart disease (Cherniss 1980; Greenwood 1979, 117–163; Holt 1982, 427–433). These problems affect both the individual and the society at large and have had a clear depressive effect on economic productivity in the United States. It has been estimated that output could be boosted by at least 10 percent if the work loss and impaired job performance attributable to mismanaged stress were eliminated (Greenwood 1979, 128–163; Ivancevich 1980, 18). As most executives surely now realize, the connection between distress and the bottom line is indeed real.

Since work plays a central role in the lives of most people, it should not be surprising that a great deal of potentially harmful stress originates in the workplace. In recent years many researchers in the area of occupational safety and health have turned their attention to the issue of occupational stress. Potential sources of stress at work are many and include irritants in the physical or social environment, problems of role or responsibility, poor job fit, inadequate rewards or support, and deadline pressure (Greenwood 1979, 103; Holt 1982, 420–427; Ivancevich 1980, 96). Some jobs tend to be more stressful than others; studies have shown, for example, that air traffic controllers are much more susceptible to stress-related diseases and behavioral problems than the average person. The same is probably true for lawyers, doctors, social workers, and salespeople (Ivancevich 1980, 171). But because tension may arise in any situation where personal needs are not being satisfied, every job has the potential for being stressful (NDACTRD 1980, 225).

When work-related stress becomes severely debilitating, the affected individual may be said to be "burned out." Burnout has been defined broadly as "a process in which individuals become exhausted by making excessive demands on energy strength" (NDACTRD 1980, 1). It is generally agreed that at the root of the problem is the individual's own overcommitment to frustrating work. The source of a burnout victim's frustration is the inability to attain high expectations set by others or, more frequently, by the individual. The entire process may take weeks, months, or years (Freudenberger 1980, 13–16; Greenwood 1979, 47; NDACTRD 1980, 1, 125). Some have said that organizations or even entire societies may burn out if they push themselves too hard (Greenwood 1979, 126).

The burnout process begins when the individual attempts to meet unmet expectations by working longer hours. Longer hours mean more exposure to the normal stress

System Dynamics Review 1 (no. 1, Summer 1985): 42–62. ISSN 0883-7066. © 1985 by the System Dynamics Society.

Jack Homer is assistant professor of systems science at the University of Southern California. His model-based doctoral dissertation at M.I.T. explored the emergence of new medical technologies. His research in system dynamics spans a variety of industrial, social, and personal applications. *Address:* Jack B. Homer, Dept. of Systems Science, Institute of Safety and Systems Management, University of Southern California, University Park, Los Angeles, CA 90089.

of work and consequently more of a drain on the individual's finite store of "adaptation energy" and less time available for recovery of that lost energy (Greenwood 1979, 31–43; Ivancevich 1980, 176; Selye 1974, 38–40). This drain of energy may, in turn, render the individual weaker and less capable of reaching his or her goals. The response to continued inadequacy of performance is to work harder, which depletes energy further (Freudenberger 1980, 5–6). In addition, the worker's growing frustration at work increases the very stressfulness of that work, so that energy is drained still more rapidly (Greenwood 1979, 47; Selye 1974, 78, 96). If the individual refuses or is unable to take time out to recover, the vicious cycle of frustration-exhaustion-dysfunction will ultimately produce chronic and severe problems which *force* him off the job, a burned-out ember of his former self (Ivancevich 1980, 96; NDACTRD 1980, 151).

The ideal context for burnout combines a workaholic personality type with a disagreeable, and therefore highly stressful, job. In general, a nonworkaholic will simply be unwilling to work unusually long and stressful hours and certainly will not seek out such a position. Workaholics or high achievers, on the other hand, create conditions of work overload for themselves and feel uneasy when they are not working (Ivancevich 1980, 177; Machlowitz 1980, 87; McClelland 1961). They compete most strongly with themselves, essentially expanding their goals whenever necessary to maintain the challenge they inherently need; current accomplishments are never quite enough (Freudenberger 1980, 49; Machlowitz 1980, 27, 122). The fact that most workaholics can push themselves this way for many years without burning out speaks well for their ability to find work that is so fulfilling as to minimize the wear and tear of long hours (Machlowitz 1980, 103, 117). But put a self-driven workaholic in a disagreeable work setting—for example, one lacking in tangible rewards—and the prescription for burnout is complete: Unmet goals become more of a burden than a motivating challenge (Freudenberger 1980, 42; Machlowitz 1980, 109; NDACTRD 1980, 128).

It is worth looking more closely at the workaholic personality. Workaholism is a way of life for perhaps 5 percent of American adults. Beneath their energetic and intense surface lies an obsessive perfectionism associated with the fear of failure or boredom (Machlowitz 1980, 6, 26–32, 41–46). Their inability to relax, compromise, seek assistance, or admit limitations often arises first in childhood, reflecting in part the internalization of family values regarding the importance of determined effort, self-reliance, and self-improvement (Freudenberger 1980, 11, 32; Machlowitz 1980, 39; NDACTRD 1980, 108). But the relatively recent phenomenon of workaholism is not attributable solely to the American work ethic, which, after all, goes back many generations. The difference in America today is that with the rapid and widespread erosion of traditions and support systems, few activities outside of the workplace continue to give the individual the sense of meaning and control that they once did. As a result, ". . . our expectations from work become disproportionate to what, in most cases, it can provide" (Freudenberger 1980, 186, 198). Since workaholics invest little of themselves outside work, they risk losing all if work itself ceases to be rewarding.

Strategies for the prevention and amelioration of burnout are as numerous as the writers who propose them. But they all seem to fall into one of three categories: (1) work less. (2) minimize the stress of work, and (3) relax more effectively when not working. Working less may involve working fewer hours per day or taking more frequent vacations (Freudenberger 1980, 158; Machlowitz 1980, 133; Selye 1974, 40). Minimizing work stress may involve shifting tasks or reducing expectations or learning through techniques such as assertiveness training and "stress inoculation" consciously to defuse and to redirect potentially stressful situations (Cameron 1982, 702; Freudenberger 1980, 158, 175; Holroyd 1982, 29; Ivancevich 1980, 230; Janis 1982, 82; NDACTRD 1980, 192; Stoyva 1982, 754). Minimizing stress also may require action by the supervisor or employer, such as clarifying goals or roles, providing more rewards and "strokes." allowing for more personal initiative, and generally improving the work climate (Ivancevich 1980, 207–214; Kiefer 1982, 6; Machlowitz 1980, 133–137, NDACTRD 1980, 192–202). More effective relaxation may involve exercise, hobbies, loafing, close friendship, meditation, biofeedback, body realignment therapy, or any of several techniques specifically designed to elicit the calm state of "passive attention" (Greenwood 1979, 175–211, 229; Freudenberger 1980, 136, 206; Ivancevich 1980, 176, 217; NDACTRD 1980, 186–191; Stoyva 1982, 749–753).

Because the sources of stress are manifold and varied, there exists no single panacea for this complex problem. But each of the preceding techniques can be helpful; in contrast, palliatives that are themselves addictive, such as drinking and gambling, clearly are not effective in managing stress and may only worsen the situation (Freudenberger 1980, 99; Greenwood 1979, 171; Maslach 1978). The prevention of burnout ultimately depends on maintaining a healthy balance of activities, giving adequate time to both active coping and rest (Freudenberger 1980, 210; Greenwood 1979, 220–221; Stoyva 1982, 746).

While the literature provides graphic descriptions of the process of burnout and offers many pieces of advice, it does little to address the question of what actually happens following burnout, that is, during and after recovery. Presumably, the individual can regain lost energy during the time off from work, but what then? If he or she returns to stressful work following recovery, without adopting measures to manage energy more effectively, another round of burnout would seem likely.

Although repeated cycles of burnout are apparently undocumented in the literature, they were, in fact, my experience from 1981 to 1983, the two years of researching and writing my doctoral thesis. Perhaps four times during this long and often lonely effort, I experienced periods of moderate burnout, the symptoms including exhaustion, confusion, anger, headaches, stomach pains, and depression. My productivity plummeted during these periods, and I would finally take time off to recover my strength. When I returned to work, I would not only feel refreshed but also have a more relaxed attitude toward work, with lower expectations for my weekly output. I could meet these new goals without working too hard, but soon I would find my expectations rising and work hours increasing, true to the workaholic profile. First, my evenings disappeared, then my weekends, and it was only a matter of time before I would start to feel tired and confused again.

Purpose and approach

The purpose of this paper is to present a dynamic model of worker burnout, consistent with both the literature and with my personal experience, which can suggest guidelines for attaining greater stability and higher productivity. The model is tested to determine whether stabilizing policies do in fact raise overall accomplishment; uncertainty on this point arises because the ways of the burnout candidate are precisely those of the high achiever. Can one actually push less and achieve more?

The next section describes the structure of the small (four-level, nineteen-equation) system dynamics model used for the subsequent analysis, first in overview and then in detail. This is followed by an investigation of model behavior: First, a description and causal-loop explanation of the base run; and second, an exploration of policies to prevent and control burnout cycles, culminating in the idea of an "optimal work limit." The paper concludes with a general strategy for fighting worker burnout, based on insights provided by the model.

Model structure

Overview

The model of worker burnout presented here focuses on the psychological dynamics underlying the problem, namely, the dynamics of workaholism. (See the Appendix for a complete model listing.) The model has four major elements: (1) accomplishments per week, the outcome measure of interest; (2) expected accomplishments per week, the outcome goal set by the worker; (3) hours worked per week, also adjusted by the worker and one determinant of the weekly accomplishment rate; and (4) energy level, which determines hourly productivity, this being the other factor determining weekly accomplishment. The energy level is depleted by the stress of work and replenished during periods of rest; thus, longer hours at work lead to faster depletion and slower recovery of energy. The frustration that comes from accomplishment falling short of expectations can also accelerate energy depletion. If current accomplishment seems adequate or nearly so, the textbook workaholic will increase expectations; the worker's goals drift downward only if they appear far too high given current output. On the other hand, he or she will tend to work longer hours if accomplishments are perceived to be inadequate; and will decrease hours only if goals are exceeded or if the energy level is so low as to make time off unavoidable.

The model does not attempt to account for the individual's basic approach to and compatibility with his work. Thus, the factors that define the workaholic personality are exogenous, as is the worker's normal level of stress. Also beyond the model's boundary are those personal and environmental factors that determine the individual's basic ability to relax. In addition, hourly output at a given energy level, determined in real life by such factors as preparation for the task at hand, native intelligence, and assistance from others, is exogenous. The model is not concerned with learning curves or organizational dynamics. This is not to say that changes in the individual or the surroundings are not important; indeed, the role of sensitivity testing is precisely to analyze the importance of various exogenous factors. Rather, the model is intended to show how the workaholic syndrome can by itself lead to burnout, without the aid of other dynamic factors. One need not examine changes in the individual's career goals, for example, in order to explain the month-to-month dynamics of burnout, though such longer-term changes may be central to an explanation of a midlife "achievement crisis" (Dabiri 1979).

Equation description

A note on parameter values: The values ascribed to constants and table functions in the baseline model are entirely based on logic and considered judgment. The numerical data of interest are simply not to be found in the literature, nor have I attempted to measure these parameters myself. The numbers were drawn primarily in an impressionistic fashion from my own experience as a victim of burnout. Note that the burnout cycle reference mode itself was presented in descriptive, not numerical, terms. Thus, lacking numerical data on both structure and behavior, attempts at historical accuracy or prediction are obviously out of the question. Still, if the endogenous structure is potent enough, one can learn much about the generic process through careful model testing. The baseline set of parameters is treated simply as a takeoff point for investigating model behavior under a variety of circumstances. The results can thereby be considered applicable to a whole spectrum of individuals and work settings.

```
AW.K = (AH.K)(HWW.K)                    A, 1

AH.K = TABLE(TAH,EL.K,0,1,,2)           A, 2
TAH = 0/.2/.4/.6/.8/1                   T, 2.1

    AW = Accomplishments per week (A-units/week)
    AH = Accomplishments per hour (A-units/week)
    HWW = Hours worked per week (hours/week)
    TAH = Table for accomplishments per hour
    EL = Energy level (0-1)
```

The worker's weekly output is the product of hourly productivity and number of hours worked per week. Output is measured in units of accomplishment (A-units), the meaning of which depends on the kind of work involved. For simplicity's sake, it has been assumed that the worker can produce at most one accomplishment per hour. Hourly productivity is determined by the worker's energy level: When one's energy level (and correspondingly, one's levels of rationality and self-esteem) is low, one is easily distracted from the task at hand and prone to erratic performance and poor decision making (Greenwood 1979, 125; Ivancevich 1980, 201; Janis 1982, 69, 99). Indeed, hourly accomplishment might be thought of as an operational definition of the energy level, implying the linear relationship depicted in Eq. 2.1. When the worker has no energy at all, he produces nothing, while at full energy he produces at maximum hourly rate.

```
HWW.K = HWW.J + (DT/TAHWW)(IHWW.J -           L, 3
  HWW.J)
HWW = HWWI                                    N, 3.1
HWWI = 40                                     C, 3.2
TAHWW = 1                                     C, 3.3

IHWW.K = MIN(LHWW,HWW.K*EELHW.K*EPAHW.K)      A, 4
LHWW = 80                                     C, 4.1
```

> HWW   = Hours worked per week (hours/week)
> HWWI = Hours worked per week, initial (hours/week)
> TAHWW = Time to adjust HWW (weeks)
> IHWW = Indicated hours worked per week (hours/week)
> LHWW = Limit on hours worked per week (hours/week)
> EELHW = Effect of energy level on hours worked
> EPAHW = Effect of perceived adequacy on hours worked

The number of hours worked per week, initialized at a standard value of 40, is adjusted by the worker toward an "indicated" value more in line with his perception of the current situation. It is assumed here that hours are flexible and the adjustment can be made within one week's time. The indicated workweek may be longer than the current workweek if output is perceived as inadequate and may be shorter if output is perceived as more than adequate or if the energy level is low (see Eqs. 5 and 6 following); but the workweek is assumed never to exceed some upper limit (or to exceed it so rarely as to be insignificant). This weekly work limit may be determined by the worker or by others and may be explicit or implicit; in any case, it represents the maximum workweek the individual is willing to put in, on a continuous basis if need be. Many people regularly work 60 to 90 hours per week (Ivancevich 1980, 17); the baseline model assumes a work limit of 80 hours per week.

```
EELHW.K = TABLE(TEELHW,EL.K,0,1,.2)           A, 5
TEELHW = 0/.4/.7/.9/1/1                        T, 5.1

EPAHW.K = TABLE(TEPAHW,PAA.K,0,1.6,.2)         A, 6
TEPAHW = 2.3/1.9/1.6/1.35/1.15/1/.9/.8/.75     T, 6.1
```

> EELHW = Effect of energy level on hours worked
> TEELHW = Table for EELHW
> EL   = Energy level
> EPAHW = Effect of perceived adequacy on hours worked
> TEPAHW = Table for EPAHW
> PAA   = Perceived adequacy of accomplishment

Figure 1 shows the "time off" effect that a low energy level may have on hours worked. The function becomes steep only in the region of lower energy, reflecting the workaholic's natural reluctance to break away from work unless the situation is desperate. Workaholics feel guilty and anxious about leaving work and so tend to skip or shortchange vacation time (Machlowitz 1980, 93–99). But *should the worker's*

Fig. 1. Table for effect of energy level on hours worked

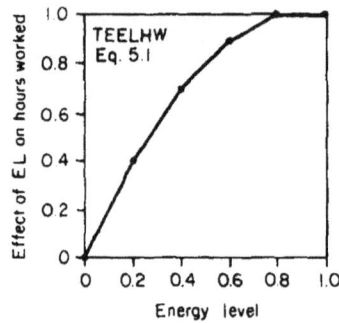

Fig. 2. Table for effect of perceived adequacy of accomplishment on hours worked

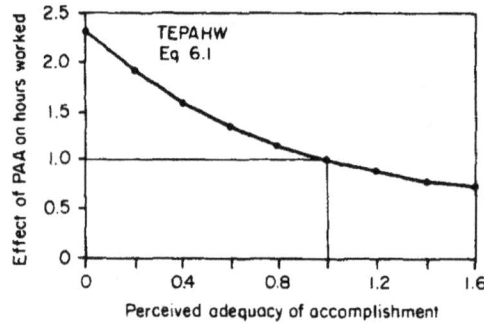

energy drop toward zero, a vacation will become a matter of necessity rather than choice.

Figure 2 depicts the "work harder" response to performance that falls short of one's goals; studies have shown that the further the goal is, the harder people will work to achieve it (Welford 1973). Experience and the logic of symmetry suggest that achievement exceeding one's goals tends to call forth a more relaxed attitude with fewer hours worked. The concave function used here represents a response roughly proportional to the perceived degree of inadequacy or surplus for values of input in the normal range, but considerably less than proportional in the region of very low adequacy. The latter reflects a natural resistance to radical increases in the workweek over a short period of time.

```
PAA.K = PAW.K/XAW.K                          A, 7

PAW.K = PAW.J + (DT/TPAW)(AW.J - PAW.J)      L, 8
PAW = AW                                      N, 8.1
TPAW = 1                                      C, 8.2

    PAA = Perceived adequacy of accomplishment
    PAW = Perceived  accomplishments  per  week  (A-units/
          week)
    XAW = Expected accomplishments per week (A-units/week)
    AW = Accomplishments per week (A-units/week)
    TPAW = Time  to  perceive  accomplishments  per  week
          (weeks)
```

The worker's satisfaction with his performance has both psychological and behavioral consequences and reflects a comparison of perceived accomplishment with expected accomplishment. The dimensionless ratio measure used here seems a reasonable approximation to the informal calculus done in real life. The worker assesses his output rate by averaging it over some recent time period; the smoothing time of one week used here reflects the attentive self-observation typical of a high achiever.

```
XAW.K = XAW.J + (DT)(XAW.J*FCXAW.J)          L, 9
XAW = XAWI                                    N, 9,1
XAWI = 40                                     C, 9,2

FCXAW.K = BFCX + FCXPA.K                      A, 10
BFCX = ,1                                     C, 10,1

FCXPA.K = TABLE(TFCXPA,PAA.K,0,1,6,,2)        A, 11
TFCXPA = -,7/-,5/-,35/-,2/-,1/0/,1/,25/,4     T, 11,1
```

```
XAW = Expected accomplishments per week (A-units/week)
XAWI = XAW, initial (A-units/week)
FCXAW = Fractional change in XAW (1/week)
BFCX = Bias for fractional change in expectations
       (1/week)
FCXPA = Fractional change in expectations from per-
        ceived adequacy (1/week)
TFCXPA = Table for FCXPA
PAA = Perceived adequacy of accomplishment
```

The worker's expectation for weekly output, initialized to equal actual output, is adjusted up or down in response to the perceived adequacy of his performance. The high achiever pushes to expand his goals once they have been met. This is represented in the model by a bias causing the increase of expected output whenever perceived adequacy is neutral (PAA = 1). When accomplishment is more than just satisfactory, the worker will feel encouraged to expand his goals even faster than this bias, as shown in Figure 3. Conversely, when output is inadequate, the individual will nat-

Fig. 3. Table for fractional change in expectations from perceived adequacy of accomplishment

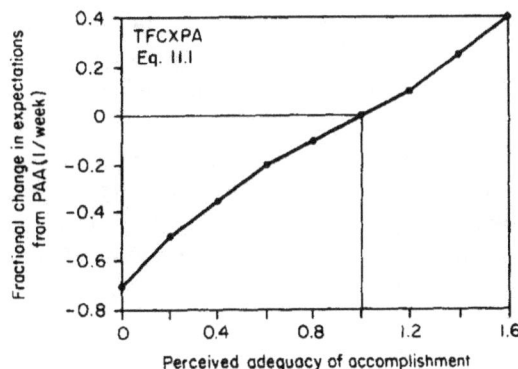

urally be tempted to draw back his goals somewhat to avoid undue frustration.[1] In the baseline model this reaction overcomes the upward bias and causes an actual shrinking of expectations whenever perceived adequacy is less than 0.8.[2] Figure 3 indicates that this response becomes stronger as the worker's dissatisfaction becomes more acute.

```
EL.K = EL.J + (DT)(ER.JK - ED.JK)              L, 12
EL = ELI                                       N, 12.1
ELI = 1                                        C, 12.2

ER.KL = (ERN)(EHWER.K)(EHEFR.K)                R, 13
ERN = .3                                       C, 13.1

EHWER.K = TABLE(TEHWER,HWW.K,0,120,20)         A, 14
TEHWER = 1.3/1.2/1/.7/.5/.35/.25               T, 14.1

EHEFR.K = TABHL(TEHEFR,EL.K,.8,1,.05)          A, 15
TEHEFR = 1/.9/.7/.4/0                          T, 15.1
```

```
    EL = Energy level (0-1)
    ELI = Energy level, initial (0-1)
    ER = Energy recovery (1/week)
    ED = Energy depletion (1/week)
    ERN = Energy recovery normal (1/week)
    EHWER = Effect of hours worked on energy recovery
    TEHWER = Table for EHWER
    HWW = Hours worked per week (hours/week)
    EHEFR = Effect of high energy on further recovery
    TEHEFR = Table for EHEFR
```

The individual's energy level, initialized at its maximum value of 1, is affected by rates of depletion and recovery. The state of low energy known as exhaustion or fatigue is generally associated with other psychological problems, which may include irritability, sadness, detachment, disorientation, and low self-esteem (Freudenberger 1980, 17–18; Holt 1982, 427–433).

Energy is recovered during periods of leisure, relaxation, and of course, sleep. The recovery rate is normalized at a point where the individual is working 40 hours per week and has an energy level lower than 0.8; the recovery rate in this situation is 30 percent per week. But a longer workweek leaves less time for recovery and therefore a slower rate of recovery, as indicated in Figure 4 (Breznitz 1982, 5; Ivancevich 1980, 176). This effect becomes proportionately greater as the workweek increases to cut into evenings, weekends, and even late nights. When the energy level exceeds 0.8, the "effect of high energy on further recovery" acts to suppress the recovery rate somewhat, simply because not much lost energy remains to be recovered. This effect becomes stronger as the energy level approaches 1, at which point the recovery rate must equal zero.

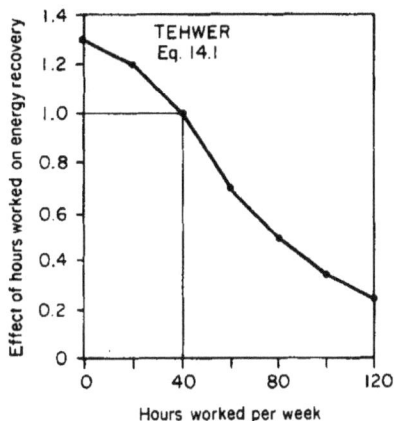

Fig. 4. Table for effect of hours worked on energy recovery

```
ED.KL = (EDN)(EPAED.K)(EHWED.K)(ELEFD.K)        R, 16
EDN = .06                                        C, 16.1

EPAED.K = TABLE(TEPAED,PAA.K,0,1.6,.2)           A, 17
TEPAED = 5/4/3.1/2.3/1.6/1/.6/.4/.3              T, 17.1

EHWED.K = TABLE(TEHWED,HWW.K,0,120,20)           A, 18)
TEHWED = .3/.6/1/1.5/2/2.5/3                     T, 18.1

ELEFD.K = TABHL(TELEFD,EL.K,0,.2,.05)            A, 19
TELEFD = 0/.4/.7/.9/1                            T, 19.1
```

```
    ED = Energy depletion (1/week)
    EDN = Energy depletion normal (1/week)
    EPAED = Effect of perceived adequacy on energy deple-
            tion
    TEPAED = Table for EPAED
    PAA = Perceived adequacy of accomplishment
    EHWED = Effect of hours worked on energy depletion
    TEHWED = Table for EHWED
    HWW = Hours worked per week (hours/week)
    ELEFD = Effect of low energy on further depletion
    TELEFD = Table for ELEFD
    EL = Energy level (0-1)
```

Energy is depleted as the result of repeated exposure to stress. The depletion rate is normalized at a point where the perceived adequacy of accomplishment is neutral, the workweek is 40 hours, and the energy level is greater than 0.2; the depletion rate in this situation is only 6 percent per week. But should the work become frustrating or the hours at work increase, the normal stress of work can be greatly compounded, as illustrated in Figures 5 and 6 (Greenwood 1979, 42, 47; Howard 1965; Ivancevich 1980, 77, 176; Lazarus 1979; NDACTRD 1980, 1; Selye 1974, 96).

Fig. 5. Table for effect of perceived adequacy of accomplishment on energy depletion

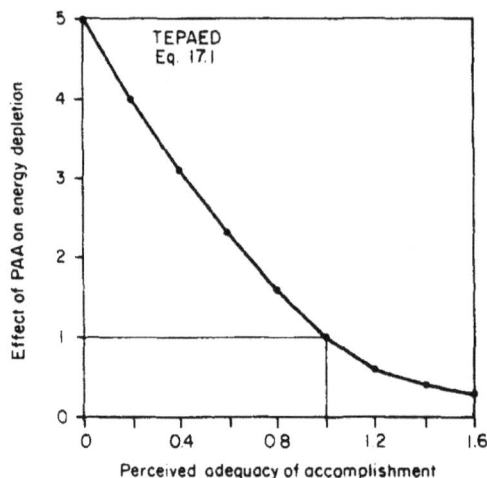

Fig. 6. Table for effect of hours worked on energy depletion

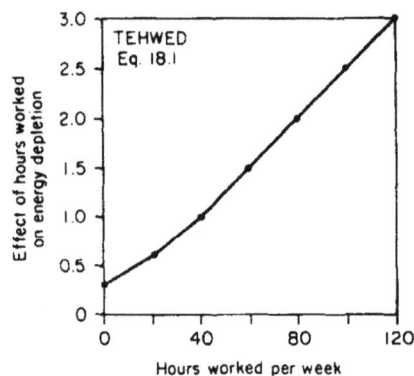

Frustration has been defined as the result of experiencing "undue delay in the fulfillment of a desired goal" (Greenwood 1979, 85), which may be interpreted here as a perceived inadequacy of accomplishment. Figure 5 indicates that as frustration increases, so too will its draining effect on energy. Figure 6 indicates that as the workweek increases, so will the exposure to work-related stressors, again resulting in faster depletion of energy. But note that energy depletion occurs even when the individual is not working at all, as a result of frustration, guilt, or boredom (Machlowitz 1980, 114; Selye 1974, 73; Weiman 1977). After all, "Ability brings with it the need to use that ability" (Albert Szent-Györgi, quoted in Selye 1974, 73). When the energy level falls below 0.2, the "effect of low energy on further depletion" acts to suppress the depletion rate somewhat, because not much energy remains to be lost. This effect becomes stronger as energy falls to zero, at which point the depletion rate must also equal zero.

Model behavior

Description and explanation of baseline behavior

The model's baseline behavior is presented in Figures 7 and 8, over a 75-week time horizon. During this time, the weekly accomplishment rate rises three times, only to fall precipitately along with the energy level as part of a self-sustaining burnout cycle. The observed limit cycle has a period of 30 weeks, one-third of this period being the decline phase during which accomplishment drops to one-sixth of its peak value.

The burnout cycle may be understood most clearly by referring to the causal-loop diagram presented in Figure 9. Initially, the perceived adequacy of accomplishment is relatively high, leading to increased expectations via loop 2. Rising expectations keep the worker somewhat dissatisfied with his output, which drives the "work

Fig. 7. Base run: accomplishment and hours worked per week

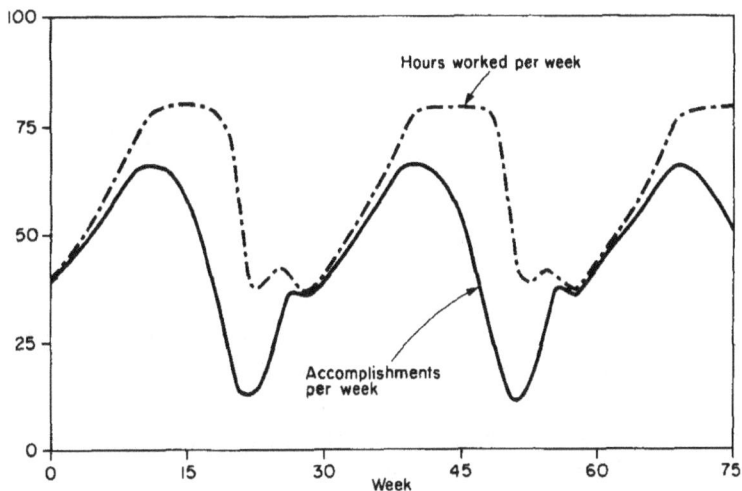

Fig. 8. Base run: energy level and perceived adequacy of accomplishment

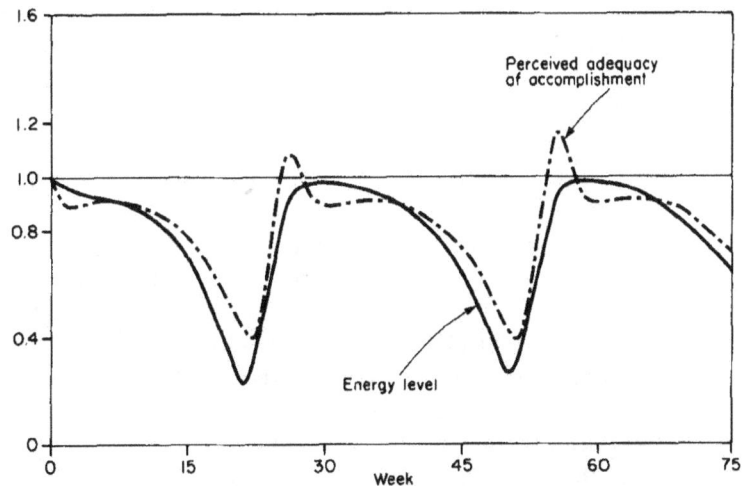

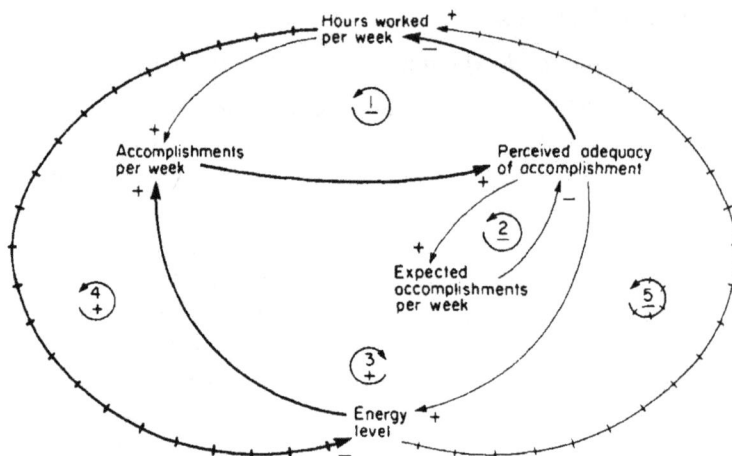

Fig. 9. Causal-loop structure underlying burnout cycles

harder" response of loop 1. Weekly accomplishment rises accordingly and continues to do so as loops 1 and 2 combine to "bootstrap" the workweek upward toward its limit.

But as the hours worked increase, so does the stress of work, causing the energy level to decline. Falling energy puts a damper on output, leading to greater dissatisfaction and the beginnings of frustration. Dissatisfaction causes the individual to continue working hard, while frustration makes work even more stressful, both of which further speed the drain of energy. If the vicious cycles just described, seen in Figure 9 as positive loops 3 and 4, grow strong enough relative to the individual's ability to recover energy during nonwork periods, then a collapse of energy and output like that seen in the base run will result.

Recovery from burnout is made possible by less work and reduced expectations. When energy falls to a low enough level, the individual finally breaks free of the addictive "work harder" response and reduces hours at work; in the base run, hours worked per week fall rapidly from their peak of 80 to a trough of about 40.[3] This "time off" response finally stems the decline of energy, as suggested by loop 5, and indeed allows lost energy to be reclaimed. While the workweek has been reduced, so have achievement expectations. Loop 2 now counters dissatisfaction by bringing output goals down to a level more in keeping with the exhausted worker's depressed capability.

The final stage of the cycle comes when modest expectations and rising energy enable the worker to achieve at a rate that is satisfying without having to put in a lot of hours to do so. As seen in the base run, the perceived adequacy of accomplishment actually bounces back to exceed the neutral level of 1, at a time when the individual is working a relatively short workweek. But this euphoric period of "coasting" is short-lived, lasting only a couple of weeks (as corroborated by my own experience). Why? Because satisfaction is unnatural for workaholics and only encourages them to expand their goals and begin the cycle once again.[4]

14

Searching for stability

In seeking to stabilize the limit cycle of burnout, it is instructive to examine first whether and under what conditions a stable equilibrium can be found. As it turns out, one can state the conditions for stable equilibrium with a single inequality comparing the forces affecting energy depletion with those affecting energy recovery. Interestingly, such an equilibrium has the individual working steadily at his or her work limit. Taking the duration effects on depletion (TEHWED, Eq. 18.1) and recovery (TEHWER, Eq. 14.1) and the frustration effect on depletion (TEPAED, Eq. 17.1) as givens, stability can be said to be threatened by (1) a high depletion normal (EDN, Eq. 16.1); (2) a low recovery normal (ERN, Eq. 13.1); (3) a high upper limit on hours worked (LHWW, Eq. 4.1); (4) a large bias to expand expectations (BFCX, Eq. 10.1); and (5) downwardly inflexible expectations (TFCXPA, Eq. 11.1). On the other hand, stability is independent of the steepness of both the "work harder" (TEPAHW, Eq. 6.1) and "time off" (TEELHW, Eq. 5.1) functions affecting hours worked per week.[5]

Model tests verify that these five factors can be adjusted to stabilize the burnout cycle. The results of one such test, in which the energy depletion normal is reduced by 20 percent, from 0.06 per week to 0.048 per week, are shown in Figure 10. Although a limit cycle is still observed, its amplitude is considerably less than that of the base run (weekly accomplishment falls by less than 60 percent, compared with more than 80 percent in the base run), and its period is considerably longer (39 weeks versus 30). Another noticeable difference is that the decline phase now accounts for more than two-thirds of the entire cycle, compared with one-third in the base run. In effect, the vicious cycles causing burnout have been rendered less vicious, so that the decline is a slower and milder one, and one requiring less time off for recovery. As a result, weekly accomplishment averaged over the full 75 weeks of output is raised by 17 percent relative to the base run.[6]

Fig. 10. Energy depletion normal reduced by 20 percent

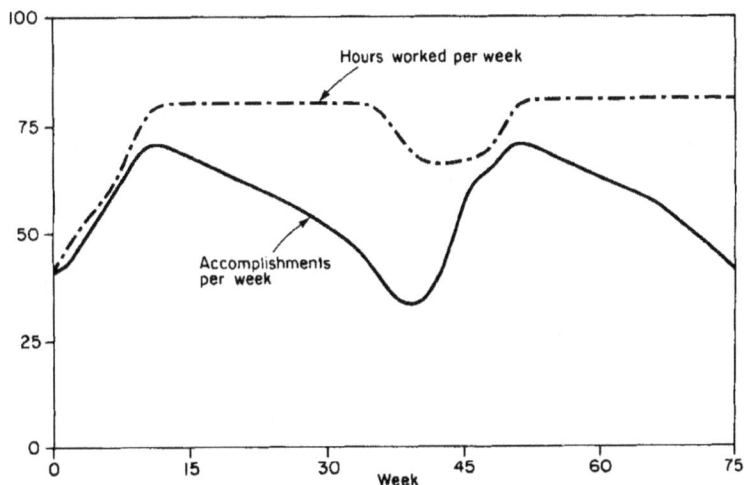

Fig. 11. Steeper effect of energy level on hours worked

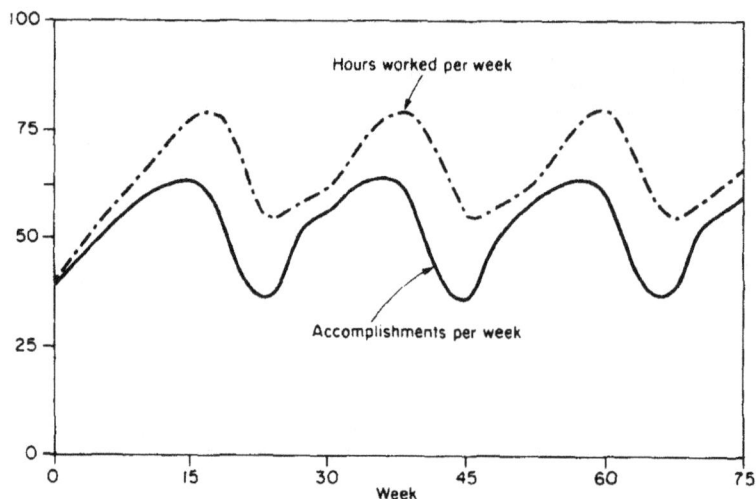

The results of this test are similar to the results of all those tests in which the described stability condition is closer to being satisfied. The limit cycle becomes smaller in amplitude, longer in period, more drawn out in the decline phase, and faster in the recovery phase. The average workweek increases, as does average weekly accomplishment. If the stability condition is actually satisfied, the limit cycle is replaced by a critically damped oscillation, with significantly higher average weekly accomplishment over the 75 weeks. For example, by reducing the energy depletion normal to 0.04 instead of 0.048 (a 33 percent reduction from the baseline instead of 20 percent), average weekly accomplishment is raised by 42 percent relative to the base run (instead of 17 percent) and a consistent 80-hour workweek is maintained. Thus, stabilizing policies do have the effect of increasing overall accomplishment, permitting the individual to work longer hours and to spend less time recuperating from exhaustion.

If the "time off" response to exhaustion does not affect the condition for stable equilibrium, one might wonder what effect this reasonable-sounding policy does have on behavior. Figure 11 shows the results of a run in which the individual is less reluctant than in the base run to break away from work during periods of reduced energy; in other words, the "time off" function falls more steeply than that seen in Figure 1.[7] As in the previous test, the limit cycle's amplitude is significantly reduced relative to the base run, a direct result of acting earlier to escape the vicious cycles of collapse. But this earlier response also leads to a cycle of shorter period, though similar in shape to the base run's cycle. The general impact is to shrink and speed up the cycle, rather than to stabilize it. In terms of overall output, though, this impact is beneficial; average weekly accomplishment is 12 percent higher in this test run than in the base run.

While a policy to reduce work hours in response to lower energy cannot stabilize the burnout cycle, a policy that reduces the limit on hours worked (LHWW) can; recall that this limit enters the condition for stable equilibrium. The essential differ-

Fig. 12. Three tests
with reduced limit
on hours worked per
week

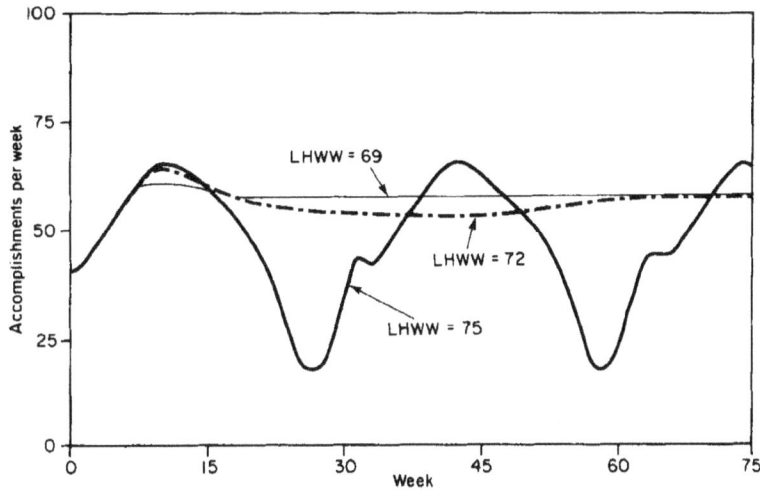

ence between these two workweek policies is that the former is reactive, responding only when a problem has already surfaced, while the latter is proactive and preventive in nature. Also, as a practical matter, it may be easier to implement a shorter work limit (perhaps, as Machlowitz (1980, 133) suggests, by locking the office doors after certain hours) than to fight the workaholic's natural reluctance to take time off to relieve fatigue. Figure 12 shows the results of three test runs in which the work limit has been reduced from its baseline value of 80. When LHWW = 75, a limit cycle is still produced, though it has a smaller amplitude and longer period than the base run cycle. When LHWW = 72, the limit cycle is replaced by a damped oscillatory mode that achieves stable equilibrium by week 60, after an initial overshoot and protracted undershoot. When LHWW = 69, the behavior is even more stable, consisting of a single small overshoot and virtual equilibrium soon after week 15. In terms of average accomplishment, stability again wins out; the first run improves on the base run by only 1 percent, the second run by 20 percent, and the third by 23 percent.

Figure 13 offers a broader view of the effect of the work limit (LHWW) on overall results.[8] Three 75-week summary statistics—average accomplishments per week to date, average hours worked per week to date, and average accomplishments per hour to date—are graphed against values of LHWW ranging from 50 to 90.[9] These graphs show clearly the difference between stability and instability, as far as total output and hours worked are concerned. Within the region of stability (where LHWW ≤ 72.2 hours per week),[10] a longer work limit means more hours worked (since HWW = LHWW in equilibrium) but also lower hourly productivity (since longer hours reduce the energy level). Because the depressive effect of longer hours on average energy, small at first, accelerates as the region of instability is approached, a point of maximum average weekly accomplishment exists; given all other baseline parameter values, this "optimal work limit" is approximately 69 hours per week.

A tradeoff of hours and energy no longer exists once the region of instability is entered. A longer limit on hours worked does not increase the average number of

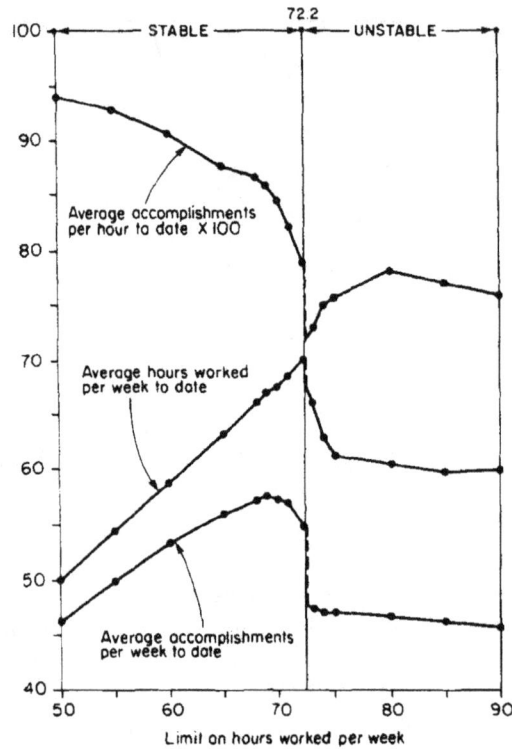

Fig. 13. 75-week summary statistics as affected by work limit

hours worked in this region. Instead, its effect is to increase the amplitude and decrease the period of the limit cycle; that is, to destabilize the behavior further. In fact, changes in the work limit have little effect on any of the cumulative measures shown in Figure 13. It is only in the transition from stability to instability that major differences appear; both average accomplishment and average hours worked drop significantly when the burnout cycle is introduced. Burnout both reduces available energy and requires more time off from work. The advantages of stability are underscored by the following sort of comparison: As shown in Figure 13, the individual can accomplish more on the average with a work limit of 55 hours per week than with a work limit of 75 hours per week, while actually working fewer hours per week on the average.

In noting that the optimal work limit is found within the region of stability but not far from the critical point of transition to instability, one is led to an important conclusion: The more advantageous the other parameters affecting stability are, the higher the *critical* work limit will be, so the higher the *optimal* work limit will be as well. In concrete terms, this means that if the individual can (1) reduce stress at work (by finding ways to make the basic tasks more enjoyable, or by adopting a less pushy or more flexible approach to setting goals), or (2) relax more effectively when not working, then he will be able to work longer hours and accomplish more without risking burnout.

Conclusion

Probably the most effective first step in attempting to fight a complex problem like burnout is to understand its dynamic source. Burnout is not caused by a stressful work environment alone but, more importantly, by the individual's workaholic response to that environment. As many psychiatrists and counselors would agree, real progress begins only when the client has understood clearly his or her own role in creating the problems and has accepted the responsibility to adopt healthier behavior. I can personally attest to the powerful impact of (1) seeing my own burnout cycles reproduced by a computer model, (2) adjusting behavioral parameters until a stable solution was found, and (3) realizing then that my problem was not inevitable but, to a large degree, a product of my own work habits. It was particularly enlightening to discover that even from the point of view of productivity, psychological stability is preferable to instability.

The model suggests not only that stability is preferable to instability but that it is generally attainable, even in a normally stressful work setting, through proper adjustment of one's work limit.[11] Chronic feelings of moderate tiredness or irritability at work must be met directly by a firm commitment to reduce one's maximum workweek permanently, or at least until the completion of a particularly stressful project. If the signs of incipient burnout later return even at this lower work limit, then the limit should be reduced again. It is important not to return to the old work limit, even when one is feeling rested and alert again; as the model demonstrates, vacations alone (the "time off" response) do not prevent a recurrence of burnout.

The model also demonstrates the potential benefit of reducing the normal stress at work, or relaxing better when not working, through methods such as those described at the beginning of this paper. By adopting ways that make one less vulnerable to the negative consequences of stress, one expands the region of stability described in the previous section. This means that one can work longer hours without risking burnout and so accomplish more. An increased ability to (1) anticipate and counteract potential sources of stress, (2) relax without feeling guilty, or (3) adopt more realistic and flexible goals, can do much to make life as a high achiever more enjoyable and satisfying and return to the individual a sense of control.

Appendix: Worker burnout model listing

```
        *  WORKER BURNOUT CYCLES
        NOTE BY J. B. HOMER, APRIL 1984
1    A  AW.K=(AH.K)(HWW.K)
2    A  AH.K=TABLE(TAH,EL.K,0,1,.2)
     T  TAH=0/.2/.4/.6/.8/1 A-UNITS/HR
3    L  HWW.K=HWW.J+(DT/TAHWW)(IHWW.J-HWW.J)
     N  HWW=HWWI
     C  HWWI=40 HRS/WK
     C  TAHWW=1 WK
4    A  IHWW.K=MIN(LHWW,HWW.K*EELHW.K*EPAHW.K)
     C  LHWW=80 HRS/WK
5    A  EELHW.K=TABLE(TEELHW,EL.K,0,1,.2)
     T  TEELHW=0/.4/.7/.9/1/1
6    A  EPAHW.K=TABLE(TEPAHW,PAA.K,0,1.6,.2)
     T  TEPAHW=2.3/1.9/1.6/1.35/1.15/1/.9/.8/.75
7    A  PAA.K=PAW.K/XAW.K
8    L  PAW.K=PAW.J+(DT/TPAW)(AW.J-PAW.J)
     N  PAW=AW
     C  TPAW=1 WK
9    L  XAW.K=XAW.J+(DT)(XAW.J*FCXAW.J)
     N  XAW=XAWI
     C  XAWI=40 A-UNITS/WK
10   A  FCXAW.K=BFCX+FCXPA.K
     C  BFCX=.1 PER WK
11   A  FCXPA.K=TABLE(TFCXPA,PAA.K,0,1.6,.2)
     T  TFCXPA=-.7/-.5/-.35/-.2/-.1/0/.1/.25/.4
12   L  EL.K=EL.J+(DT)(ER.JK-ED.JK)
     N  EL=ELI
     C  ELI=1
13   R  ER.KL=(ERN)(EHWER.K)(EHEFR.K)
     C  ERN=.3 PER WK
14   A  EHWER.K=TABLE(TEHWER,HWW.K,0,120,20)
     T  TEHWER=1.3/1.2/1/.7/.5/.35/.25
15   A  EHEFR.K=TABHL(TEHEFR,EL.K,.8,1,.05)
     T  TEHEFR=1/.9/.7/.4/0
16   R  ED.KL=(EDN)(EPAED.K)(EHWED.K)(ELEFD.K)
     C  EDN=.06 PER WK
17   A  EPAED.K=TABLE(TEPAED,PAA.K,0,1.6,.2)
     T  TEPAED=5/4/3.1/2.3/1.6/1/.6/.4/.3
18   A  EHWED.K=TABLE(TEHWED,HWW.K,0,120,20)
     T  TEHWED=.3/.6/1/1.5/2/2.5/3
19   A  ELEFD.K=TABHL(TELEFD,EL.K,0,.2,.05)
     T  TELEFD=0/.4/.7/.9/1
     NOTE SUMMARY STATISTICS
20   A  AVAWD.K=AD.K/(TIME.K+1E-7)
21   L  AD.K=AD.J+(DT)(AW.J)
     N  AD=AW*1E-7
22   A  AVHWWD.K=HWD.K/(TIME.K+1E-7)
23   L  HWD.K=HWD.J+(DT)(HWW.J)
     N  HWD=HWW*1E-7
24   A  AVAHD.K=AD.K/HWD.K
     NOTE  CONTROL STATEMENTS
     SPEC  DT=.25/LENGTH=75/PLTPER=1.5/PRTPER=15
     PRINT  AVAWD,AVHWWD,AVAHD
     PLOT  AW=A,HWW=H(0,100)
     PLOT  EL=E,PAA=P(0,1.6)
     RUN  BASE
```

Notes

1. This response of accommodation is known in the stress literature as syntoxic, in contrast with the catatoxic, or fighting, response of working harder to meet one's goals. Both responses are considered homeostatic, because their purpose is to reduce potentially harmful stress. See Selye (1974, 41, 47).

2. For values of PAA less than 0.8, BFCX + FCXPA < 0. Equilibrium can occur only when PAA = 0.8.

3. In this case of moderate burnout an extended full-time vacation is not needed.

4. In postscript to the explanation of the burnout cycle, it should be noted that loops 3 and 5—the "frustration" and "time off" loops— are not strictly necessary for generating the cycle. Model tests show that if the work is normally disagreeable enough (i.e., if the energy depletion normal EDN is large enough), the draining power of long hours can generate a collapse without the added stress of mounting frustration. But in the baseline model (with its relatively small value of EDN), removal of the aggravating effect of loop 3 does result in stable behavior. Other model tests show that when the individual does not respond to exhaustion by taking time off, expectations eventually fall so low that they can be satisfied even while working less. The resulting respite from work permits a recovery of energy and a renewal of higher accomplishment. But the removal of loop 5 does delay the recovery, resulting in a considerably more severe and protracted period of burnout.

5. Equilibrium requires that hours worked per week (HWW), expectations of weekly accomplishment (XAW), and the energy level (EL) all be unchanging. An equilibrium at which HWW is less than its limit (LHWW) can be shown to exist under certain circumstances, but this equilibrium is unstable in the sense that any small perturbation of certain exogenous parameters will trigger a limit cycle. The condition for a stable equilibrium requires that HWW = LHWW. Labeling EHWED as f, EHWER as g, and EPAED as h, the condition may be stated as follows:

   $$EDN * h(PAA') * f(LHWW) \leq ERN * g(LHWW)$$

   where $FCXPA(PAA') = -BFCX$

   In the baseline model the left-hand side of the inequality equals 0.19, while the right-hand side equals 0.15: The condition for stable equilibrium is not met and a limit cycle results.

6. Average accomplishments per week to date, or AVAWD, increases from the baseline value of 47.0 to 55.1.

7. In the test run, TEELHW = 0/.3/.5/.7/.9/1, compared to the base run's TEELHW = 0/.4/.7/.9/1/1.

8. This figure summarizes the results of a number of model tests, like those in Figure 12, in which the work limit was altered. Recall that LHWW = 80 in the base run.

9. DYNAMO equations for these three summary statistics may be found in the Appendix, listed as AVAWD, AVHWWD, and AVAHD, respectively. AVAWD is computed by integrating accomplishments per week (AW) over time and then dividing by the total time elapsed. AVHWWD is similarly computed by intergrating hours worked per week (HWW) and then dividing by time elapsed. AVAHD is found by dividing cumulative accomplishments (accomplishments to date, AD) by cumulative hours worked (hours worked to date, HWD).

10. The analytic stability condition is satisfied in this region. When LHWW = 72.2, the left-hand (depletion) side equals the right-hand (recovery) side.

11. The person-environment fit may conceivably be so poor that the optimal work limit is less than a standard workweek or even nonexistent. In such cases, an individual in search of full-time work is best advised to find it elsewhere. Why *court disaster*?

References

Breznitz, S., and L. Goldberger. 1982. Stress Research at a Crossroads. In *Handbook of Stress*, ed. L. Goldberger and S. Breznitz. New York: Free Press.

Cameron, R., and D. Meichenbaum. 1982. The Nature of Effective Coping and the Treatment of Stress-Related Problems: A Cognitive-Behavioral Perspective. In *Handbook of Stress.*

Cherniss, C. 1980. *Staff Burnout: Job Stress in the Human Services.* Beverly Hills, Calif.: Sage.

Dabiri, H. E. 1979. Dynamics of Human Development: Achievement Crisis. Sloan School Working Paper WP-1043-79. Sloan School of Management, M.I.T., Cambridge, Mass.

Freudenberger, H. J., and G. Richelson. 1980. *Burn-Out: The High Cost of High Achievement.* Garden City, N.Y.: Anchor Press.

Greenwood, J. W., III, and J. W. Greenwood, Jr. 1979. *Managing Executive Stress.* New York: Wiley.

Holroyd, K. A., and R. S. Lazarus. 1982. Stress, Coping, and Somatic Adaptation. In *Handbook of Stress.*

Holt, R. R. 1982. Occupational Stress. In *Handbook of Stress.*

Howard, A., and R. A. Scott. 1965. A Proposed Framework for the Analysis of Stress in the Human Organism. *Behavioral Science* 10 (2):141–160.

Ivancevich, J. M., and M. T. Matteson. 1980. *Stress and Work.* Glenview, Ill.: Scott, Foresman.

Janis, I. L. 1982. Decision-Making Under Stress. In *Handbook of Stress.*

Kiefer, C., and P. M. Senge. 1982. Metanoic Organizations in the Transition to a Sustainable Society. System Dynamics Group Working Paper D-3360-3. Sloan School, M.I.T., Cambridge, Mass.

Lazarus, R. S. 1979. Positive Denial: The Case For Not Facing Reality. *Psychology Today* 13:44–60.

Machlowitz, M. M. 1980. *Workaholics: Living With Them, Working With Them.* Reading, Mass.: Addison-Wesley.

Maslach, C. 1978. Job Burnout: How People Cope. *Public Welfare* 36.

McClelland, D. C. 1961. *The Achieving Society.* Princeton, N.J.: Van Nostrand.

NDACTRD (National Drug Abuse Center for Training and Resource Development). 1980. *Trainer Manual: Staff Burnout.* NDACTRD Publication No. 80-00115. Washington, D. C.

Selye, H. 1974. Stress Without Distress. Philadelphia: Lippincott.

Stoyva, J., and C. Anderson. 1982. A Coping-Rest Model of Relaxation and Stress Management. In *Handbook of Stress.*

Weiman, C. G. 1977. A Study of Occupational Stressors and the Incidence of Disease/Risk. *Journal of Occupational Medicine* 19:119–122.

Welford, A. T. 1973. Stress and Performance. *Ergonomics* 16:567–580.

2

A Diffusion Model with Application to Evolving Medical Technologies (1987)

Homer J. *Technological Forecasting and Social Change,* 31(3): 197-218

Related writings

Finkelstein S, Homer J, Sondik E. Modeling the Dynamics of Decision-Making for Emerging Medical Technologies. *R & D Management* 14(3): 175-191, 1984.

Homer J. *A Dynamic Model for Analyzing the Emergence of New Medical Technologies.* PhD dissertation, MIT Sloan School of Management, March 1983.

Homer J. The Effect of Government Regulation on the Emergence of a New Medical Technology. *Proceedings of the System Dynamics Research Conference,* Rensellaerville, NY, 1981.

A Diffusion Model with Application to Evolving Medical Technologies

JACK B. HOMER

ABSTRACT

This paper reports on a system dynamics simulation model which was developed to study the diffusion of medical technologies but which should be applicable to other sorts of technologies as well. The model addresses both the adoption and the changing extent of use of an evolving, product-based technology and also endogenously accounts for changes in actual and perceived performance. Two specific medical case studies provide background to the model and are used for testing the model's ability to reproduce various aspects of historical behavior. The model's relatively complex structure is presented in the form of influence diagrams showing major flows and relationships. The paper concludes with a discussion of contributions and possible future uses and modifications of the model.

Introduction

Over the last thirty years, a number of mathematical models have been developed which attempt to capture the key features of a process known alternatively as innovation diffusion, market penetration, or technological substitution [21, 27, 28]. Employing ideas from various fields (notably, diffusion research, marketing, economics, and decision theory), these models have been applied extensively to explain observed diffusion patterns for specific products or technologies. As tools for prelaunch or early forecasting, they have seen only limited use, largely because of inherent uncertainties regarding parameter estimation. Even in the presence of such uncertainties, however, an appropriately designed model may provide powerful insights for decision-makers who are able to control certain parameters represented in the model. Several such normative diffusion models have been proposed in recent years [22, 23, 24, 27, 28].

Most current diffusion models may be seen as extensions or refinements of the classic "mixed-influence" model introduced by Bass [4]. This model consists of a single, three-parameter differential equation which expresses the rate of adoption as a fraction of remaining nonadopters. The model produces a generalized logistic curve depicting the number of adopters as growing continuously over time up to its maximum value, in either sigmoid (S-shaped) or inflectionless exponential fashion. Both of these patterns have firm empirical foundations, with the sigmoid curve in particular showing up throughout the empirical diffusion literature [27, 34].

JACK B. HOMER is Assistant Professor of Systems Science at the University of Southern California. He specializes in the field of system dynamics and is director of USC's System Dynamics Laboratory.

Address reprint requests to Professor Jack B. Homer, ISSM, Room 109, Department of Systems Science, University of Southern California, Los Angeles, CA 90089-0021.

© 1987 by Elsevier Science Publishing Co., Inc. 0040-1625/87/$03.50

Mahajan and Peterson [27] discuss several of the Bass model's underlying assumptions and review extensions that have relaxed these assumptions to make the model more realistic and useful in particular settings. For example, some extensions allow for repeat purchases by adopters or multiple stages of adoption. Despite indications of progress, however, Mahajan and Peterson conclude that "there clearly is a woeful lack of research on the relaxation of [the Bass model's] assumptions" [27, p. 54].

This paper reports on a diffusion model which was developed to study innovative medical technologies but which should be applicable to other sorts of technologies as well. Specifically, the model addresses a situation in which potential adopters face a stream of related decisions some subset of which may be best resolved by purchasing the new technology in question. The model is particularly useful for analyzing evolving technologies, where both the technology's acceptance and its extent of use are subject to change over time as a result of new performance-related information, and where performance itself is affected by changes in the technology or its application.

The need for such a flexible diffusion model became apparent from reviewing the medical technology literature and examining specific case histories. While several studies confirm the sigmoid growth pattern for medical technologies, a rise-and-decline pattern has also been documented [2, 7, 8, 10, 36]. The decline frequently occurs as a result of discouraging new evaluative evidence, a factor not addressed by previous diffusion models. Indeed, diffusion research in general has neglected to relate the volume and content of evaluative evidence to the diffusion process; nonetheless, it is clear that in medicine evaluations may affect both adoption and patient selection (i.e., extent of use) [8]. It is equally clear that evaluations of evolving medical technologies may be systematically biased or dated, and that actual performance may be highly dependent on technology, experience, and extent of use [2, 5, 9, 32]. The model discussed below explicitly includes all of these real-world factors in an integrated, testable framework.

The remainder of the paper is organized as follows: 1) the model's development is described and two case histories of evolving medical technologies are summarized; 2) the model's structure is discussed in some detail; 3) the approach taken in calibrating the model to represent the two cases is discussed; 4) the model's ability to reproduce various dimensions of the historical behaviors is demonstrated and explained; and 5) contributions and future uses and modifications of the model are discussed.

Model Background

The model presented here was developed over a three-year period. At all times, the primary goal was to provide insights into the diffusion of medical technologies capable of assisting government decision-makers in their efforts to improve the use of these technologies. Such efforts include regulation, cost reimbursement control, research funding, information dissemination, and the provision of guidelines and coordination for local control activities. The U.S. government has yet to adopt a consistent stand on risky new medical technologies, largely because the diffusion process is poorly understood and is surrounded by controversy, uncertainty, and the fear of unintended consequences [2, 9, 40].

The model was constructed and refined according to principles associated with the system dynamics approach to modeling complex feedback systems, systems which are prone to behave in unintended or unanticipated ways [13, 14, 33]. Others have applied system dynamics to the study of innovation diffusion [37, 41, 42], attesting to the appropriateness of using the methodology in this context. A system dynamics simulation model consists of a stock-and-flow structure and a set of decision functions controlling the various flows. A key principle is that all decision functions should have a firm basis in the real world (where "bounded rationality" predominates [31]) and should respond realistically under all conceivable conditions, including extreme ones; this requirement often necessitates the use of nonlinear functions. Other structures and parameters should have readily identifiable real-world counterparts as well. In general, the principles are intended to guide the development of realistic models with endogenous structure rich enough to reproduce observed patterns of behavior and to suggest other possible behaviors, yet parsimonious and aggregate enough to be understandable and broadly applicable.

Model development occurred in two phases and included more than two dozen major structural revisions. The first phase (1980–1981) focused on the diffusion of percutaneous transluminal coronary angioplasty (PTCA), an innovative catheter-based technology for the treatment of coronary artery disease and an alternative to open-heart bypass surgery [25]. This study was performed under the auspices of the National Heart, Lung, and Blood Institute, and came at a time when PTCA was new and evolving, with its future far from certain. The PTCA study laid much of the groundwork for the current model and included some suggestive public policy analysis [11, 18]. Unfortunately, too little historical data on PTCA existed at the time of the study for strict behavioral model validation. Nonetheless, the PTCA model was able to generate the previously documented sigmoid and rise-and-decline patterns when appropriately calibrated. It also predicted a third pattern, which may be described as rise-decline-and-rebound, under certain seemingly reasonable conditions [18].

The second, longer phase of model development (1981–1983) focused on the two case studies discussed below, the implantable cardiac pacemaker and the antibiotic drug clindamycin. The selection of these two technologies was based on a number of considerations [19]. First, for several reasons it was decided that model development should initially focus on therapeutic drugs and devices, rather than diagnostic technologies or fixed hospital equipment. Second, the two technologies had distinctly different characteristics and histories and so would test the model's breadth. Third, both technologies had been introduced recently enough to make interviews with physicians and manufacturers a useful source of supplementary information; but both technologies had also been in clinical use long enough to reveal clear patterns of diffusion.[1] Fourth, both technologies had received substantial coverage in medical journals, which served as a rich source of detailed, real-world information. In summary, the two case studies did much to strengthen the model and ground it further in reality.

PACEMAKER CASE HISTORY

The implantable cardiac pacemaker is a sophisticated electronic device for the correction of chronic abnormal heart rhythms (arrhythmias) that can cause death or significant impairment of health. Figure 1 charts annual U.S. pacemaker implantations (or, equivalently, unit sales) between 1960, when the first successful clinical implant was performed, and 1984.[2] The implantation rate grew in essentially sigmoid fashion through 1981, but then abruptly declined. This rate includes both "initial" implants and the "replacement" implants performed when implanted units near end-of-battery or show signs of impending

---

[1]The two case histories were originally tracked through 1980. This paper presents recently gathered information updating the cases through 1984.

[2]Figure 1 reflects an updating of data presented previously [19, p. 65]. These new data were provided by Medtronic, Inc., and come from a pacemaker industry survey performed by the firm of L.F. Rothschild. Unterberg, and Towbin (LFRUT), in October 1984.

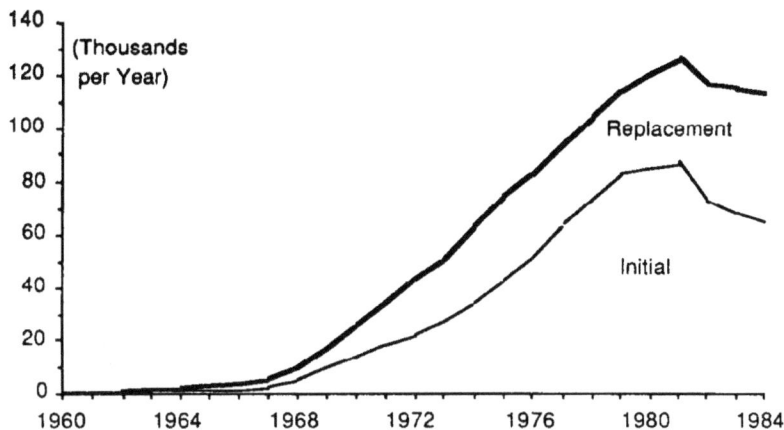

Fig. 1. Annual pacemaker implantations.

or actual failure. From 1970 to 1980, product improvements increased the average life of pacemakers from two years to about ten years [19]. The pattern of total implants essentially reflects a similar, but more accentuated, pattern of initial implants.[3]

Behind the growth of initial implants is much more than the usual story of adoption told in the diffusion literature. True, the 1960s were a period when pacemakers gained initially slow but then snowballing acceptance. But by about 1971, pacing was almost universally accepted among cardiologists as the modality of choice for a specific subset of patients with highly symptomatic and life-threatening arrhythmias. The solid growth of the next ten years therefore had little to do with adoption, but rather was almost exclusively the result of widening patient selection criteria. Indeed, selection criteria had begun to widen in the late 1960s, as technical developments made implantation and pacing much more safe and reliable and a worthwhile risk for more patients. During the 1970s and early 1980s, manufacturers developed increasingly sophisticated pacemakers, which led to the pacing of patients with less severe or more subtle arrhythmias, patients often at little risk of sudden death or lacking major symptoms [19].

Between 1981 and 1984, the initial implant rate dropped by about 25%, most of this drop occurring within the first year. In 1982, a few highly visible reports [1, 17, 45] seemed to confirm the claim made by some that patient selection criteria had become too broad and that peer review or other institutional incentives to narrow the use of pacemakers might be required [6, 44].[4] By 1984, peer review of every pacemaker implant was required in most if not all U.S. hospitals as a prerequisite for Medicare reimbursement. This appears to have caused physicians to cut back significantly the extent to which they use pacemakers, particularly when symptoms such as fainting or dizziness are absent or minor.[5]

---

[3]Initial implant data for the 1980s are based on the LFRUT total implant data and a small case-flow model used for deriving missing portions of certain time series [19, App. P1]. For this derivation, it was assumed that the average life of pacemakers remained at ten years during the 1981–1984 period.

[4]A subsequent in-depth study [38] disputed the Health Research Group's [17] methodology and findings, but its authors did support the policy of peer review.

[5]Based on discussions with Dr. Michael Bilitch, director of University of Southern California's Pacemaker Center, and Mr. Geoff Ball, marketing planning manager of Medtronic, Inc.

27

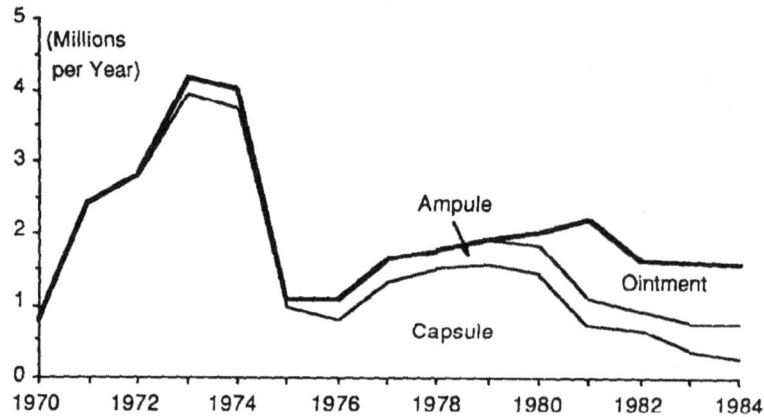

Fig. 2. Annual clindamycin orders.

CLINDAMYCIN CASE HISTORY

In 1970, the Upjohn Company received Food and Drug Administration approval to market their new antibiotic clindamycin in the U.S. under the trademark Cleocin. Figure 2 charts combined annual orders (unit sales) for the three manufactured forms of clindamycin (capsule, ampule, and ointment) between 1970 and 1984.[6] The pattern through 1981 may be described as rise-decline-and-rebound; a smaller secondary decline occurred during 1981–1982 and was followed by apparent stability.

Clindamycin is one of a group of drugs known as broad-and-medium-spectrum (BMS) antibiotics, which include ampicillin, tetracycline, erythromycin, and other commonly prescribed antibiotics. Though individual BMS antibiotics may come and go, their combined use in the U.S. appears to be quite stable at about 100 million orders per year [19]. At first, clindamycin was used primarily as a second- or third-line drug for infections not requiring hospitalization. In 1972, the injectable ampule form of clindamycin was introduced for use against severe infections. In the same year, a prominent study [3] demonstrated the unique value of clindamycin in fighting anaerobic infections, which can be particularly hard to control and are often life-threatening. These events helped to convince remaining skeptics to adopt the new drug and led to the widening of patient selection criteria.

In 1974, another prominent study was published [43], but this time the news was bad: In a hospital setting, the drug was apparently capable of causing high incidence rates of diarrhea and an untreatable, potentially fatal form of colitis. These unexpectedly high incidence rates seemed to be confirmed by a number of reports and warnings subsequently issued in 1975. In that year, most physicians either abandoned use of the drug or limited its use to severe infections for which no acceptable alternative was available [19].

Confidence in clindamycin was much restored during 1976 and 1977 as the medical community came to understand, through increased experience and a thorough review of the evidence [15], that the adverse side effects were neither as frequent nor as uncon-

---

[6]Figure 2 reflects an updating of data presented previously [19, p. 157]. Data for the entire time series were provided by IMS America, Ltd. and consist of national estimates of Cleocin use extracted from their National Disease and Therapeutic Index (NDTI).

trollable as they had previously appeared. However, the drug (in its oral form) was still clearly more risky than comparable BMS antibiotics for nonsevere infections, discouraging use in these settings and limiting the magnitude of the rebound. And yet the patient selection criteria did rewiden during the late 1970s. It was reported in 1976 that clindamycin capsules dissolved in a simple vehicle and applied topically worked remarkably well against acne and with minimal risk of serious side effects. Dermatologists took increasingly to the use of topical clindamycin for the entire spectrum of acne conditions, and by 1980, when Upjohn introduced its Cleocin-T ointment, 60% of all orders for clindamycin were designated for topical use [19].

Total orders for clindamycin abruptly declined 25% during 1981–1982, Cleocin-T taking the brunt of the drop-off, and remained approximately level through 1984. The year 1982 was apparently a big one for the introduction and promotion of new topical antibiotics, particularly topical erythromycins, which compete directly with topical clindamycin.[7] Indeed, the combined use of all topical antibiotics increased rather than decreased during the early 1980s.[8]

Model Structure

Figure 3 presents an overview of the model described in greater detail below. The model endogenously generates a variety of decisions and outcomes that can directly or indirectly affect the diffusion of an evolving, product-based technology. Potential purchasers of the technology make a number of crucial decisions in the model, including most of those related to use and evaluation. The publishing media also play a role in the evaluation process. Suppliers of the product provide support for diffusion in the form of promotional marketing and product modification. The three functional areas of use, evaluation, and support interact with one another in a number of ways.[9] For example, purchases ("use") provide suppliers with the sales revenue needed to fund further promotional marketing ("support"), which may, in turn, lead to more purchases.

Figure 3 also indicates that beyond the model's explanatory boundary lie various exogenous factors that can affect the technology's diffusion. These include attributes of the technology and of its users, evaluators, and supporters that remain essentially fixed throughout the diffusion process. The exogenous factors also include variable "environmental" influences beyond the control of the model's actors, such as the number of required new purchase decisions, the standard of performance determined by alternative technologies, and government policy affecting the technology. Note that price and the time to replacement purchase have also been considered exogenous: While these may, in particular cases, be affected by factors such as product modification, the evidence collected during model development suggested no generalizable way to model their dynamics.

The current model contains some 150 endogenous equations, including 24 state variables and delay functions, plus 90 exogenous constants and nonlinear look-up functions. Needless to say, the model will not be presented here on an equation-by-equation

---

[7]Based on discussions with Dr. Ronald Reisner, chairman of the Division of Dermatology at the University of California, Los Angeles, and Dr. Richard Homer, dermatologist in private practice.

[8]NDTI summary data on "Topical Dermatologic Antiinfectives," 1980–1984, were provided by IMS America, Ltd.

[9]This basic framework is in line with that of Zaltman and Stiff [46], who view product diffusion as the final result of interactions among "adopter units" and between those units and the product's suppliers.

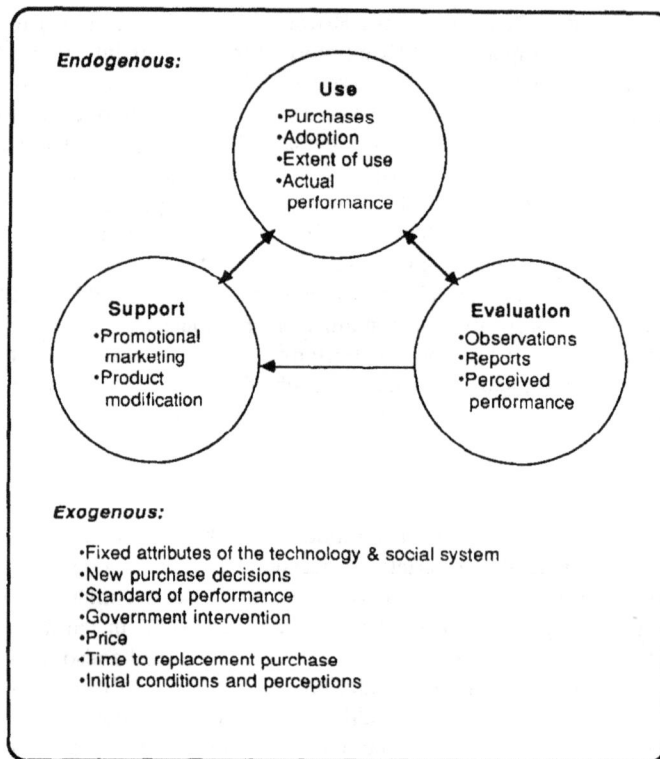

Fig. 3. Model overview.

basis.[10] Nonetheless, it is possible to present the essence of the model in the form of somewhat simplified influence diagrams which focus on the model's major flows and relationships. These diagrams, which appear as Figures 4–10, employ some of the flow diagramming symbols used in system dynamics [33].

Figure 4 shows how purchases (or unit sales) of the new product are determined in the model. Purchases may be of two types: new and replacement. New purchases reflect the selection of some fraction of the stream of new purchase decisions in favor of the new technology rather than alternative approaches. This selection fraction is found by multiplying 1) the fraction of potential purchasers who buy the product at all ("purchaser fraction") and 2) the fraction of purchase decisions in which these purchasers specifically favor the new technology ("extent of use").

Replacement purchases represent purchases to replace products that are depleted, that is, products discarded for reason of wear-and-tear. Note that these are not the same as products discarded because the original need for the product has terminated; as, for example, in the death of a pacemaker implant recipient. Note also that replacement is not automatic following product depletion: In the event that extent of use has narrowed

---

[10]Full documentation of the current model is available from the author upon request. The final version of the model prior to its recent update has been previously documented and described [19].

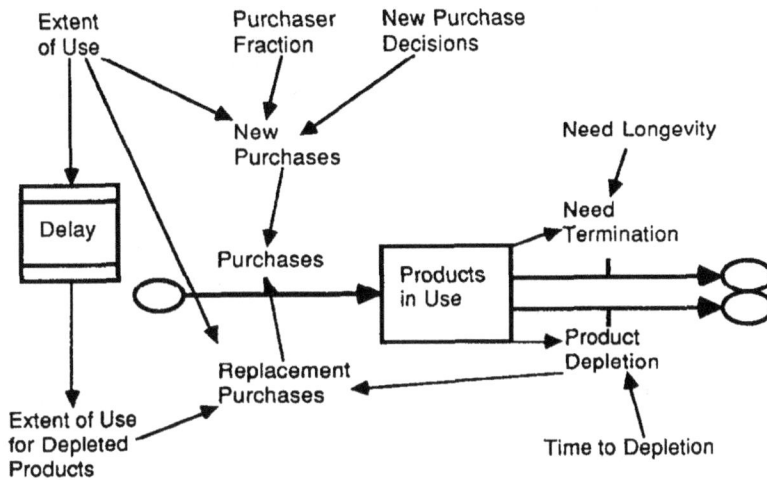

Fig. 4. Purchases.

in the time since original purchase, replacements will be indicated for only some fraction of depleted products.

Figure 5 shows the model's representation of the adoption process. Unless product availability is restricted (as a result of government policy or problems of supply), the purchaser fraction will be identical to the fraction of potential purchasers who wish to purchase the new product ("accepter fraction"). The accepter fraction, in turn, is increased when nonaccepters decide to accept the new product and decreased when accepters decide to abandon it. Both acceptance and abandonment rates will be affected when perceptions of the technology's overall expected relative performance change significantly.

Acceptance is enhanced when information on the new technology is not only encouraging but plentiful enough to break down natural resistances. The model's "volume of information" term sums the effects of personal, professional, and commerical channels of communication in keeping potential accepters aware of the new technology and its advantages. As the purchaser fraction increases, so does the important word-of-mouth

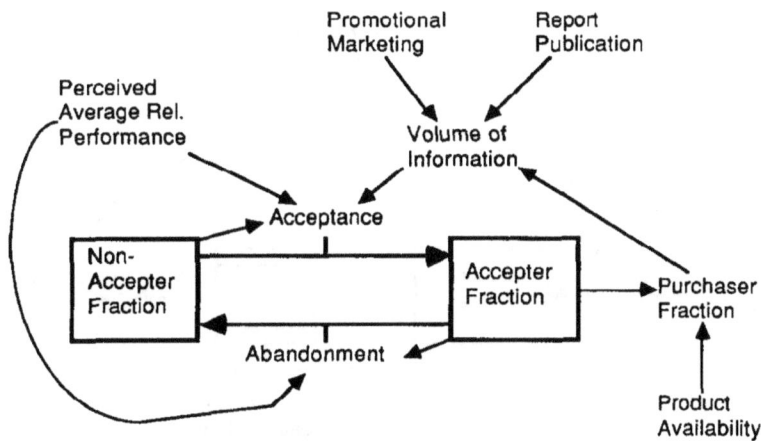

Fig. 5. Adoption.

influence. The model's non-word-of-mouth influences include promotional marketing and the publication of reports on the new technology. Increases in either of these factors can increase the volume of influential information, but both impacts are also assumed to have diminishing returns to scale [19, 26].

Figure 6 illustrates the determination of extent of use. Purchasers seek to apply the new technology in only those types of cases for which the expected performance meets or exceeds the existing standard, that is, for which "relative performance" is nonnegative. Evaluations (observations and reports) of past cases provide the primary source of information on expected performance by type of case. If it appears that the expected relative performance for marginal cases (cases on the periphery of use, where the technology's advantage is least certain) has been negative, then purchasers will narrow the extent of use so as to eliminate these cases from consideration. If, on the other hand, this perceived marginal relative performance is positive, then purchasers will tend to widen the extent of use so as to include some of those cases that had formerly been beyond the periphery of accepted use. Thus, extent of use will continue to be adjusted until perceived marginal relative performance equals zero, or until extent of use can be narrowed or widened no further.

Because the evaluation process takes time to complete, evaluations sometimes fail to reflect the impact of the latest product modifications on performance; this "moving target" situation can be particularly problematic when modification is rapid or evaluation is slow [9, 32]. When evaluations do not reflect the latest modifications, purchasers may extend the product's use beyond the point indicated by those evaluations. Some fraction of modifications may result in readily apparent performance gains. These "self-evident product improvements" lead purchasers to widen extent of use prior to full evaluation.

Figure 7 illustrates the determination of actual performance. Two quantities, average (overall) performance and marginal (peripheral) performance, are computed to summarize the full distribution of outcomes. The concept of performance is represented in the model as an explicit comparison of expected benefits with expected risks. These expected outcomes are determined by three fractors: extent of use, user skill, and product capability. Increases in skill or product capability relative to a given extent of use will result in improved performance. Similarly, narrowing of the extent of use relative to given levels

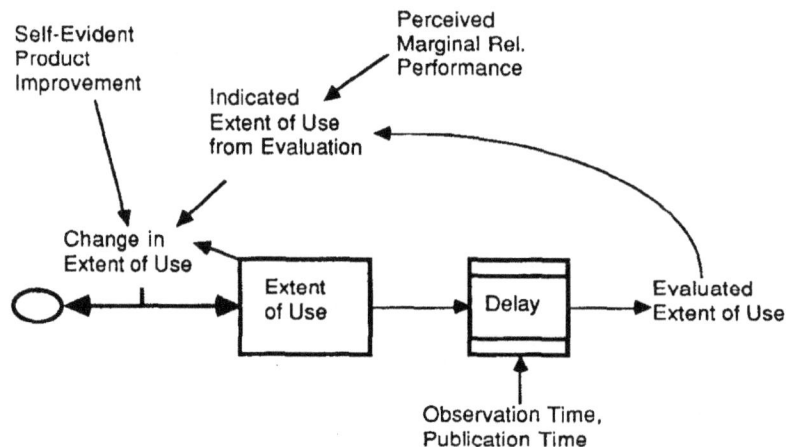

Fig. 6. Extent of use.

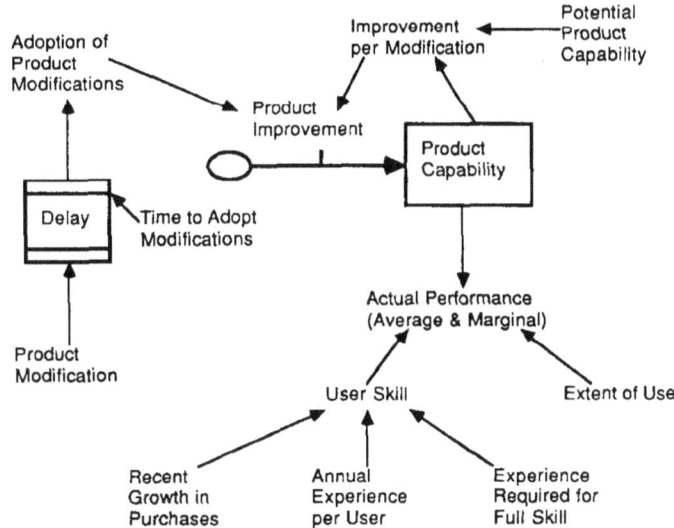

Fig. 7. Actual performance.

of skill and product capability will result in an improved distribution of performance. The model computes both the accumulation and depreciation of user experience over time and the average level of skill implied by this experience. The supply of hands-on users (not to be confused with purchasers, though overlap is possible) is assumed to adjust rapidly to the demand created by purchases; thus, rapid growth in purchases will lead to an influx of inexperienced users. Such an influx will reduce average skill, unless a typical user's annual experience well exceeds the experience required for full skill.

Product capability is an index-scaled measure representing the technology's functional value or "figure of merit" when applied by fully-skilled users. Product capability will be increased when purchasers adopt recent product modifications which succeed in improving the product's value.[11] (Though modifications require some time to be adopted into general use, this process will generally be faster and more automatic than acceptance of the basic technology [35].) The average degree of improvement per modification declines as the product's ultimate potential is approached and equals zero when that potential is achieved. This diminishing-returns formulation is similar to that used by Floyd in a well-known model for forecasting technological figures of merit [12].

Figure 8 portrays the generation of evaluative evidence in the form of observations and reports. As long as the technology is evolving and uncertainty remains about its optimal use, every purchase becomes a subject for observation and assessment. Observations remain current for a period of time which may or may not be equal to the life of the product but will be at least as long as the time perceived by purchasers as necessary for observation of all possible outcomes, including delayed complications. (In medicine, this period is commonly known as the "follow-up time".)[12]

---

[11]Product improvements (for given values of skill and extent of use) increase benefits and/or decrease risks. The nonlinear relationships between product capability and the components of performance are considered fixed attributes of the technology.

[12]Based on the medical context, the model's observation time is endogenous: It grows longer when unforeseen complications consistently start appearing beyond the time period formerly thought sufficient to track all complications.

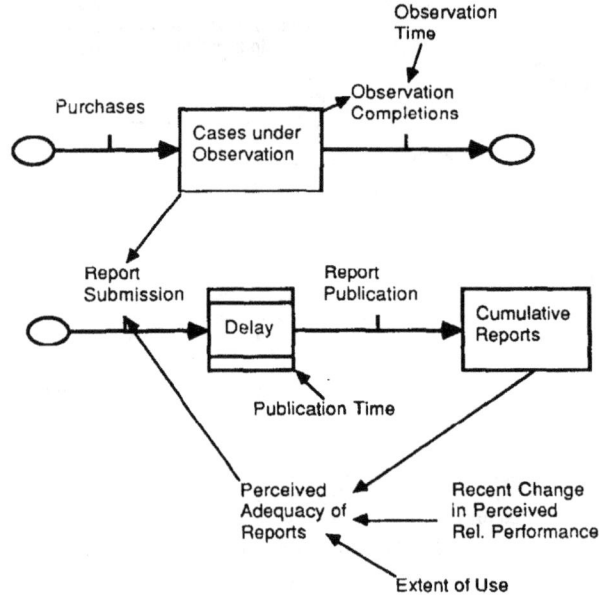

Fig. 8. Observations and reports.

Some fraction of cases under observation become the subject of evaluative reports published in journals and other influential media; a delay separates the submission and publication of these reports. The fraction of cases reported is inversely related to the perceived adequacy of existing (cumulative) reports: Both submitters and publishers of reports will tend to turn their attentions to other topics as the technology's capabilities and shortcomings become well-established. But uncertainty regarding the technology's performance may actually increase at times, spurring renewed interest in reports. One possible source of increased uncertainty is a widening of the extent of use, which implies application of the technology to additional types of cases not previously evaluated. A second possible source of uncertainty, if not controversy, is new evaluation-based perceptions of performance which contradict or depart significantly from prior perceptions.

Figure 9 illustrates the determination of perceived performance and perceived relative performance; the latter simply compares perceived performance with the exogenous standard of performance. The model structure for perceived performance is replicated for computation of both average and marginal quantities.

Perceived performance may differ from actual performance due to built-in delays or inaccuracies in evaluation. Observations are more timely, responding sooner to changes in actual performance, than reports which must wait for publication.[13] However, observations have much less capability than reports of bringing newly discovered complications to the general attention of fellow potential purchasers: Media reports may enable even a small number of complications to come to widespread attention speedily while local observations generally cannot. (On the other hand, such reports may have a tendency initially to exaggerate the risk of complications, as happened with clindamycin.)

---

[13]Reported performance may become significantly dated (more so than the publication time alone would suggest) during periods when needed reports are not submitted due to a misguided perception that existing reports are adequate. The possibility of such a "trap of complacency" has been demonstrated using the present model [19, Chap. 8].

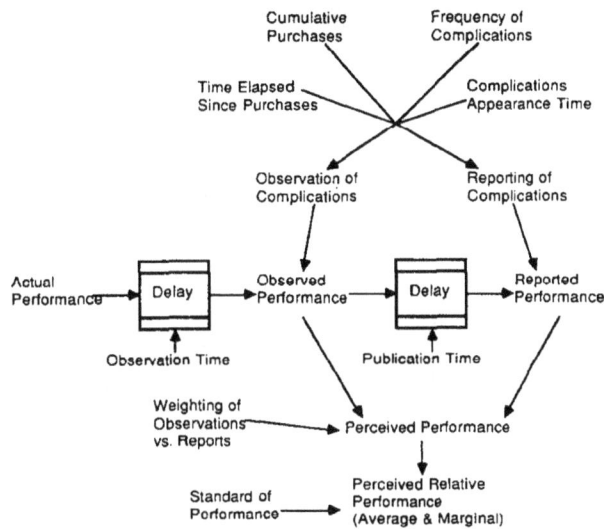

Fig. 9. Perceived performance.

The discovery and accurate assessment of significant complications may take a long time, either because the complications are relatively infrequent or because they are simply slow to appear. Infrequent complications start to appear as more purchases are made. Delayed complications start to appear as the time elapsed since purchase (in a sufficient number of cases) grows to exceed the appearance delay time for those complications.

Perceived performance represents a synthesis, a weighted combination, of the information provided by observations and reports. It is formulated in a nonlinear fashion which effectively increases the weight on "bad news" relative to "good news" when the two sources of evaluation provide divergent information.[14] This input-sensitive weighting scheme reflects the typical risk-aversion of potential purchasers [5, 23]. The basic relative weights attached to observations and reports will depend on the degree to which potential purchasers feel that performance of the new technology can be judged in a reliable way on the basis of locally-obtained evidence without reference to the national media.

Figure 10 illustrates the determination of promotional marketing and product modification efforts by suppliers of the new product. Both types of effort require time from initial planning to final implementation, with product modifications sometimes requiring several years for successful completion [30]. The planned level of effort in both cases is typically budgeted as a given fraction of the new product's sales revenue [19, 39]; the model also allows for short-term "kick-off" efforts supported by an initial pool of available funds.

The fraction of sales revenue allocated for promotional marketing may decrease below its traditional value if suppliers perceive that the fraction of accepters is already

---

[14]$PP = (OP^{w_o})(RP^{w_r})$, where PP is perceived performance, OP is observed performance (OP > 0), RP is reported performance (RP > 0), w_o is the weight on observations ($0 < w_o < 1$), and w_r is the weight on reports ($w_r = 1 - w_o$). The more disparate the inputs OP and RP are, the closer PP will be to the lower of the two values.

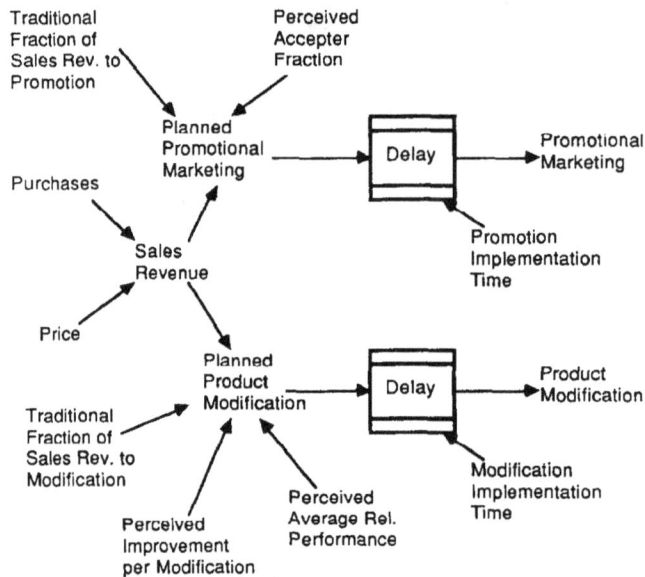

Fig. 10. Promotional marketing and product modification.

high enough to cause promotion to be less cost-effective than desired. Similarly, the fraction of sales revenue allocated for product modification may decrease below its traditional value if the perceived improvement per modification (or alternatively, the expected increase in extent of use per modification) falls below its desired level. On the other hand, the modification budget fraction may increase above normal to support attempts to upgrade performance perceived to be relatively weak.

Model Calibration

The model was calibrated to represent the pacemaker and clindamycin cases using an approach based on system dynamics modeling principles. These principles suggest that model validity is enhanced when parameters have been estimated as much as possible on the basis of disaggregate, process-oriented information about the system, rather than on the basis of aggregate time-series data [16]. To the extent parameter values are estimated simultaneously and with the sole aim of duplicating historical behavior, one runs the risk in a nonlinear feedback model of establishing a parameter set that works well within that historical context only. The whole-model or full-information "best fit" approach may well produce a flawed set of parameter estimates which compensate for each other's faults during the historical period but produce spurious results when looking at future possibilities or in analyzing policy alternatives [20].

Disaggregate, technology-specific information is usually plentiful but is often scattered or unavailable in the desired numerical form. Thorough searches of the pacemaker and clindamycin literatures did provide useful detailed data on the evaluation, performance (benefits and risks), and other aspects of these technologies. These data were formally analyzed to derive specific parameter values. Interviews with physicians and company representatives provided much detailed, process-oriented information in areas often not covered by the literature [19]. Physicians were asked to describe their experiences with regard to adoption, patient selection, factors affecting performance, and the evaluation of performance. Company representatives were asked to describe their experiences with regard to promotional marketing and product modification efforts. Others have noted the practical usefulness of survey or judgmental data in parameterizing diffusion models [22, 29].

Though there were a number of parameters for which technology-specific data were not available for one or both of the two case studies, it was nonetheless possible in many of these cases to develop estimates on the basis of logic or knowledge gained from general experience. For example, there were some parameters for which pertinent data were available for only one of the technologies, but where the nature of the parameter in question suggested that its estimate should be the same in the two cases.

Although it is desirable to estimate parameter values without reference to aggregate time-series data, this may often be difficult in actual practice where time and resource constraints limit the effort that can be devoted to information gathering. Thus, there may be some model parameters for which the techniques described above are insufficient to establish satisfactory estimates; in the pacemaker and clindamycin cases, fewer than 20% of the parameters were of this sort. In such situations, a carefully circumscribed use of partial-model testing can be useful for estimating parameter values [20]. This procedure involves simulating the behavior of a piece of model structure—which may be as small a single equation—in response to historical input data and for comparison with historical output data. The uncertain parameter or parameters of interest are then adjusted (within whatever logical restrictions might apply, such as nonnegativity) until an acceptable or best possible fit is obtained.

Model Behavior

PACEMAKER CASE

After calibrating the model to represent the pacemaker case, including initial conditions for the year 1960, a thirty-year "base run" simulation was produced. Various aspects of the simulated behavior are presented graphically in Figures 11–14 for comparison with the corresponding actual time series data.[15] The model clearly does a good job of reproducing the historical patterns of behavior along all dimensions of comparison. Based on model output presented here, as well as other dimensions of simulated behavior and sensitivity test results not shown here, a simplified reinterpretation of pacemaker diffusion emerges. This new telling of the story serves primarily to clarify the role of key endogenous feedback loops operating within a unique set of exogenous boundary conditions. Figure 15 illustrates these key loops in the form of a system dynamics causal-loop diagram [33]. The diagram's five loops are numbered for convenient reference in the following discussion.

During the early and mid-1960s, nearly all growth in pacemaker purchases (Figure 11) was due to the effect of word-of-mouth in generating increasing acceptance (Figure 12) of the technology (see loop 1). Extent of use (Figure 13) remained quite limited as long as the pacemaker was perceived to have no marginal advantage over alternative treatments. By the late 1960s, however, product modifications—funded by growing sales revenue—led to a dramatic reduction in risk and greatly boosted perceived performance. This added fuel to the technology's acceptance (see loop 2), which was complete by the

---

[15]Details regarding construction of the historical time series presented in Figures 11–14 (pacemaker) and Figures 16–19 (clindamycin) and the various contributing data sources have been previously presented [19]. Various time scales have been used for presentation purposes.

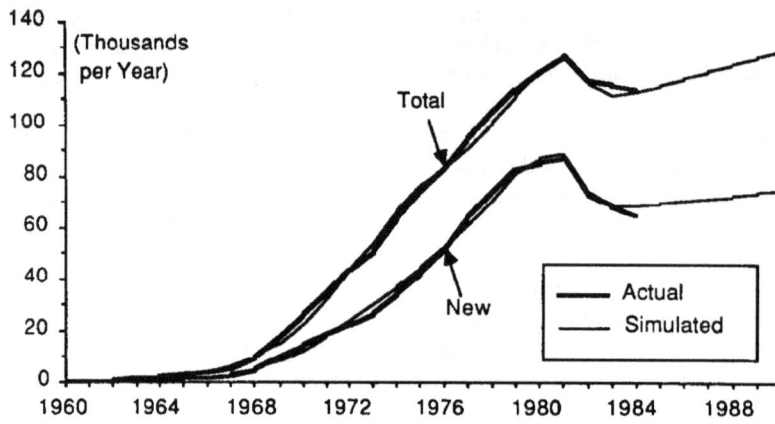

Fig. 11. Pacemaker "purchases" behavior.

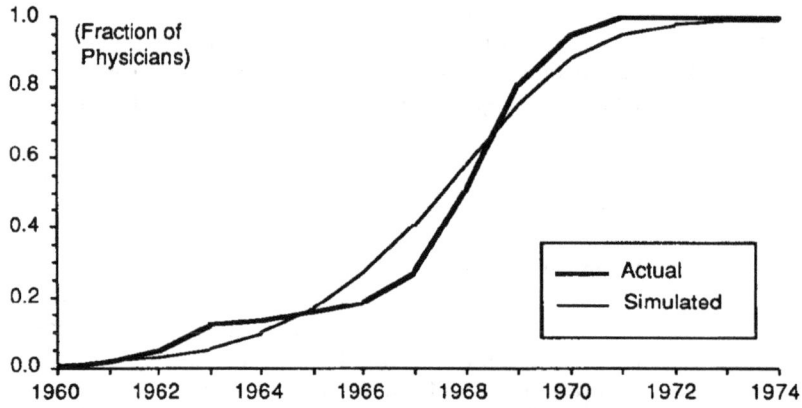

Fig. 12. Pacemaker "purchaser fraction" behavior.

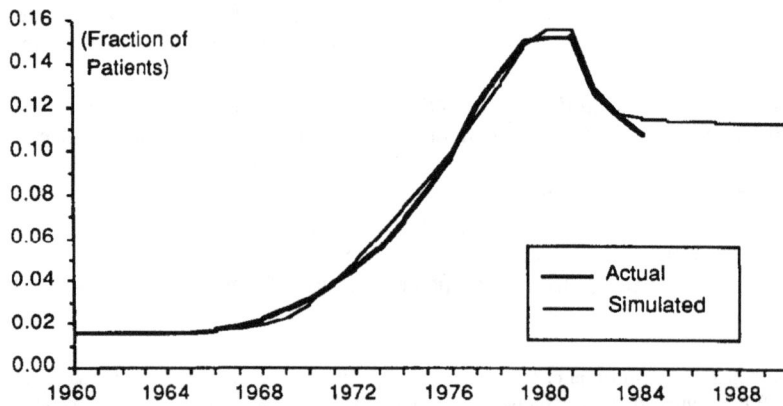

Fig. 13. Pacemaker "extent of use" behavior.

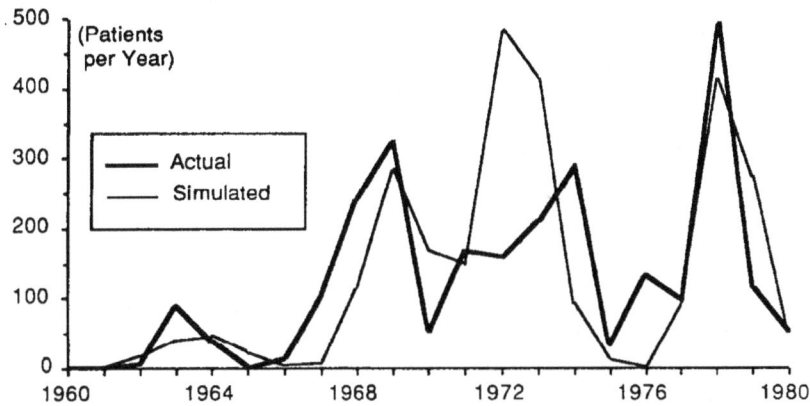

Fig. 14. Pacemaker "reports" behavior.

early 1970s. More importantly, it triggered an accelerating increase in extent of use, as well as growth in new purchases, sales revenue, and product modification efforts. The resulting product improvements, in turn, formed the basis for further increases in extent of use (see loop 3) during the 1970s.

By the mid-1970s, however, increases in extent of use were starting to outstrip diminishing product improvements (see loop 4), and the pacemaker's superiority on the margin began to slip quickly. Even according to the standards of the time, extent of use should have stopped increasing around 1977; but due to the inherent delays in evaluation and perception, it continued to increase through the end of the decade. Thus, in the early 1980s, extent of use would have narrowed somewhat in any case (see loop 5), as perceptions caught up with reality. But this narrowing was hastened and accentuated in 1982 by a significant increase in the standard of performance.

The model predicts that pacemaker purchases should grow continuously throughout

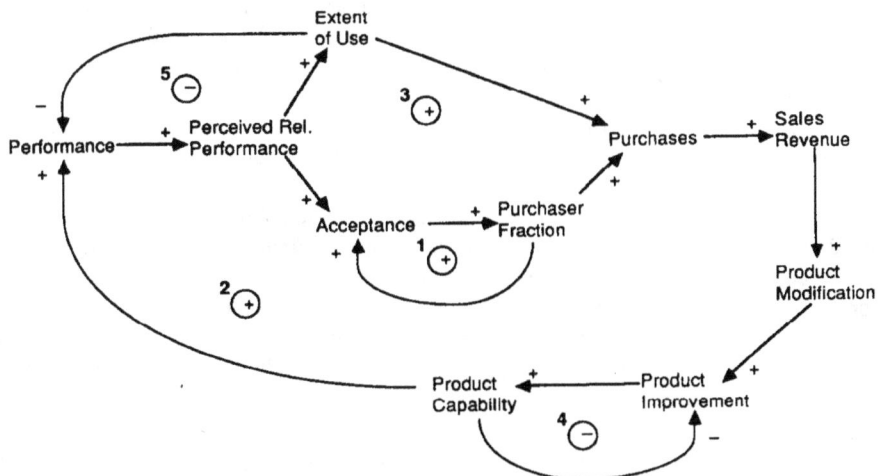

Fig. 15. Key feedback loops for pacemaker diffusion.

the late 1980s, despite lack of growth in either the purchaser fraction or the extent of use. The reasons for this are two: 1) resumed growth in new purchases and 2) continued growth in replacement purchases. New purchases resume their growth (following the stabilization of extent of use) because of the continued aging of the population which leads to a steadily increasing incidence of cardiac arrhythmias requiring attention ("new purchase decisions").[16] Replacement purchases, on the other hand, grow to reflect former growth in new purchases: With a time to replacement (i.e., battery life) of about ten years, growth in replacement purchases during the late 1980s reflects the growth in new purchases during the late 1970s.

In response to accelerating growth in extent of use, the feedback structure for reports shown in Figure 8 endogenously generates an oscillation quite similar to the historical pattern seen in Figure 14. The reporting system is apparently subject to periods of excess followed by relative quiet on a given topic, a result of built-in delays and the competitive, often redundant nature of the media. Although the oscillating pattern of pacemaker reports is interesting, it appears to have had no effect on other dimensions of pacing: Pacemaker prescribers have generally given little weight to published reports relative to local observations in their decision-making [19].

CLINDAMYCIN CASE

A fifteen-year "base run" for clindamycin was produced after calibrating the model appropriately and specifying initial conditions for the year 1970. Various aspects of the simulated behavior are presented in Figures 16–19 for comparison with actual time series. The model again does quite a good job of reproducing the historical patterns of behavior, though these patterns may seem unusual. The model tells a simplified story of clindamycin diffusion quite different from that of pacemaker diffusion. The differences between the two simulated cases come from different sets of exogenous conditions and, consequently, differences in the relative strengths of the model's feedback loops. Figure 20 presents a causal-loop diagram containing the five loops that the model suggests were key to the diffusion of clindamycin.

Most of the rapid growth in clindamycin purchases (Figure 16) during 1970–1973 came from corresponding growth in the purchaser fraction (Figure 17). Extent of use (Figure 18) also grew somewhat during this period, as a result of the drug's initially strong perceived relative performance. The rapid acceptance of clindamycin was generated primarily by two factors, promotional marketing (see loop 1) and word-of-mouth (see loop 2). Promotional marketing (Figure 19) appears to have started with a "kick-off" but was soon fully funded by growing sales revenue.

With the accumulation of cases, it was inevitable that clindamycin's severe but infrequent complications should finally be observed and reported (see loop 3). Perceived relative performance dropped sharply as a result of discouraging reports, leading to rapid declines during 1974–1975 in both acceptance and extent of use. (Sales revenue also dropped sharply as a result, forcing a cutback in promotional marketing.) But the first reports of complications were based on incomplete evidence and substantially exaggerated the problem, as revealed by subsequent reports. This reassessment of risks (see loop 4) along with the narrowed extent of use (see loop 5) served to bolster substantially the perceived relative performance of clindamycin. In response, both the purchaser fraction

---

[16]The incidence rate per million among the 55-and-older segment of the population appears to have been quite stable throughout the 1960–1980 period. It was assumed, conservatively, that this group would continue to grow at its historical 1.8% per year rate throughout the 1980s [19].

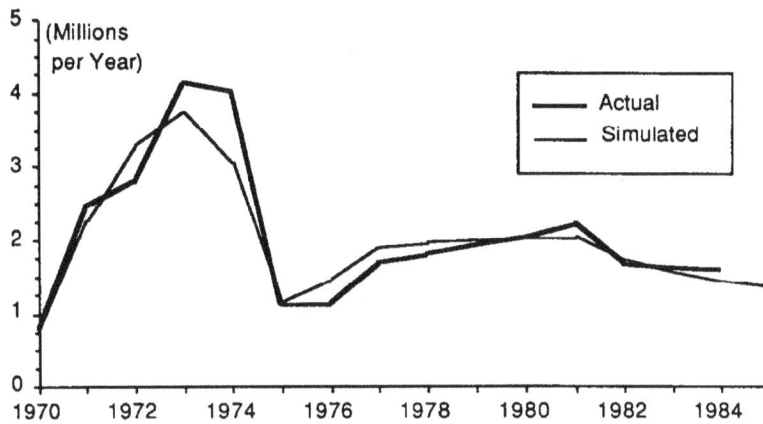
Fig. 16. Clindamycin "purchases" behavior.

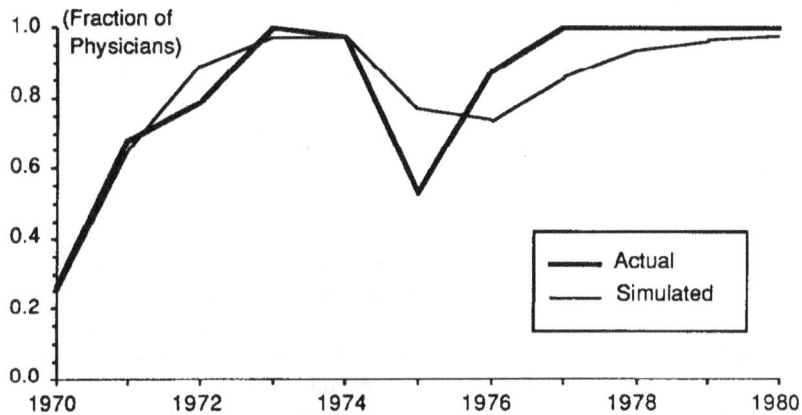
Fig. 17. Clindamycin "purchaser fraction" behavior.

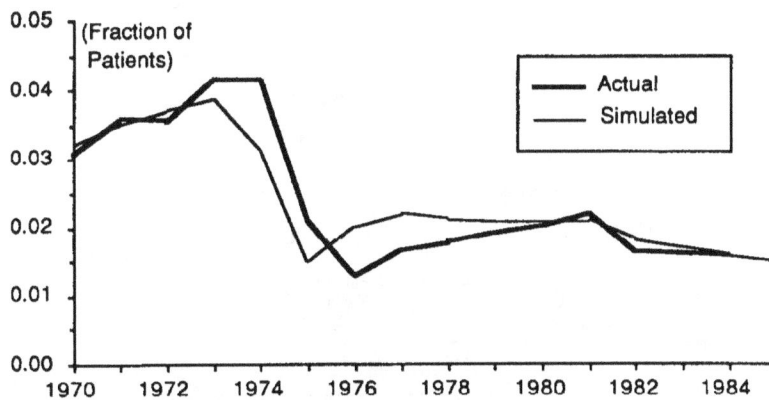
Fig. 18. Clindamycin "extent of use" behavior.

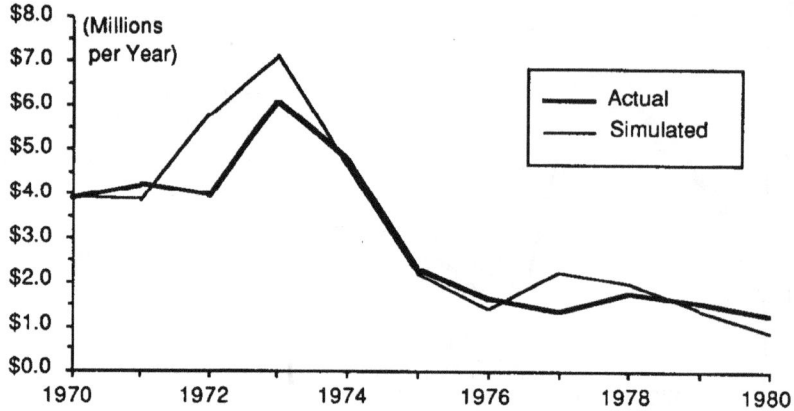

Fig. 19. Clindamycin "promotional marketing" behavior.

and extent of use rebounded quickly and then stabilized; correspondingly, the purchase rate rose sharply at first and then adjusted more slowly through 1981.

In 1982, the standard of performance for clindamycin increased significantly, leading to an immediate narrowing of extent of use followed again by relative stability. The model predicts continued stability into the foreseeable future, with the possibility of a bit more narrowing of use.

Conclusion

The multiloop diffusion model discussed in this paper was developed for the purpose of analyzing the complex dynamics of evolving technologies. Although the case studies and much of the literature which aided in model development were medical in nature, the model's generic structure suggests that it may prove useful for analyzing other sorts of evolving technologies as well. For example, it would be interesting to apply the model to study the diffusion of alternative energy sources, an area in which the Department of Energy has previously supported diffusion modeling efforts [22, 23]. Alternative energy

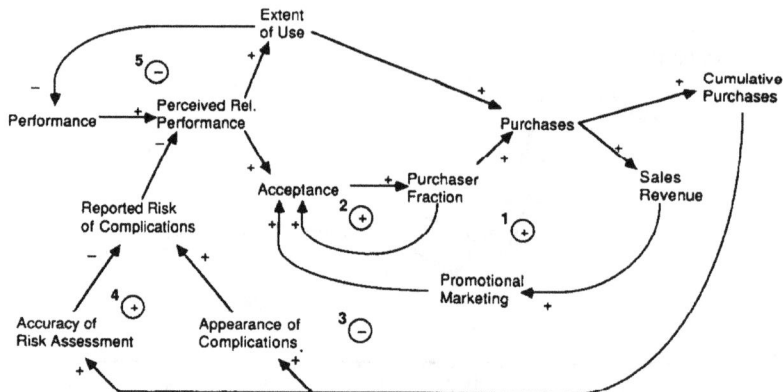

Fig. 20. Key feedback loops for clindamycin diffusion.

sources, like new medical technologies, appear to have intrinsic uncertainties that involve more than the rate of adoption.

The model explicitly relates within its endogenous structure a number of potentially important elements considered exogenous or not considered at all by diffusion models which focus solely on the process of adoption. These elements include extent of use, product modification, promotional marketing, product capability, user skill, actual performance (benefits and risks), and the evaluation and perception of performance. A model such as this which integrates a wide range of real-world concerns can be a powerful tool for policy analysis. Although there was not space in this paper to demonstrate the model's usefulness as a normative tool, the model has proved capable of generating original policy insights [11, 18, 19]. Much work remains to be done in the area of policy analysis and design using the current model.

Although the current model is quite inclusive as diffusion models go, it (like any other model) has limitations born of its particular focus and course of development. These limitations naturally suggest ways in which the model may be modified. For example, in the current model—based on medical technologies with acceptable cost to patients—price is exogenous and affects only sales revenue. But there may be situations in which price should be endogenized or modeled as an explicit part of the potential purchaser's decision-making calculus. The model might also be usefully modified to examine diffusion in the presence of an actively competitive technology which responds to the new technology or diffusion of a pair of new technologies whose use is interdependent.

References

1. ABC News 20/20, Pacemakers: Pumping Gold, American Broadcasting Companies, New York, 9 September 1982.
2. Banta, H. D., Behney, C. J., and Willems, J. S., *Toward Rational Technology in Medicine*, Springer, New York, 1981.
3. Bartlett, J. G., Sutter, V. L., and Finegold, S. M., Treatment of Anaerobic Infections with Lincomycin and Clindamycin, *New England Journal of Medicine* 287, 1006–1010 (1972).
4. Bass, F. M., A New Product Growth Model for Consumer Durables, *Management Science* 15, 215–227 (1969).
5. Blume, S., Aspects of the Dynamics of Medical Technologies, in *Research on Research: Proceedings of the European Symposium*. E. Heinekken and others, eds., Academy of Finland, Helsinki, 1980.
6. Chokshi, A. B., Friedman, H. S., and others, Impact of Peer Review in Reduction of Permanent Pacemaker Implantations, *Journal of the American Medical Association* 246(7), 754–757 (1981).
7. Coleman, J., Katz, E., and Menzel, H., *Medical Innovation: A Diffusion Study*, Bobbs-Merrill, Indianapolis, 1966.
8. Fineberg, H. V., Effects of Clinical Evaluation on the Diffusion of Medical Technology, in *Assessing Medical Technologies*, Committee for Evaluating Medical Technologies in Clinical Use, Division of Health Sciences Policy, Institute of Medicine, eds., National Academy Press, Washington, DC, 1985.
9. Fineberg, H. V., and Hiatt, H. H., Evaluation of Medical Practices: The Case for Technology Assessment, *New England Journal of Medicine* 301, 1086–1091 (1979).
10. Finkelstein, S. N., and Gilbert, D. L., Scientific Evidence and the Abandonment of Medical Technology: A Study of Eight Drugs, Working Paper 1419-83, Sloan School of Management, MIT, Cambridge, MA, 1983.
11. Finkelstein, S. N., Homer, J. B., and Sondik, E. J., Modelling the Dynamics of Decision-Making for Emerging Medical Technologies, *R & D Management* 14(3), 175–191 (1984).
12. Floyd, A. L., A Methodology for Trend-Forecasting of Figures of Merit, in *Technological Forecasting for Industry and Government*. J. R. Bright, ed., Prentice-Hall, Englewood Cliffs, NJ, 1968.
13. Forrester, J. W., *Industrial Dynamics*, MIT Press, Cambridge, MA, 1961.
14. Forrester, J. W., and Senge, P. M., Tests for Building Confidence in System Dynamics Models, *TIMS Studies in Management Sciences (System Dynamics)* 14, 209–228 (1980).
15. Friedman, G. D., Gerard, M. J., and Ury, H. K., Clindamycin and Diarrhea, *Journal of the American Medical Association* 236, 2498–2500 (1976).

16. Graham, A. L., Parameter Formulation and Estimation in System Dynamics Models, in *The System Dynamics Method: The Proceedings of the 1976 International Conference on System Dynamics*. J. Randers and L. K. Ervik, eds., Resource Policy Group, Oslo, 1977.

17. Greenberg, A., Kowey, P. R., Bargmann, E., and Wolfe, S. M., Permanent Pacemakers in Maryland, report by Health Research Group, Washington, DC, 7 July 1982.

18. Homer, J. B., The Effect of Government Regulation on the Emergence of a New Medical Technology, in Proceedings of the 1981 International System Dynamics Conference, Rensselaerville, NY, 1981.

19. Homer, J. B., A Dynamic Model for Analyzing the Emergence of New Medical Technologies, Ph.D. Dissertation, Sloan School of Management, MIT, Cambridge, MA, 1983.

20. Homer, J. B., Partial-Model Testing as a Validation Tool for System Dynamics, in Proceedings of the 1983 International System Dynamics Conference, Chestnut Hill, MA, 1983.

21. Hurter, A. P., and Rubenstein, A. H., Market Penetration by New Innovations: The Technological Literature, *Technological Forecasting and Social Change* 11, 197–221 (1978).

22. Kalish, S., and Lilien, G. L., Applications of Innovation Diffusion Models in Marketing, in *Innovation Diffusion Models of New Product Acceptance*. V. Mahajan and Y. Wind, eds., Ballinger, Cambridge, MA, 1986.

23. Kalish, S., and Lilien, G. L., A Market Entry Timing Model for New Technologies, *Management Science* 32, 194–205 (1986).

24. Kalish, S., and Sen, S. K., Diffusion Models and the Marketing Mix for Single Products, in *Innovation Diffusion Models of New Product Acceptance*. V. Mahajan and Y. Wind, eds., Ballinger, Cambridge, MA, 1986.

25. Kleinmann, A. J., Projecting the Impact of a Radical New Health Technology, M.S. thesis, Sloan School of Management, MIT, Cambridge, MA, 1980.

26. Little, J. D. C., Aggregate Advertising Models: The State of the Art, *Operations Research* 27, 629–677 (1979).

27. Mahajan, V., and Peterson, R.A., *Models for Innovation Diffusion*, Sage, Beverly Hills, CA, 1985.

28. Mahajan, V., and Wind, Y., Innovation Diffusion Models of New Product Acceptance: A Reexamination, in *Innovation Diffusion Models of New Product Acceptance*. V. Mahajan and Y. Wind, eds., Ballinger, Cambridge, MA, 1986.

29. Mahajan, V., Mason, C. H., and Srinivasan, V., An Evaluation of Estimation Procedures for New Product Diffusion Models, in *Innovation Diffusion Models of New Product Acceptance*. V. Mahajan and Y. Wind, eds., Ballinger, Cambridge, MA, 1986.

30. Marquis, D., and Meyers, S., *Successful Industrial Innovation*, National Science Foundation, Washington, DC, 1969.

31. Morecroft, J. D. W., System Dynamics: Portraying Bounded Rationality, *Omega* 11(2), 131–142 (1983).

32. National Academy of Sciences, A Study of the Diffusion of Equipment-Embodied Technology, Government Printing Office, Washington, DC, 1979.

33. Richardson, G., and Pugh, A., III, *System Dynamics Modeling with DYNAMO*, MIT Press, Cambridge, MA, 1981.

34. Rogers, E. M., *Diffusion of Innovations*, The Free Press, New York, 1983.

35. Rosenberg, N., Factors Affecting the Payoff to Technological Innovation, Manuscript DDC-0812N, Stanford University, Stanford, CA, 1977.

36. Russell, L. B., The Diffusion of Hospital Technologies: Some Econometric Evidence, *Journal of Human Resources* 12, 482–501 (1977).

37. Sanatani, S., Market Penetration of New Products in Segmented Populations: A System Dynamics Simulation with Fuzzy Sets, *Technological Forecasting and Social Change* 19, 313–329 (1981).

38. Scherlis, L., and Dembo, D. H., Problems in Health Data Analysis: The Maryland Permanent Pacemaker Experience in 1979 and 1980, *American Journal of Cardiology* 51, 131–136 (1983).

39. Schmookler, J., Economic Sources of Inventive Activity, *Journal of Economic History* 22, 1–20 (1962).

40. Schroeder, S. A., and Showstack, J. A., The Dynamics of Medical Technology Use: Analysis and Policy Options, in *Medical Technology: The Culprit Behind Health Care Costs? Proceedings of the 1977 Sun Valley Forum on National Health*. S. H. Altman and R. Blendon, eds., PHS-79-3216, Government Printing Office, Washington, DC, 1979.

41. Sharif, M. N., and Kabir, C., System Dynamics Modeling for Forecasting Multilevel Technological Substitution, *Technological Forecasting and Social Change*, 9, 89–112 (1976).

42. Sharif, M. N., and Ramanathan, K., Polynomial Innovation Diffusion Models, *Technological Forecasting and Social Change* 21, 301–323 (1982).

43. Tedesco, J. F., Barton, R. W., Alpers, D. H., Clindamycin-Associated Colitis: A Prospective Study, *Annals of Internal Medicine* 81, 429–433 (1974).

44. Tullis, J. L. L., Apocalypse Maybe: Buy Medtronic—The Domestic Pacemaker Market, 1981–1985 Outlook, Morgan Stanley Investment Research, New York, 1981.
45. United States Senate (Special Committee on Aging), Fraud, Waste, and Abuse in the Medicare Pacemaker Industry, Government Printing Office, Washington, DC, September 1982.
46. Zaltman, G., and Stiff, R., Theories of Diffusion, in *Theoretical Perspectives in Consumer Behavior*, S. Ward and T. Robertson, eds., Prentice-Hall, Englewood Cliffs, NJ, 1972.

Received 13 October 1986; revised 4 December 1986.

3

A Model of HIV Transmission Through Needle Sharing (1991)

Homer J, St. Clair C. *Interfaces*, 21(3): 26-49

A Model of HIV Transmission through Needle Sharing

JACK B. HOMER
Institute of Safety and Systems Management
University of Southern California
Los Angeles, California 90089

CHRISTIAN L. ST. CLAIR
County of Los Angeles Department of Health Services
Los Angeles, California 90012

Intravenous drug users (IVDUs) are a high-risk group for human immunodeficiency virus (HIV) infection because of the common practice of needle sharing. A mathematical model simulates the spread of HIV infection and HIV-related death through a population of IVDUs. Special attention is given to the movement of needles between noninfectious and infectious states. The model has several input parameters that may be adjusted to represent local population characteristics and policy interventions. Use of the model for policy analysis is illustrated with a series of simulations examining the potential benefits of a needle-cleaning campaign.

A new mathematical model of human immunodeficiency virus (HIV) transmission dynamics focuses on needle sharing among intravenous drug users (IVDUs). It combines a basic population structure with a set of variables related to the risk of infection, making it possible to simulate the multiple year consequences of various key input factors, including those that may be affected by public policy interventions. Proposed interventions include needle-cleaning and needle-exchange programs, increased resources for drug treatment, expanded education regarding HIV transmission, and expanded HIV blood testing [Institute of Medicine 1986].

We decided to focus on the IVDU population for two distinct reasons. First, the incidence of HIV-associated acquired immune deficiency syndrome (AIDS) in

Copyright © 1991, The Institute of Management Sciences
0091-2102/91/2103/0026$01 25
This paper was refereed

HEALTH CARE—EPIDEMIOLOGY
SIMULATION—APPLICATIONS

INTERFACES 21: 3 May–June 1991 (pp. 26–49)

IVDUs is growing substantially in the US and threatens to affect the general population. By 1986, 17 percent of newly reported AIDS cases in the US were attributable to IV drug use [Institute of Medicine 1986]; and by 1987, IVDUs were estimated to comprise 20 percent of the HIV-infected population [CDC 1987]. Although homosexual males still account for the majority of AIDS cases, the rate of new infections is declining among that group, while it may still be growing among IVDUs in some areas. IVDUs thus currently represent a key target group for behavioral change efforts. IVDUs also present a threat well beyond their own numbers, because they are the primary source of both heterosexual transmission and parent-to-child transmission of HIV [Institute of Medicine 1986].

The second reason for focusing on IVDUs is that HIV transmission through needle sharing has received considerably less attention from mathematical modelers than has sexual transmission. While several published modeling studies have focused on sexual transmission [Anderson 1988; Kaplan 1989b; May and Anderson 1987], it was only rather recently that a model focusing on IVDUs and needle sharing by Kaplan [1989a] appeared in print. While the Kaplan model is an important contribution, we found a few of its key simplifying assumptions to be overly restrictive or possibly unrealistic. We therefore set out to develop an alternative model, based both on the literature and on discussions with clinicians and other experts.

For example, one key assumption of the Kaplan model that we wished to relax is that needles circulate through the IVDU population entirely at random, from one use to the next, in the manner of an open "shooting gallery." Although this may be a reasonable approximation for certain areas, such as parts of New York City, it does not describe well the situation in other areas where needles are primarily shared within small, relatively closed friendship groups or cliques. In two recent studies, two-thirds to three-quarters of sharers reported sharing only with friends, relatives, or sexual partners [Magura et al. 1989; Mascola et al. 1989]. This difference in the extent of unrestricted or open needle sharing, which is explicitly represented in our model, may help to explain the large variations in HIV seroprevalence among IVDUs that have been observed from one urban area to another [Chaisson, Moss, et al. 1987; Institute of Medicine 1986]. Studies during the years 1984 to 1987 found rates of seroprevalence among IVDUs of 50 to 60 percent in New York City, northern New Jersey, and Puerto Rico, 20 to 30 percent in several cities in the Northeast (including Boston, Baltimore, and New Haven), and generally less than 20 percent in the rest of the country [CDC 1987].

Model Assumptions

Our model focuses on the dynamic consequences of needle sharing among IVDUs in terms of HIV infection and HIV-related death within a given geographic area. For the sake of clarity and parsimony, we have made the following assumptions regarding the population and the transmission process:

(1) Only IVDUs who inject frequently enough to be considered addicts may become infected. The definition of an addicted injector varies by drug, with heroin

and methedrine addicts injecting daily, while cocaine injectors tend to concentrate their weekly use during weekend binge periods [Murphy 1987]. Addicts who cut back to no more than occasional use, without subsequent relapse to addiction, are considered to have ceased addictive use.

(2) From a needle-sharing standpoint, the total addicted IVDU population in a given area may be considered a collection of small cliques, including those that consist of only a single individual. The IVDU population may then be categorized into "sharers," individuals whose cliques use needles that have been used by other cliques, and "nonsharers," individuals whose cliques never do so. This dichotomy of sharers and nonsharers is the model's sole representation of heterogeneity among addicted IVDUs in regard to needle-sharing behavior.

(3) The nonsharer population is fixed in size and the sharer population is initially constant as well, with inflows (representing initiation of addictive IV drug use and migration into the area) equal to outflows (representing cessation of addictive IV drug use, death, and migration out of the area). However, all HIV infections are assumed ultimately to be fatal, and HIV-related deaths increase outflow from the sharer population.

(4) Needle sharing is the only route of HIV transmission represented in our model. Sexual transmission is excluded because needle sharing is the predominant route of transmission within the IVDU population [Institute of Medicine 1986], and also because the model does not address transmission outside of the IVDU population.

(5) Each clique of IVDUs is identifiable, at any given point in time, as either HIV seropositive or HIV seronegative. That is, we assume that all members of any particular clique share the same HIV infection status. One implication is that all sharers may be at risk for HIV infection through needle sharing, while nonsharers face no such risk.

(6) An uninfected sharer risks infection, with a fixed probability, whenever he or she injects with an infectious needle not effectively cleaned prior to use. The probability that an uninfected sharer will encounter an infectious needle, in turn, depends upon a number of factors, including the openness of sharing among cliques. If sharing is entirely open, as in an idealized shooting gallery, then this probability may become large. If sharing is quite closed, rarely crossing from one clique to another, then this probability may remain close to zero.

(7) A stock of used needles is maintained in proportion to the sharer population. These needles are disposable and are discarded (typically due to breakage,

---

Two-thirds to three-quarters of sharers reported sharing only with friends, relatives, or sexual partners.

---

blocking, or dulling [Murphy 1987]) after an average number of uses, and new needles are started as necessary. A needle becomes infectious if it is used by an infected sharer. It remains infectious until it is deliberately and effectively cleaned prior to use, or until it "self-disinfects" due to natural inactivation of the virus. It may then become reinfected and may go through the cycle of disinfection and reinfection any number of times until it is finally discarded.

Model Structure

A complete equation listing of our model is presented in the appendix. Population movements are expressed on a per-year basis but conceptually take place in continuous time. Although the system of differential equations cannot be solved analytically, its solution can be approximated closely by simulating the corresponding difference equations using Euler integration with a time interval of one month ($\frac{1}{12}$ year); we used this computation interval for all simulations of the model discussed here.

The model starts at time zero with an initial total IVDU population and an initial sharer fraction of that population, which together give the initial sharer and nonsharer populations. As time progresses, the total IVDU population changes only when the population of sharers changes, while the nonsharer population remains fixed. The population of sharers is subdivided into two stocks, uninfected sharers and infected sharers, initialized so that the stock of infected sharers is only a tiny fraction (0.01 percent) of all sharers at time zero. We have also modeled a stock of infected former sharers, individuals who are no longer a part of the active IVDU population but who do account for a proportion of all HIV-related deaths in the area traceable to needle sharing. For purposes of comparing alternative simulations, the model includes a running total of cumulative HIV deaths, as well as cumulative HIV infections.

The model's basic stock-and-flow structure is diagrammed in Figure 1. New entries or initiates to the uninfected sharer population eventually exit this stock either (1) by quitting addictive IV drug use, (2) through "standard loss," namely, out-migration or non-HIV death, or (3) by becoming infected. Those who become infected eventually exit the infected sharer stock either (1) by quitting addictive IV drug use, (2) through standard loss, or (3) through HIV death. Finally, those infected

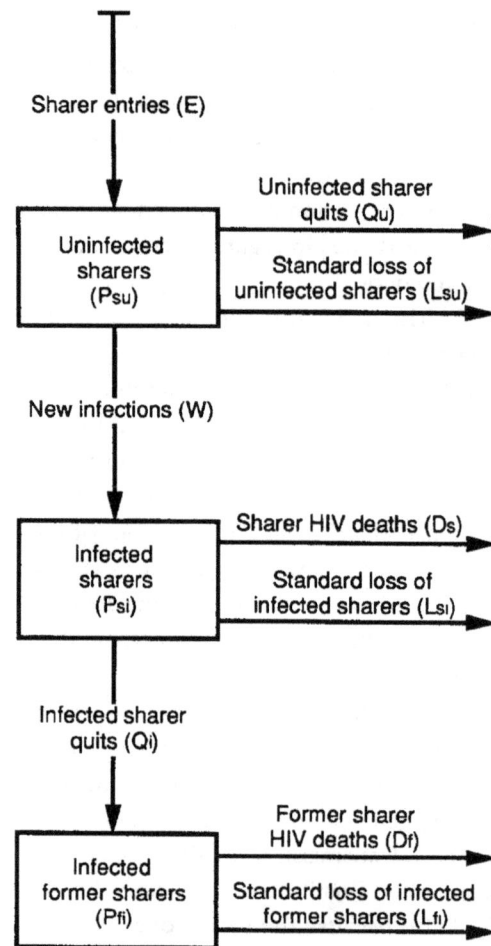

Figure 1: Sharer population stocks and flows. Flows expressed in people per year.

sharers who have quit addictive IV drug use eventually exit the infected former sharer stock either (1) through standard loss or (2) through HIV death.

In regard to new entries, our baseline assumption is that this flow is constant over time and equal to the initial outflow of sharers (quits plus standard losses), so that the population of sharers will remain close to its initial value as long as the number of HIV deaths remains small. This baseline assumption may be altered to represent alternative scenarios for sharer entry, perhaps as the result of policy intervention. For example, a primary objection to needle-cleaning or needle-exchange programs has been that, by reducing the risk of needle sharing, such programs may cause more people to be attracted to IV drug use and needle sharing [Institute of Medicine 1986]. Alternatively, expanded AIDS education may be advocated based on an expectation that it will cause new entries to decrease.

The flows of standard loss (out-migration plus non-HIV death) occur according to an annual rate or fraction which is assumed identical for all three of the population stocks in Figure 1. The two flows of quits also occur according to annual rates which may differ in value. These quit rates may be altered to represent certain policy interventions. For example, an increase in drug treatment resources might succeed in increasing both of the quit rates to a similar degree. On the other hand, a policy of expanded HIV testing might affect the quit rate for infected sharers differently than it affects the quit rate for uninfected sharers. Also, policy interventions aside, these quit rates may be altered to represent the im-

pact of increased knowledge and fear of AIDS; in a survey of jailed IVDUs in New York City, 23 percent reported having stopped sharing so as to avoid AIDS [Des Jarlais, Tross, and Friedman 1987].

We calculate the two HIV death flows by first computing total HIV deaths and then allocating them proportionally to sharers and former sharers. This simple allocation scheme assumes that the infection

---

In a survey of jailed IVDUs in New York City, 23 percent reported having stopped sharing so as to avoid AIDS.

---

progresses toward death at the same rate within the two infected populations. (This assumption may only be approximate, since infected former sharers may, on the average, be closer to death than infected current sharers.) The total HIV death flow is formulated as a third-order exponential delay [Meadows 1974] of the transmission of new infections, with a delay time parameter representing the number of years from transmission to death. The use of a full pass-through delay reflects our assumption that HIV infection leads inevitably to death. The particular function selected is characterized by a pulse-response which is equivalent to an Erlang (or gamma) type 3 distribution [Johnson and Kotz 1970]. The Erlang distribution, along with the Weibull distribution, has previously been used to estimate the incubation period of HIV in transfusion-associated AIDS cases [Anderson 1988]. Also, the fact that HIV infection and AIDS progress through a sequence of relatively well-defined stages [Institute of Medicine 1986], rather than being an undifferentiated, first-order-type process, would seem to support our choice of a higher-order delay function.

The annualized flow of new infections is formulated as a daily flow multiplied by 365 days per year. Daily infections are found, in turn, by multiplying the population of uninfected sharers by their daily risk of infection. The daily risk to uninfected sharers is derived from basic principles of probability, given the infection risk per needle use and the number of uses per sharer per day. While the daily frequency of use is an input parameter, the risk per use is itself a function of other parameters.

The equation for risk per use says that an uninfected sharer may, with a given probability, become infected with HIV by using an infectious needle contaminated with active HIV without first cleaning it effectively. (By effective cleaning we mean actual decontamination or sterilization of the needle. Although IVDUs commonly rinse their equipment with water prior to use, this is not an effective means of cleaning; in contrast, proper rinsing with household bleach is extremely effective [Resnick et al. 1986].) The probability of an uninfected sharer using a needle which is infectious (prior to possible cleaning) is expressed as the product of two factors. The first factor represents the probability that the needle has been used previously, as opposed to being newly started; this reuse fraction is determined by the average number of uses per needle prior to discard. The second factor represents the probability that a previously used needle currently in the possession of an uninfected sharer is

infectious.

The infectious fraction of uninfected sharers' used needles (denoted ϕ) is itself a function of several other factors, and we discuss its derivation below. Before getting to the derivation, however, it is instructive to note certain general properties of ϕ: first, it increases with the infected fraction (or seroprevalence) of sharers. Consequently, an increase in seroprevalence implies an increase in the risk of infection, which, in the model, may lead to a further or faster increase in seroprevalence. A self-reinforcing feedback loop therefore exists in the model, which is typical of epidemics in general. Second, ϕ also increases with the number of uses per needle and with a parameter representing the openness of sharing among cliques, as one might expect. Third, it decreases with the effective cleaning fraction and with the relative frequency of natural self-disinfection of contaminated needles (specifically, the average time between uses of a needle divided by the average time required for self-disinfection).

Analysis of Shared-Needle Flows

The expression for ϕ was derived through a separate analysis of equations that describe a system of shared-needle flows (appendix) whose basic structure is diagrammed in Figure 2. In particular, this expression is obtained by assuming that the needle system is in equilibrium—with all four used-needle stocks pictured in Figure 2 unchanging over time—and solving for a ratio that expresses the infectious fraction of uninfected sharers' needles.

Although the 15 flows and four stocks pictured in Figure 2 are subject to continuous perturbations due to changes in the

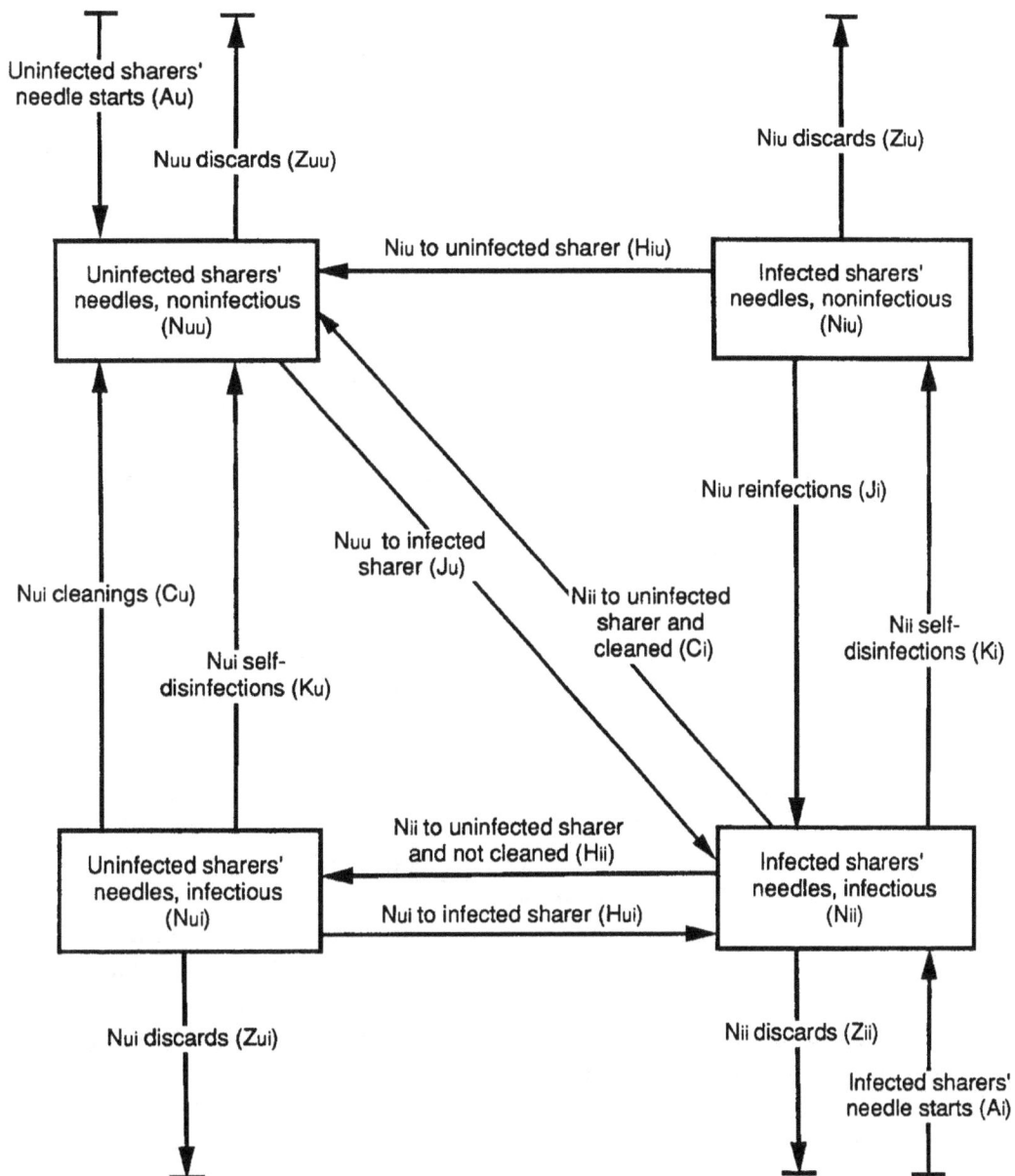

Figure 2: Shared-needle stocks and flows. Flows expressed in needles per day.

sharer population and its seroprevalence, simulations of this needle system (performed using a computation interval of $\frac{1}{4}$ day) suggest that it adjusts fully to such perturbations in two or three weeks' time at most. This short adjustment time is determined primarily by the average lifetime of a needle before it is discarded and by the time that it takes users to obtain new needles to replace discards, both of which we estimate at less than a week. Since the adjustment is such a rapid one relative to the population model's time horizon of years, it appears reasonable to make the simplifying assumption that the infectious fraction of uninfected sharers' needles is equal to its indicated equilibrium value at any point in time.

The system of shared-needle flows, as we have formulated it, requires some explanation. The total stock of sharers' used needles is equal to the number of sharers

---

The IVDU population of New York City has been estimated at 200,000.

---

multiplied by the average stock per sharer, and it is subdivided into four separate categories: at any given point in time, a needle may be either infectious or noninfectious and may be in the possession of either a clique of uninfected sharers or a clique of infected sharers. Needles enter the stock of used needles after their initial use—that is, through fresh needle starts—and leave it through discards. Total discards per day are determined by dividing the total number of needle uses per day (equal to the number of sharers multiplied by their daily frequency of use) by the average number of uses per needle; each of the four used needle categories is then allocated its proper fraction of these discards. Total needle starts are, in our assumed equilibrium, equal to total discards and are divided proportionally between infected and uninfected sharers; those needles started by infected sharers immediately become infectious, while needles

started by uninfected sharers are initially noninfectious.

After a needle has been started and before it is discarded, it may move among the four needle categories, doing so as a result of reuse or as a result of natural self-disinfection. Infectious needles may self-disinfect between uses and require some average time period to do so; there is one self-disinfection flow for infectious needles in the possession of infected cliques and one for those in the possession of uninfected cliques.

Every time a needle is reused, it either remains within the clique of its previous use, or it moves to another clique. Cross-clique sharing is assumed to occur with a fixed probability per needle use (the openness of sharing) that applies equally to all sharer cliques. When a needle is shared from one clique to another, its probability of going to a clique of infected sharers is simply equal to the fraction of such sharers, while its probability of going to a clique of uninfected sharers equals one minus this fraction.

The shared-needle system includes seven reuse-related flows. Two of these flows comprise changes in category that may occur when a needle remains with the same clique or is shared to another clique having the same infection status. They include the effective cleaning of infectious needles in the possession of uninfected cliques (C_u); and the reinfection of noninfectious (self-disinfected) needles in the possession of infected cliques (J_i). The system's five other reuse-related flows represent movements of needles from uninfected sharer cliques to infected sharer cliques, or vice versa. Moving from an uninfected clique to an infected clique, a non-infectious needle immediately becomes infectious (J_u), while an infectious needle remains infectious (H_{ut}). Moving from an infected clique to an uninfected clique, a noninfectious needle remains noninfectious (H_{tu}), while an infectious needle either becomes noninfectious via effective cleaning (C_t) or otherwise remains infectious (H_n).

Parameter Values

We assigned the model's 14 input parameters baseline values (appendix) based on a combination of published figures, personal communications with researchers and clinicians, and educated guesswork. New York City, with its large IVDU population and a fairly well-documented history of HIV spread among IVDUs, was taken as the target area for model calibration and behavioral validation. We attempted to employ data specific to New York City or the urban Northeast wherever possible, but due to the unavailability of such data in many cases, we also employed data based on studies and observations from other locales. Each parameter is presented below, in the order of its appearance in the appendix, with its assumed baseline value and explanatory citations.

The initial IVDU population is 200,000: The IVDU population of New York City has been estimated at 200,000 [Kaplan 1989a]. One may compare this large figure with the estimated 10,000 to 12,000 IVDUs in San Francisco [Chaisson, Moss, et al. 1987].

The initial sharer fraction of the IVDU population is 0.6: Surveys of IVDUs have found that 40 to 89 percent report some needle sharing during the past month or year [OTA 1988; Chaisson, Bacchetti, et al. 1989; Magura et al. 1989]. More specific to the definition used in our model, a current survey in Los Angeles [UCLA 1990] finds 47 percent of IVDUs reporting that they shared needles, at least on occasion, with individuals outside their usual small group. We have boosted this figure to 60 percent to account for the reported 50 to 60 percent seroprevalence among IVDUs in New York City noted previously.

The factor for sharer entry is 1: As stated above, we assume that sharer entries remain equal to the initial sharer outflow of quits and standard losses.

The infection risk per infectious needle use is 0.01: A seroconversion rate of 0.6 percent has been found among health care workers accidentally exposed to HIV through needles and other sharp objects [CDC 1988]. We have increased this figure somewhat to account for the greater risk that would seem to apply to intentional intravenous injection by IVDUs, who may require more than one needle stick to find a usable vein.

The number of uses per sharer per day is 2.5: Most heroin and methedrine addicts inject two to three times per day [Murphy 1987]. Cocaine injectors, on the other hand, were found in one study [Gawin and Kleber 1985] to average 1.6 "runs" (binges) per week, with the average run lasting 6.9 hours, and with one to three injections per hour; multiplying these numbers gives 11 to 33 injections per week.

The effective cleaning fraction of shared uses is 0.1: A study in San Francisco [Becker and Joseph 1988] found that, in 1985 (prior to an intensive needle-cleaning campaign in 1986), only six percent of needle sharers reported sterilizing their needles with bleach.

The number of uses per needle before discard is 15: In a Los Angeles-based study [UCLA 1990], needle sharers reported a wide range of uses per needle, with a median of eight and a mean of 20; however, if one omits the three percent of respondents who reported more than 100 uses per needle, a less skewed distribution is obtained which has a mean of 15.

The cross-clique sharing probability per use, or openness of sharing, is 0.2: We are aware of no attempts to estimate this parameter, which represents the fraction of the time that sharers (those who share at least occasionally outside of the clique) do, in fact, share outside of the clique. Moreover, it may be expected to vary widely from one urban area to another. After assigning all other parameter values, we adjusted this parameter until the model generated an exponential growth rate in HIV-related deaths similar to that observed for AIDS cases in the urban Northeast US during the early 1980s [Kaplan 1989a].

The needle stock per sharer is 0.5: This parameter value may vary from one area to another. The only pertinent data we have found come from a Los Angeles-based study [UCLA 1990], in which needle sharing cliques of one to five individuals are reported, with a mean of 2.3. We assume that a typical clique of two individuals has one used needle in its possession on the average.

The needle self-disinfection time is two days: Active HIV survives for up to three days in dried material at room temperature [Resnick et al. 1986].

The quit rate for uninfected sharers is 0.11 per year: IVDUs may quit drug use either following drug treatment or without the benefit of treatment. The rate of quitting may vary widely with such factors as age and criminality and is the subject of continuing discussion [Anglin et al. 1986]. About 11 percent of IVDUs in the US receive drug treatment each year [DHHS 1990], and 30 to 50 percent of these subsequently relapse to addictive use [Anglin and Hser 1990]; thus, drug treatment by itself may account for IVDU quits on the order of five to eight percent per year. In contrast, a California-based study [Biernacki 1986] of opiate addicts (largely Caucasian and reporting few criminal arrests) who quit without treatment found a mean duration of addiction of less than six years, an average quit rate of over 17 percent per year. However, it appears that the majority of IVDUs do eventually receive treatment [Iguchi 1990]; and we assume that only 25 percent of needle sharers quit without ever having been treated. Thus, we estimate quits without treatment to be four to five percent of IVDUs per year.

The quit rate for infected sharers is 0.13 per year: The responses of IVDUs to the news of being HIV infected vary widely; some individuals sense a new urgency in their lives and become more open to addiction treatment, while others become more hopeless and resistant to treatment [Young 1990]. We assume that the quitting rate for infected sharers aware of their infection is, on the average, 50 percent greater than that for uninfected sharers, but that only one-third of infected sharers are aware that they are infected.

The standard loss rate for population is 0.015 per year: Standard loss is defined to include both non-HIV death and out-migration; non-HIV deaths may occur due to drug overdose, natural causes, or violence. A study of arrested narcotics addicts in California [Anglin, Hser, et al. 1988] provides data on a single population over the course of nearly 30 years, with transitions including addiction, incarceration, treatment, death, and loss from the study. From these data, we estimate an average annual death rate of one percent and an average annual study loss rate of about one-half percent. Several other studies [for example, Vaillant 1973; Concool et al. 1979], mostly preceding the AIDS era, have examined mortality in narcotics addicts and have generally found rates in the range of one to two percent per year.

The time from HIV transmission to death is nine years: this period consists of the mean incubation time from HIV transmission to clinical manifestation, plus the mean survival time from clinical manifestation to death. Estimates of the mean incubation time appear to have settled around seven to eight years, independent of the transmission mode [Anderson 1988; CDC 1987]. We assume an average survival time of one to two years, which is typical for victims who lack effective treatment [Ward, Hardy, and Drotman 1987] and whose health is compromised by intravenous drug use [Rothenberg et al. 1987]. Survival times may be increased with the drug AZT (also known as zidovudine or ZDV), which has proved effective in slowing progression of the disease in both AIDS and AIDS-related complex [Cimons 1989].

Baseline Model Behavior

The model's baseline behavior patterns are graphed in Figures 3 to 5 over a time horizon of 16 years; this time horizon is sufficient to see the system move well toward ultimate equilibrium. With an assumed initial sharer seroprevalence of 0.01 percent, the model starts with only 12 infected sharers out of a population of 120,000 needle sharers and 200,000 IVDUs in total.

Figure 3 shows three key population categories and the annual incidence of new infections. New infections increase rather little during the first year, but then grow during the second year, reach a peak of about 100,000 per year, and decline during the third and fourth years. By the fifth year, new infections have declined to about 14,000 per year, and they remain nearly constant at that level thereafter. The concentration of new infections during the second and third years results in a rapid rise in the number of infected sharers during those years. The population of infected sharers peaks in the fourth year and declines steadily thereafter due to HIV deaths and quits. The total infected population, which includes both current and former sharers, continues to grow until the seventh year before declining due to HIV deaths. Former sharers grow in number and account for about half of the infected population from the twelfth year onward.

Figure 4 shows three key population fractions and the infectious fraction of uninfected sharers' used needles. We have previously described how infected sharers and infectious needles may act to reinforce each other's growth. As a result, by the fourth year the infected fraction of sharers exceeds 90 percent and the infectious fraction of uninfected sharers' used needles

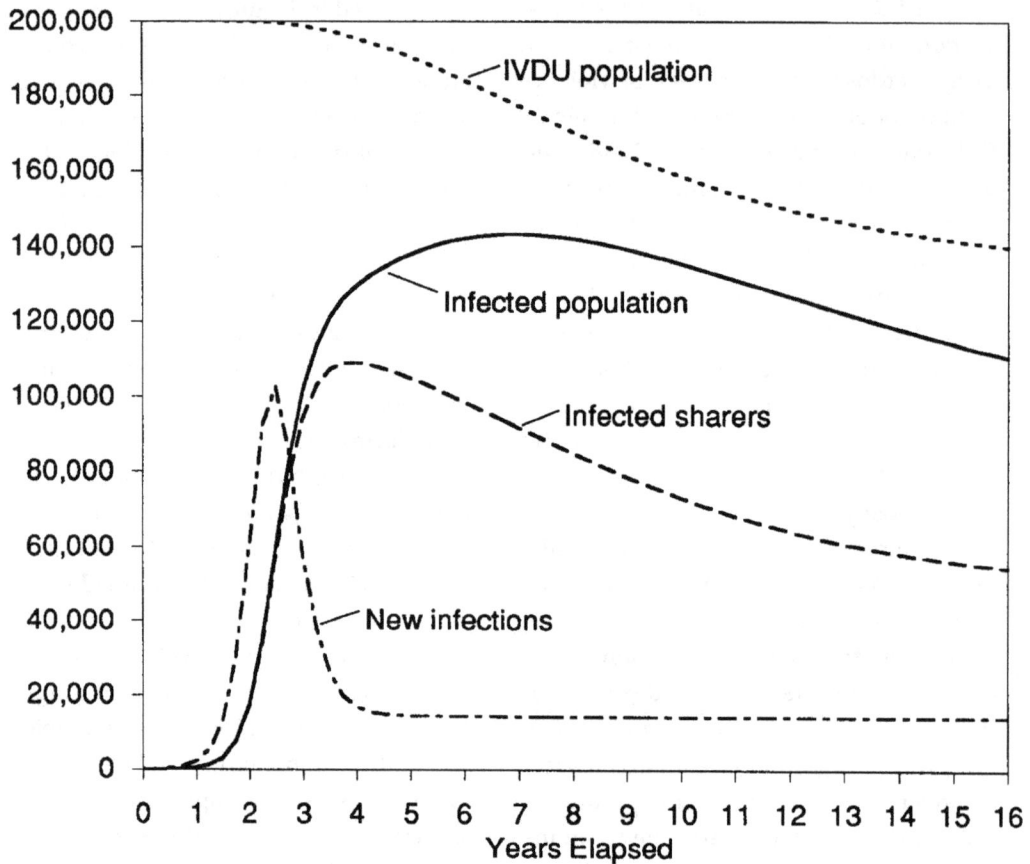

Figure 3: Baseline behavior patterns for population categories and new infections per year.

exceeds 33 percent; and these high levels are maintained for the remainder of the simulation. Since relatively few HIV deaths have occurred by the fourth year, the sharer fraction of IVDUs is still near its initial value of 60 percent; as a result, the infected fraction of total IVDUs is able to grow to as much as 56 percent in the baseline simulation. But as HIV deaths cause the sharer population to decline, the loss becomes reflected in reductions in the sharer fraction of IVDUs and the infected fraction of IVDUs. By the 16th year, a net loss of 60,000 sharers has occurred, the sharer fraction is reduced to 43 percent,

and the infected fraction of IVDUs has declined to 39 percent.

Figure 5 shows two categories of HIV deaths, along with a rescaled graph of new infections for comparison. Total HIV deaths, which represent the sum of sharer HIV deaths and former sharer HIV deaths, follow a course which is dictated by our use of a third-order exponential delay and an average transmission-to-death time of nine years. Following the growth of new infections during the second and third years, HIV deaths climb most rapidly during the fourth through the ninth years and reach their peak of about 17,000 in the

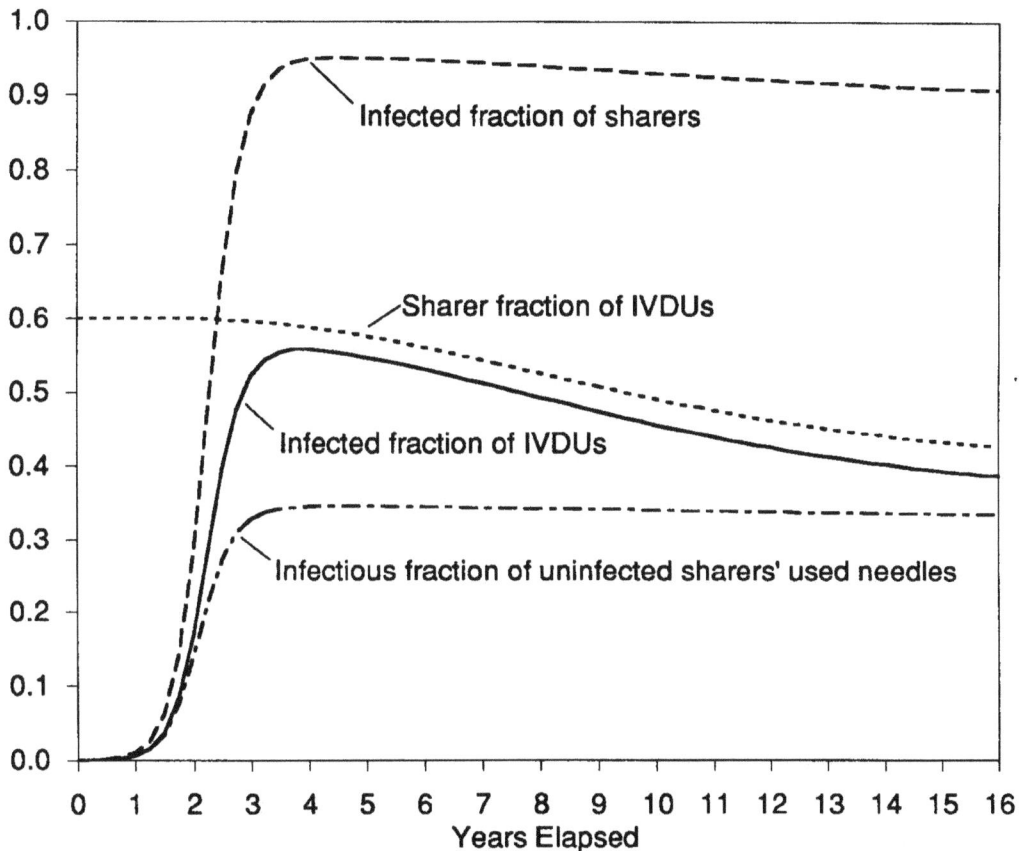

Figure 4: Baseline behavior patterns for population fractions and the infectious fraction of un-infected sharers' used needles.

twelfth year, nine years after the peak in new infections. HIV deaths then begin a gradual decline, converging toward the number of new infections because of our assumption that every HIV infection leads ultimately to death. From Figure 5, it is also apparent that former sharers grow to comprise a larger and larger fraction of HIV deaths; indeed, former sharers account for about half of all HIV deaths from the twelfth year onward.

Analysis of a Policy

The use of a dynamic model such as ours for policy analysis typically involves the modification of input parameter values, including both those parameters directly related to a given policy and those parameters whose values may differ from one setting or population to another. Although simulation studies have often found policy results to be insensitive to questions of calibration [Richardson and Pugh 1981], such as the exact size or composition of a population, such insensitivity is not always the case and should be investigated. Even for a model of limited size, such as ours, many simulations may be required to thoroughly analyze a given policy. Therefore, we seek in this paper only to give an illustration of such policy analysis.

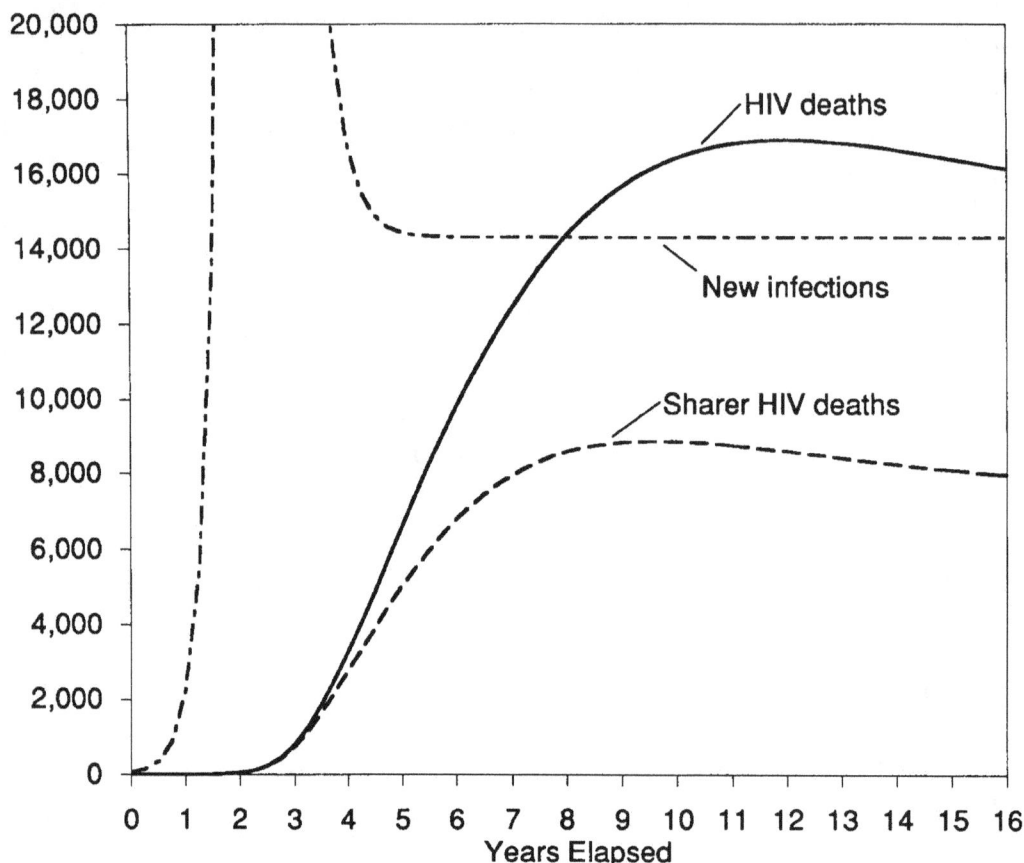

Figure 5: Baseline behavior patterns for HIV deaths and new infections per year.

We have performed a series of simulations to investigate the potential benefits of a campaign to increase the effective cleaning of shared needles. In San Francisco, an intensive campaign throughout 1986 by public health agencies helped to increase the fraction of sharers who reported sterilizing their needles with bleach from six percent to 47 percent [Becker and Joseph 1988]. Our simulations were designed to address the following questions regarding the ultimate effectiveness of such a needle-cleaning campaign in reducing HIV infections and deaths:

(1) How important is the magnitude of the campaign?

(2) How important is the timeliness of the campaign?

(3) How might the openness of needle sharing affect the impacts of the campaign's magnitude and timeliness?

(4) If the campaign results in a greater attraction to needle sharing, to what extent might its benefits be undermined?

Table 1 presents a summary of results from the baseline simulation and from twelve simulations of alternative policy scenarios. Figures 6 to 8 show, for most of these scenarios, the patterns of HIV deaths over time compared to the baseline pat-

Scenario and Assumptions	Cum Infections	At 16th Year Cum Deaths	IVDUs
1. Baseline ($c = .1, \rho = .2, \epsilon = 1$)	280,535	170,110	139,991
2. $c = .3$ throughout	213,210	93,220	154,006
3. $c = .5$ throughout	7,838	867	199,271
4. $c = .5$ from 2nd year	182,930	94,216	160,977
5. $c = .5$ from 4th year	222,115	145,434	152,420
6. $\rho = .1$	244,654	129,276	146,211
7. $\rho = .1; c = .5$ from 3rd year	26,231	11,877	194,722
8. $\rho = .1; c = .5$ from 4th year	84,371	44,095	181,753
9. $\rho = .05$	173,105	48,135	168,536
10. $\rho = .05; c = .5$ from 4th year	1,109	754	199,751
11. $\rho = .05; c = .5$ from 8th year	31,472	15,322	192,766
12. $c = .5$ and $\epsilon = 1.5$ from 2nd year	231,801	107,925	201,658
13. $c = .5$ and $\epsilon = 2$ from 2nd year	278,908	120,505	242,965

Table 1: Scenarios and summary statistics for the analysis of a needle-cleaning campaign. Scenarios differ according to the effective cleaning fraction (c), the openness of sharing (ρ), the factor for sharer entry (ϵ), and the timing of changes c and ϵ.

tern. Each scenario involves changes in one or two of the following parameters: the effective cleaning fraction (c), the openness of sharing (ρ), and the factor for sharer entry (ϵ). For each scenario in Table 1, we report three summary (16th year) statistics, namely, cumulative HIV infections, cumulative HIV deaths, and the IVDU population. Not surprisingly, a reduction in cumulative deaths is associated with a reduction in cumulative infections. Also, with the exception of the last two scenarios (where sharer entry has been boosted), a larger IVDU population is associated with a reduction in cumulative deaths; that is, with fewer deaths more IVDUs remain in the population.

Scenarios 2 through 5 (Table 1 and Figure 6) address the first two questions above regarding the importance of the needle-cleaning campaign's magnitude and timeliness. Each one of these scenarios involves a lasting stepwise increase in the cleaning fraction (c), attributable to the campaign, above its baseline value of 10 percent; the four scenarios differ only according to the magnitude and timing of this increase. In Scenario 2, c is increased to 30 percent for all 16 years of the simulation. In Scenario 3, c is again increased for the entire simulation, but to a greater value of 50 percent. In Scenario 4, c is again increased to 50 percent, but not until the second year, a time at which new infections are on the rise in the baseline simulation. In Scenario 5, c is increased to 50 percent, but this time not until the fourth year, by which time the epidemic burst of new infections is essentially complete.

An examination of the results from Scenarios 2 through 5 makes clear the critical importance of both the magnitude and the timeliness of the cleaning campaign. While the moderate increase in cleaning of Scenario 2 does result in a reduction of cumulative infections and deaths, it does so

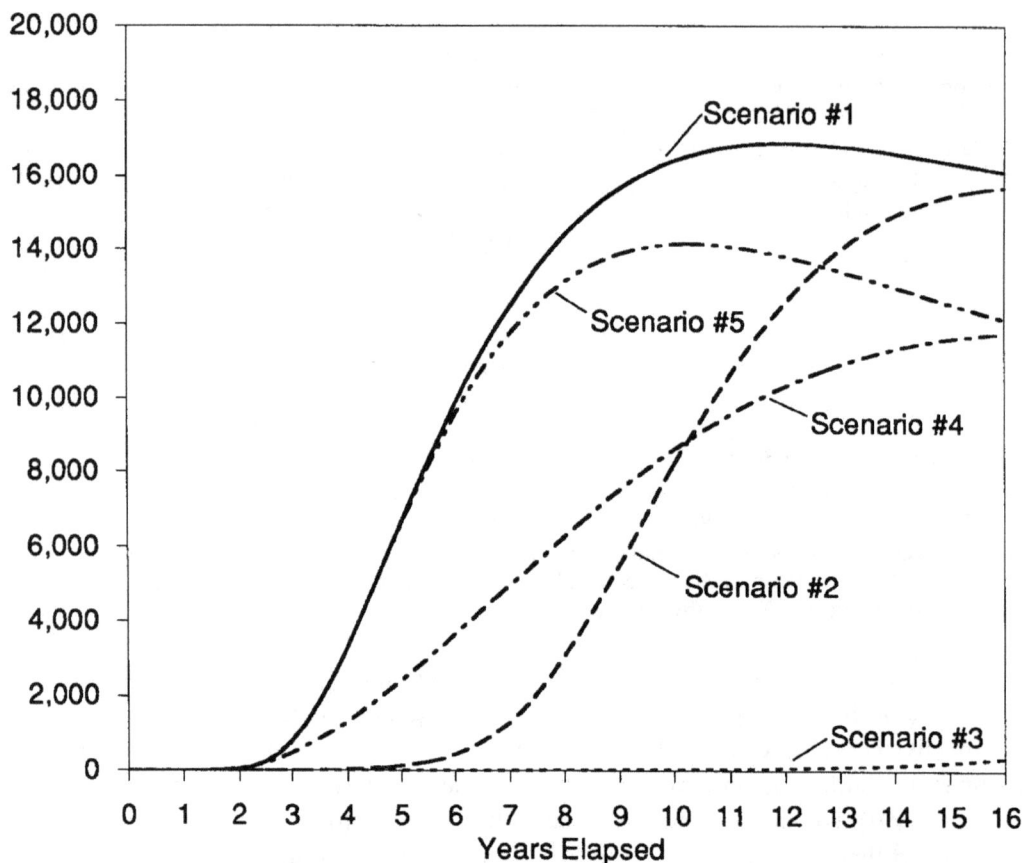

Figure 6: HIV deaths per year under alternative scenarios (Table 1), examining the impacts of magnitude and timing on the results of a cleaning campaign. The cleaning fraction is increased throughout #2 and #3, from the second year in #4, and from the fourth year in #5.

by delaying rather than actually preventing the rapid transmission of HIV. In contrast, the larger campaign of Scenario 3 is sufficient to effectively curtail the epidemic throughout the 16-year time horizon. However, when, in Scenario 4, this same larger campaign is delayed until the second year (by which time 17,000 infections have occurred), the cumulative results are little better than those of Scenario 2. And when, in Scenario 5, the larger campaign is delayed until the fourth year (by which time 132,000 infections have occurred), the re-

sults are decidedly worse than the moderate but very timely campaign of Scenario 2. In summary, to be truly effective—at least in our baseline setting—the needle-cleaning campaign should be sufficiently large in its direct impact and must be implemented before the spread of HIV is already well under way.

To what extent is this conclusion generalizable to sharer populations whose pattern of needle sharing may be less open than that assumed in our baseline? Scenarios 6 through 11 (Table 1 and Figure 7) ad-

May–June 1991

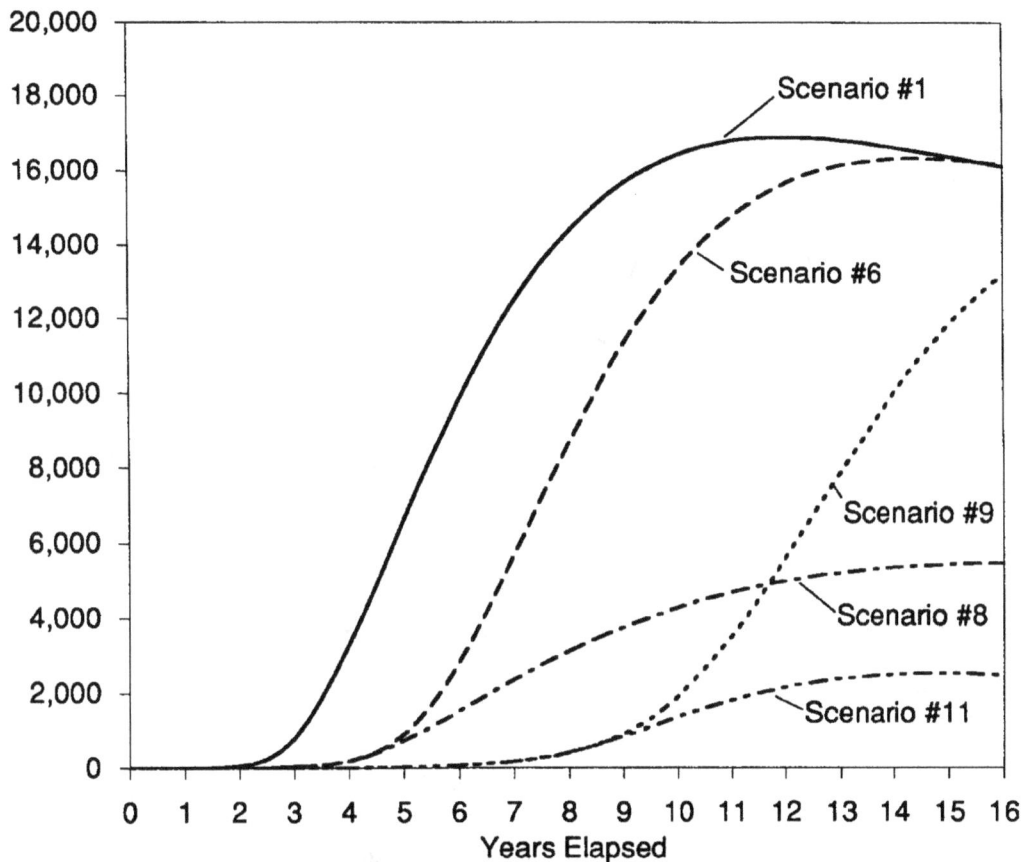

Figure 7: HIV deaths per year under alternative scenarios (Table 1), examining the impact of the openness of sharing on the results of a cleaning campaign. The openness of sharing is reduced in #6, #8, #9, and #11. The cleaning fraction is increased from the fourth year in #8 and from the eighth year in #11.

dress this question. Each one of these scenarios involves a reduction in the openness of sharing (ρ) below its baseline value of 20 percent for the entire simulation: In Scenarios 6, 7, and 8, ρ is reduced to 10 percent, while in Scenarios 9, 10, and 11, ρ is reduced to five percent. In Scenarios 6 and 9, no needle-cleaning campaign is assumed, while in Scenarios 7, 8, 10, and 11, the cleaning fraction is increased to 50 percent at different points in time.

Scenarios 6 and 9 demonstrate that when the openness of sharing is reduced,

the period of very rapid HIV transmission is delayed though not eliminated. (In other simulations we have found that the epidemic is permanently avoided only when the openness of sharing is extremely small, on the order of two percent or less.) Under such conditions, it appears that the timeliness of the needle-cleaning campaign becomes a less pressing issue than it is in more openly sharing populations. In Scenarios 8 and 11, the needle-cleaning campaign is initiated when new infections are on the rise and something on the order of

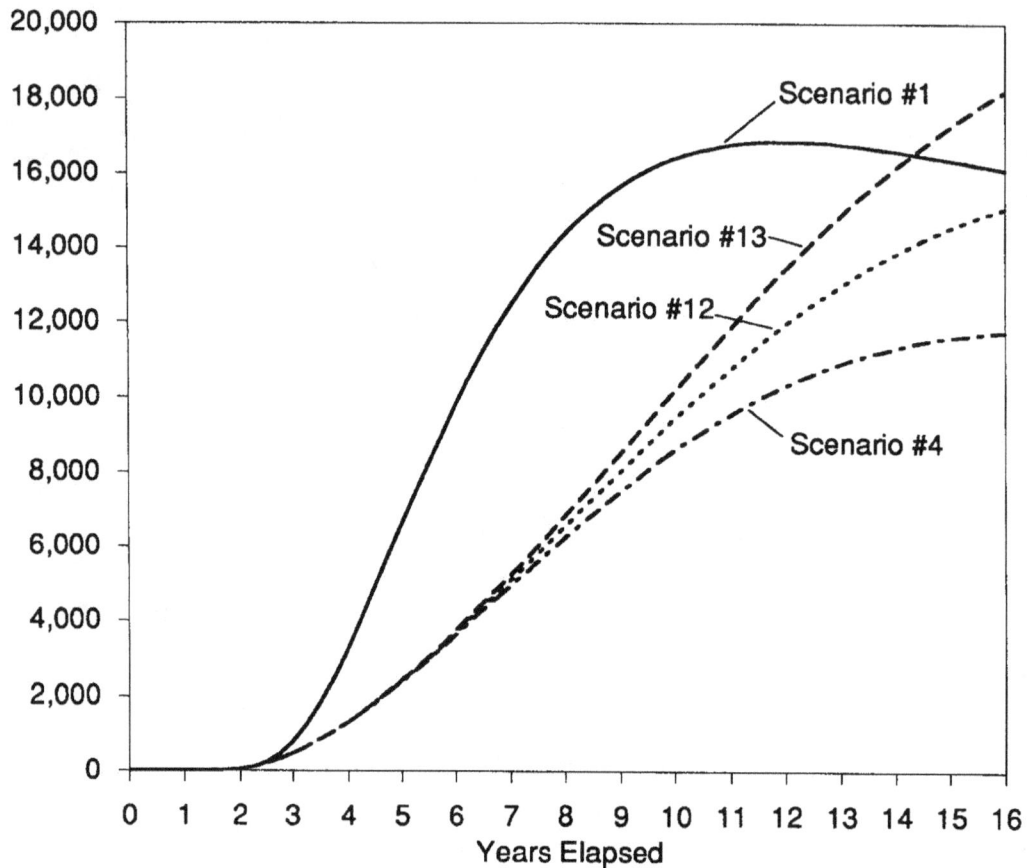

Figure 8: HIV deaths per year under alternative scenarios (Table 1), examining the impact of relative sharer entry on the results of a cleaning campaign. The cleaning fraction is increased from the second year in #4, #12, and #13. The factor for sharer entry is increased from the second year in #12 and #13.

20,000 have already occurred; these conditions are similar to those seen previously in Scenario 4. These early epidemic conditions are increasingly delayed as the openness of sharing is reduced, thereby giving policymakers more time to react. Furthermore, a comparison of Scenarios 4, 8, and 11 (Table 1) suggests that the efficacy of a needle-cleaning campaign implemented during the growth spurt is greater when the openness of sharing is reduced, even though the campaign is implemented at a later point in time. Not only does reduced openness of sharing delay the epidemic, but it also slows it down, making it less virulent and more responsive to intervention.

To what extent might the potential benefits of a needle-cleaning campaign be undermined by a resulting increase in the inflow of needle sharers? Scenarios 12 and 13 (Table 1 and Figure 8) address this question, making what seem to be rather pessimistic assumptions for the sake of argument. (We have found no actual evidence to suggest that a needle-cleaning

May–June 1991

campaign will in fact increase the attraction of IV drug use, although this has been a popular objection to such a policy.) These scenarios take as a starting point the assumptions of Scenario 4, in which a cleaning campaign implemented in the second year increases the cleaning fraction from 10 percent to 50 percent. They then additionally assert that the campaign will have the adverse consequence of increasing the inflow of new sharers (thereby increasing the total IVDU population): In Scenario 12, sharer entries are increased (from a baseline value of 15,000 per year) by 50 percent starting in the second year (ϵ is increased to 1.5), while in Scenario 13, they are increased by 100 percent (ϵ is increased to two).

The results of Scenarios 12 and 13 are open to different interpretations. From the standpoint of trying to reduce HIV infections and deaths over the span of 16 years, the needle-cleaning campaign remains attractive even under the strongly pessimistic assumptions of Scenario 13. Figure 8 does indicate that, under Scenario 13, the incidence of HIV deaths eventually grows to exceed that of the baseline scenario, but the crossover does not occur until after the 14th year. In essence, many years are required for the accumulation of new sharers finally to overcome, in terms of absolute numbers, the immediate and lasting reduction of infection risk from increased needle cleaning. On the other hand, one may argue that the risks of IV drug use go beyond HIV infection to include various drains on personal and social resources, such as those resulting from criminal activity, underemployment, and general ill health. From this standpoint, the needle-

cleaning campaign may logically be opposed if one finds that the increase in IVDUs evident in Scenario 12 or Scenario 13 has costs that exceed the benefits of reducing HIV infections and deaths.

Conclusion

Our model is intended to capture the essential features of HIV transmission through needle sharing within a population of IVDUs. This mode of transmission is conceptually more complex than sexual transmission, because it involves an intermediating agent in the form of infectious needles. We have found that one must therefore consider both population flows and needle flows when modeling this phenomenon. Because the adjustment of the needle stock takes place over days or weeks, compared with the multiple-year dynamics of the sharer population, one may model the infectious fraction of uninfected sharers' used needles as an instantaneously adjusting function of other parameters, without tracking needle flows per se. However, the correct formulation of this function still requires a detailed analysis of needle flows, albeit in equilibrium.

In addition to paying attention to realistic details of cause and effect, we have attempted to gather the best information available for calibrating the model. For some of the model's input parameters, published data are sparse or lacking, which has led us to rely upon a certain amount of expert judgment and guesswork. We are hopeful that models such as ours will help to identify those parameters, such as the openness of sharing and the used needle stock per sharer, most in need of data collection and estimation efforts, from the standpoint of making more accurate projections for public policy decisions.

We have presented a set of simulations that illustrate the use of the model for understanding basic behavior and for policy analysis. In these simulations, the infected population grows exponentially at first, but eventually reaches a peak and declines gradually thereafter. The exponential growth is the result of a self-reinforcing feedback loop connecting infected sharers and infectious needles, while the decline is primarily the result of HIV deaths, which typically occur after a long delay following transmission of the virus. HIV deaths may affect both current and former needle sharers; indeed, the simulations suggest that HIV deaths among former sharers may increase over time to account for a large proportion of total HIV deaths attributable to needle sharing in a given area.

In regard to policy, our analysis of a needle-cleaning campaign suggests that the epidemic can be curtailed if the campaign has sufficient direct impact on the cleaning fraction, and if it is implemented prior to the peak in new infections. But the analysis also suggests that the timeliness of policy implementation may be a more pressing issue for certain populations than for others. For example, timeliness becomes more critical to the policy's effectiveness when the population's characteristic pattern of sharing is sufficiently open. Also, our analysis suggests that the needle-cleaning campaign may be defensible, over a relatively generous time frame, even if one assumes that it will attract more individuals into addictive IV drug use. This finding is related to the fact that the benefits of the needle-cleaning campaign (in terms of reducing HIV transmission) begin immediately, while the disadvantages build only gradually over time.

Our model may be used to examine policy interventions other than a needle-cleaning campaign. For example, a needle exchange policy may be simulated by appropriately reducing the value of the parameter representing the average number of uses per needle. With appropriate changes in other parameter values, one may also simulate policies that effectively increase the rate of quitting (for example, through expanded drug treatment), that slow the inflow of new sharers (for example, through targeted educational programs), or that extend the time from HIV transmission to death (for example, through increased availability of AZT).

Although we believe that our model's structure, parameter values, and dynamic behavior are credible, and that the model is sufficiently flexible to consider a variety of populations and policies, it should be recognized that the model is a simplified representation of a rather complex reality. Consequently, the model may not be adequate for analyzing certain policy questions, such as those related to different approaches to drug treatment, different approaches to law enforcement and legal supervision, or different approaches to preparing the health care system for the increasing burden of HIV patients. Nonetheless, we must urge policymakers to rely upon the best models and sources of information currently available, despite the inevitable limitations, so that they may consider carefully the future implications of their actions and act both appropriately and expeditiously.

May–June 1991

HOMER, ST. CLAIR

Acknowledgments

We acknowledge the contribution of two anonymous referees who provided detailed comments on an earlier version of this paper. We are also indebted to the UCLA Drug Abuse Research Group for providing us with a variety of documents, research data, and valuable contacts with other researchers and clinicians. Finally, we wish to thank those personal contacts who gave us their observations for the record.

APPENDIX

Definitions of variables in the population model (flows expressed in people per year):

P_{su} = Uninfected sharers;
P_{si} = Infected sharers;
P_{fi} = Infected former sharers;
$CumD$ = Cumulative HIV deaths;
$CumW$ = Cumulative infections;
P_s = Sharer population;
P = IVDU population;
F_{ts} = Infected fraction of sharers;
F_{ip} = Infected fraction of IVDUs;
E = Sharer entries;
W = New infections;
ω = Infection risk per day;
β = Infection risk per use;
ϕ = Infectious fraction of uninfected sharers' used needles;
Q_u = Uninfected sharer quits;
Q_i = Infected sharer quits;
L_{su} = Standard loss of uninfected sharers;
L_{si} = Standard loss of infected sharers;
L_{fi} = Standard loss of infected former sharers;
D_s = Sharer HIV deaths;
D_f = Former sharer HIV deaths; and
D = HIV deaths.

Definitions of variables in the system of shared needles (flows expressed in needles per day):

N_{uu} = Uninfected sharers' needles, noninfectious;
N_{ui} = Uninfected sharers' needles, infectious;
N_{ii} = Infected sharers' needles, infectious;
N_{iu} = Infected sharers' needles, noninfectious;
N = Sharers' used needle stock;
Z = Sharers' used needle discards;
Z_{uu} = Uninfected sharers' noninfectious needle discards;
Z_{ui} = Uninfected sharers' infectious needle discards;
Z_{ii} = Infected sharers' infectious needle discards;
Z_{iu} = Infected sharers' noninfectious needle discards;
A_u = Uninfected sharers' needle starts;
A_i = Infected sharers' needle starts;
R = Sharers' needle reuses;
C_u = Uninfected sharers' infectious needle cleanings;
C_i = Infectious needles moved to uninfected sharer and cleaned;
K_u = Uninfected sharers' needle self-disinfections;
K_i = Infected sharers' needle self-disinfections;
J_u = Noninfectious needles moved to infected sharer;
J_i = Infected sharers' needle reinfections;
H_{ui} = Infectious needles moved to infected sharer;
H_{ii} = Infectious needles moved to uninfected sharer and not cleaned; and
H_{iu} = Noninfectious needles moved to uninfected sharer.

Definitions of input parameters:

P_0 = Initial IVDU population;
F_{sp0} = Initial sharer fraction of IVDU population;
ϵ = Factor for sharer entry (relative to initial sharer outflow);
α = Infection risk per infectious needle use;
ν = Uses per sharer per day;
c = Effective cleaning fraction of shared uses;

NEEDLE SHARING

μ = Uses per needle before discard;

ρ = Cross-clique sharing probability per use (openness of sharing);

n = Used needle stock per sharer;

T_k = Needle self-disinfection time (days);

ψ_u = Quit rate for uninfected sharers (1/year);

ψ_i = Quit rate for infected sharers (1/year);

λ = Standard loss rate for population (1/year); and

T_d = Time from HIV transmission to death (years).

The population model:

$$dP_{su}/dt = E - W - Q_u - L_{su}, \qquad (1.1)$$

$$dP_{si}/dt = W - Q_i - D_s - L_{si}, \qquad (1.2)$$

$$dP_{fi}/dt = Q_i - D_f - L_{fi}, \qquad (1.3)$$

$$dCumD/dt = D, \qquad (1.4)$$

$$dCumW/dt = W, \qquad (1.5)$$

$$P_s = P_{su} + P_{si}, \qquad (1.6)$$

$$P = P_s + P_0[1 - F_{sp0}], \qquad (1.7)$$

$$F_{is} = P_{si}/P_s, \qquad (1.8)$$

$$F_{ip} = P_{si}/P, \qquad (1.9)$$

$$E = \epsilon[\psi_u + \lambda]F_{sp0}P_0, \qquad (1.10)$$

$$W = 365\omega P_{su}, \qquad (1.11)$$

$$\omega = 1 - [1 - \beta]^\nu, \qquad (1.12)$$

$$\beta = \alpha[1 - c][1 - [1/\mu]]\phi, \qquad (1.13)$$

$$\phi = \frac{F_{is}\rho[1 - c][1 - [1/\mu]]}{[1 + [[n/\nu]/T_k]]\{[F_{is}\rho[1 - c] + c] \atop \times [1 - [1/\mu]] + [[n/\nu]/T_k] + [1/\mu]\}}, \qquad (1.14)$$

$$Q_u = \psi_u P_{su}, \qquad (1.15)$$

$$Q_i = \psi_i P_{si}, \qquad (1.16)$$

$$L_{su} = \lambda P_{su}, \qquad (1.17)$$

$$L_{si} = \lambda P_{si}, \qquad (1.18)$$

$$L_{fi} = \lambda P_{fi}, \qquad (1.19)$$

$$D_s = DP_{si}/[P_{si} + P_{fi}], \qquad (1.20)$$

$$D_f = DP_{fi}/[P_{si} + P_{fi}], \qquad (1.21)$$

$$D = \text{3rd-order Delay } (W, T_d). \qquad (1.22)$$

The system of shared needles:

$$dN_{uu}/dt = A_u + C_u + C_i + K_u + H_{iu} - J_u - Z_{uu}, \qquad (2.1)$$

$$dN_{ui}/dt = H_{ii} - C_u - K_u - H_{ui} - Z_{ui}, \qquad (2.2)$$

$$dN_{ii}/dt = A_i + J_i + J_u + H_{ui} - C_i - K_i - H_{ii} - Z_{ii}, \qquad (2.3)$$

$$dN_{iu}/dt = K_i - J_i - H_{iu} - Z_{iu}, \qquad (2.4)$$

$$N = N_{uu} + N_{ui} + N_{ii} + N_{iu} = nP_s, \qquad (2.5)$$

$$Z = \nu P_s/\mu, \qquad (2.6)$$

$$Z_{uu} = ZN_{uu}/N, \qquad (2.7)$$

$$Z_{ui} = ZN_{ui}/N, \qquad (2.8)$$

$$Z_{ii} = ZN_{ii}/N, \qquad (2.9)$$

$$Z_{iu} = ZN_{iu}/N, \qquad (2.10)$$

$$A_u = Z[1 - F_{is}], \qquad (2.11)$$

$$A_i = ZF_{is}, \qquad (2.12)$$

$$R = \nu P_s - Z = \nu P_s[1 - [1/\mu]], \qquad (2.13)$$

$$C_u = R[N_{ui}/N][1 - \rho F_{is}]c, \qquad (2.14)$$

$$C_i = R[N_{ii}/N]\rho[1 - F_{is}]c, \qquad (2.15)$$

$$K_u = N_{ui}/T_k, \qquad (2.16)$$

$$K_i = N_{ii}/T_k, \qquad (2.17)$$

$$J_u = R[N_{uu}/N]\rho F_{is}, \qquad (2.18)$$

$$J_i = R[N_{iu}/N][1 - \rho[1 - F_{is}]], \qquad (2.19)$$

$$H_{ui} = R[N_{ui}/N]\rho F_{is}, \qquad (2.20)$$

$$H_{ii} = R[N_{ii}/N]\rho[1 - F_{is}][1 - c], \qquad (2.21)$$

$$H_{iu} = R[N_{iu}/N]\rho[1 - F_{is}]. \qquad (2.22)$$

The input parameters' baseline values:

$P_0 = 200{,}000$, $F_{sp0} = 0.6$, $\epsilon = 1$, $\alpha = 0.01$, $\nu = 2.5$, $c = 0.1$, $\mu = 15$, $\rho = 0.2$, $n = 0.5$, $T_k = 2$, $\psi_u = 0.11$, $\psi_i = 0.13$, $\lambda = 0.015$, $T_d = 9$.

References

Anderson, Roy M. 1988, "The role of mathematical models in the study of HIV transmission and the epidemiology of AIDS," *Journal of Acquired Immune Deficiency Syndromes*, Vol. 1, No. 3, pp. 241–256.

Anglin, M. Douglas; Brecht, M. L.; Woodward, J. Arthur; and Bonett, Douglas G. 1986, "An empirical study of maturing out: Conditional factors," *The International Journal of the Addictions*, Vol. 21, No. 2, pp. 233–246.

Anglin, M. Douglas and Hser, Yih-Ing 1990, "Treatment of drug abuse," in *Crime and Justice: An Annual Review of Research*, Volume 13, eds. M. Tonry and J. Q. Wilson, University of Chicago Press, Chicago, Illinois.

Anglin, M. Douglas; Hser, Yih-Ing; Booth, Mary; Speckart, George R.; McCarthy, William J.; Ryan, Timothy; and Powers, Keiko 1988, "The natural history of narcotic addiction: A 25 year follow-up," final report under NIDA Grant R01-DA-03425.

Becker, Marshall H. and Joseph, Jill G. 1988, "AIDS and behavioral change to reduce risk," *American Journal of Public Health*, Vol. 78, No. 4 (April), pp. 394–409.

Biernacki, Patrick 1986, *Pathways from Heroin Addiction: Recovery Without Treatment*, Temple University Press, Philadelphia, Pennsylvania.

Centers for Disease Control (CDC) 1987, "Human immunodeficiency virus infection in the United States: A review of current knowledge," *Morbidity and Mortality Weekly Report*, Vol. 36, No. S-6 (December 18).

Centers for Disease Control (CDC) 1988, *Morbidity and Mortality Weekly Report*, Vol. 37, No. 15 (April).

Chaisson, R. E.; Bacchetti, P.; Osmond, D.; Brodie, B.; Sande, M. A.; and Moss, A. R. 1989, "Cocaine use and HIV infection in intravenous drug users in San Francisco," *Journal of the American Medical Association*, Vol. 261, No. 4 (January 27), pp. 561–565.

Chaisson, R. E.; Moss, A. R.; Onishi, R.; Osmond, D.; and Carlson, J. R. 1987, "Human immunodeficiency virus infection in heterosexual intravenous drug users in San Francisco," *American Journal of Public Health*, Vol. 77, No. 2 (February), pp. 169–172.

Cimons, Marlene 1989, "AIDS-related complex found slowed by AZT," *Los Angeles Times*, August 4, p. 1.

Concool, B.; Smith, H.; and Stimmel, B. 1979, "Mortality rates of persons entering methadone maintenance: A seven-year study," *American Journal of Drug and Alcohol Abuse*, Vol. 6, No. 3, pp. 345–353.

Des Jarlais, D. C.; Tross, S.; and Friedman, S. R. 1987, "Behavioral change in response to AIDS," in *AIDS and Other Manifestations of HIV Infection*, eds. G. P. Wormser, E. Bottone, and R. Stahl, Noyes Publications, Park Ridge, New Jersey.

Department of Health and Human Services, Division of STD/HIV Prevention (DHHS) 1990, *Annual Report: Fiscal Year 1989*, Department of Health and Human Services, Washington, DC.

Gawin, Frank H. and Kleber, Herbert D. 1985, "Cocaine use in a treatment population: Patterns and diagnostic distinctions," in *Cocaine Use in America: Epidemiologic and Clinical Perspectives*, eds. N. J. Kozel and E. H. Adams, NIDA Research Monograph 61, US Government Printing Office, Washington, DC.

Iguchi, Martin 1990, personal communication; Dr. Iguchi is a drug abuse researcher with the Department of Psychiatry, University of Medicine and Dentistry of New Jersey, Camden, New Jersey.

Institute of Medicine 1986, *Confronting AIDS: Directions for Public Health, Health Care, and Research*, National Academy Press, Washington, DC.

Johnson, Norman L. and Kotz, Samuel 1970, *Continuous Univariate Distributions-I*, John Wiley and Sons, New York.

Kaplan, Edward H. 1989a, "Needles that kill: Modeling human immunodeficiency virus transmission via shared drug injection equipment in shooting galleries," *Reviews of Infectious Diseases*, Vol. 11, No. 2 (March–April), pp. 289–298.

Kaplan, Edward H. 1989b, "What are the risks of risky sex? Modeling the AIDS epidemic," *Operations Research*, Vol. 37, No. 2 (March–April), pp. 198–209.

Magura, S.; Grossman, J. I.; Lipton, D. S.; Siddiqi, Q.; Shapiro, J.; Marion, I.; and Amann, K. R. 1989, "Determinants of needle sharing among intravenous drug users," *American Journal of Public Health*, Vol. 79, No. 4 (April), pp. 459–462.

Mascola, L.; Leib, L.; Iwakoshi, K. A.; McAllister, D.; Siminowski, T.; Giles, M.; Run,

G.; Fannin, S. L.; and Strantz, I. H. 1989, "HIV seroprevalence in intravenous drug users: Los Angeles, California, 1986," *American Journal of Public Health*, Vol. 79, No. 1 (January), pp. 81–82.

May, Robert M. and Anderson, Roy M. 1987, "Transmission dynamics of HIV infection," *Nature*, Vol. 326 (March 12), pp. 137–142.

Meadows, Dennis L. 1974, "Delays: Exercise and supplementary notes," in *Study Notes in System Dynamics*, ed. M. R. Goodman, MIT Press, Cambridge, Massachusetts.

Murphy, Sheigla 1987, "Intravenous drug use and AIDS: Notes on the social economy of needle sharing," *Contemporary Drug Problems*, Vol. 14 (Fall), pp. 373–395.

Office of Technology Assessment (OTA) 1988, "How effective is AIDS education?" US Government Printing Office (June), Washington, DC

Resnick, L.; Veren, K.; Salahuddin, S. Z.; Tondreau, S.; and Markham, P. D. 1986, "Stability and inactivation of HTLV-III/LAV under clinical and laboratory environments," *Journal of the American Medical Association*, Vol. 255, No. 14 (April 11), pp. 1887–1891.

Richardson, George P. and Pugh, Alexander III 1981, *Introduction to System Dynamics Modeling with DYNAMO*, The MIT Press, Cambridge, Massachusetts.

Rothenberg, R.; Woelfel, M.; Stoneburner, R.; Milberg, J.; Parker, R.; and Truman, B. 1987, "Survival with the acquired immunodeficiency syndrome," *The New England Journal of Medicine*, Vol. 317, No. 21 (November 19), pp. 1297–1302.

University of California, Los Angeles, Drug Abuse Research Group (UCLA) 1990, Unpublished data developed under grant, "HIV infection and transmission risks among homosexual and heterosexual IVDUs," from National Institute on Drug Abuse (DA-055-89).

Vaillant, G. E. 1973, "A 20-year follow-up of New York narcotic addicts," *Archives of General Psychiatry*, Vol. 29, No. 2, pp. 237–241.

Ward, J. W.; Hardy, A. M.; and Drotman, D. P. 1987, "AIDS in the United States," in *AIDS and Other Manifestations of HIV Infection*, eds. G. P. Wormser, E. Bottone, and R. Stahl, Noyes Publications, Park Ridge, New Jersey.

Young, Mark 1990, personal communication;

Mr. Young is a clinician with the Substance Abuse Services of San Francisco General Hospital, San Francisco, California.

May–June 1991

4

A System Dynamics Model of National Cocaine Prevalence (1993)

Homer J. *System Dynamics Review*, 9(1): 49-78

Related writings

Homer J. A Dynamic Model of Cocaine Prevalence in the United States. In *System Dynamics*, ed. Y Barlas, *Encyclopedia of Life Support Systems* (EOLSS). Developed under auspices of UNESCO, EOLSS Publishers, Oxford, UK. 2004. Available at http://www.eolss.net.

Homer J. A System Dynamics Model for Cocaine Prevalence Estimation and Trend Projection. *Journal of Drug Issues,* 23(2): 251-279, 1993.

Homer J. Projecting the Impact of Law Enforcement on Cocaine Prevalence. *Journal of Drug Issues*, 23(2): 281-295, 1993.

Hser Y-I, Anglin MD, Wickens T, Brecht M-L, Homer J. *Techniques for the Estimation of Illicit Drug-Use Prevalence: An Overview of Relevant Issues*. National Institute of Justice Research Report NCJ 133786, May 1992.

A system dynamics model of national cocaine prevalence

Jack B. Homer

A system dynamics model reproduces a variety of national indicator data reflecting cocaine use and supply over a 15-year period and provides detailed estimates of actual underlying prevalence. Sensitivity testing clarifies the source of observed trends, such as growth in the compulsive use of crack cocaine and decline in the casual use of cocaine powder. Alternative scenarios with possible policy implications are simulated and projected for 12 years, and the results are assessed.

Jack B. Homer is a consultant with clients in business and government and an associate adjunct professor at the University of Southern California. He received a Ph.D. degree from the Sloan School of Management at M.I.T., for a thesis modeling the diffusion of new medical products and technologies. Address: 36 Covington Lane, Voorhees, NJ 08043.

The epidemic of cocaine use in the United States has proven consistently difficult to estimate and project forward in time. Self-report surveys inevitably suffer from biases related to partial coverage and inaccurate reporting, which make them unreliable as stand-alone tools for point estimation. Such surveys are suspect even as tools for trend or slope estimation, because the biases themselves are subject to change. For example, it seems likely that reporting of cocaine use has become less accurate over the last several years as the legal and social risks of such use have increased. Because of the inherent drawbacks of self-report surveys for estimating illicit drug use, researchers have often brought other indicators into play as well, such as data on drug-related morbidity and mortality, arrests and seizures, and price and purity. Although any single indicator provides only a partial, often indirect, view of the situation, the hope exists that taken together they may lead to more reliable estimates and projections (Hser et al. 1992).

Regardless of the specific approach taken, the synthesis of multiple indicators for making inferences about prevalence requires some sort of modeling of cause-and-effect relations. Models may be characterized as formal or informal, simple or complex, static or dynamic, and narrow or broad in scope. Formal mathematical models are unambiguous in their assumptions and produce results reliably consistent with these assumptions, clear advantages not shared by qualitative mental models. On the other hand, the human mind does capture realistic details and dynamics that may be overlooked by mathematical models, a fact appreciated by government decision makers who often reach their own conclusions rather than rely on mathematical models. This article starts from the premise that decision makers confronted by a phenomenon as dynamically complex as illicit drug use should have available formal models that are not only unambiguous and reliable but also as realistic in detail and at least as broad in scope as their own mental models. It is this premise that has led to the development of the system dynamics simulation model presented here.

This model builds upon previous system dynamics models of illicit drug use and economics. The "Persistent Poppy" model of Levin, Roberts, and Hirsch (1975) examined heroin use in New York City and was used for policy analysis rather than prevalence estimation. Although this rather complex model contains many plausible relations, even seeking to account for changes in budget allocations for law enforcement and treatment, it was developed at a time when the numerical data needed for its calibration and validation were lacking. More

This work was supported by grants from the National Institute of Justice to the UCLA Drug Abuse Research Center (87–IJ–CX–0042, 90–IJ–CX–0014). My thanks go to the other members of the grant team, Douglas Anglin, Mary-Lynn Brecht, Yih-Ing Hser, and Thomas Wickens, for their valuable contributions.

System Dynamics Review Vol. 9, no. 1 (Winter 1993): 49–78   *Received November 1991*
© 1993 by John Wiley & Sons, Ltd. 0883–7066/93/010001–30$20.00   *Accepted June 1992*

recent, simpler models developed by Gardiner and Shreckengost (1985; 1987; Shreckengost 1984; 1985) address drug supply and demand on a national level, with specific application to heroin and cocaine. These models have been used primarily to make inferences about drug import quantities rather than prevalence of use, and were in fact acknowledged to lack robustness for estimating the number of users (Shreckengost 1984).

Data sources and trends

During the course of model development, a variety of national indicator data have contributed to enriching and calibrating the model's structure and reducing the degree of uncertainty in model assumptions and estimates. The key longitudinal data used for these purposes, drawn from various government publications and data bases and covering the 1976–1990 period, are presented in Table 1. They include self-reported use data from the National Household Survey (NHS), the High School Senior Survey (HSSS), and the Drug Use Forecasting (DUF) project; morbidity and mortality data from the Drug Abuse Warning Network (DAWN); arrest and incarceration data from the Uniform Crime Reports (UCR) and the Offender-Based Transaction Statistics (OBTS); and data on seizure volumes, price, and purity from the Drug Enforcement Administration's (DEA) NNICC Reports and STRIDE data base. Two other useful sources of data, not shown in Table 1, include (1) annual felony prison rate data for California from that state's *Sentencing Practices Quarterly* (JCC 1990) publication, for the period 1977–1989, and (2) monthly consumer price index data for the nation's urban areas for the entire 1976–1990 period.

Some of the salient trends observed in these indicator data are presented in Figures 1 through 5. Figure 1 shows how, for both the general population (NHS) and high school seniors (HSSS), self-reported past month cocaine use prevalence grew rapidly in the late 1970s, was relatively stable during the early 1980s, and then fell rapidly in the late 1980s. Self-reported past month marijuana use, graphed in Figure 2, also displays a rise-and-fall pattern, but the timing and magnitude are clearly different from cocaine's: the marijuana figures grow only modestly in the late 1970s, then decline throughout both the early and the late 1980s. Noting that the HSSS shows a pattern for self-reported use of "any illicit drug" nearly identical to that for marijuana use, one may conclude from these two figures that self-reported cocaine use has not simply tracked illicit drug use in general but has followed its own unique path.

Figures 3 and 4 present time series on the negative consequences of cocaine use: Figure 3 presents DAWN data on emergency room (ER) and medical examiner (ME) mentions, while Figure 4 presents UCR data on drug-law felony

Table 1. National indicator data for modeling cocaine prevalence

National Household Survey (NHS) from National Institute on Drug Abuse (NIDA), reports for 1976, 1977, 1979, 1982, 1985, 1988, 1990:
- Past month, past year, and lifetime use of cocaine and marijuana
- Past month, past year, and lifetime use of crack (1988 and 1990 only)
- Weekly use of cocaine (1985, 1988, and 1990 only)

High School Senior Survey (HSSS) from NIDA, annual reports 1976–1990:
- Past month, past year, and lifetime use of cocaine, marijuana, and "any illicit drug"
- Past month, past year, and lifetime use of crack (1987–1990 only)
- Attitudes, beliefs, and social milieu for cocaine

Drug Abuse Warning Network (DAWN) from NIDA, annual and semiannual reports 1976–1989:
- Emergency room mentions of cocaine, totals reported by quarter
- Medical examiner mentions of cocaine

Drug Use Forecasting (DUF) project from National Institute of Justice (NIJ), quarterly interview data base 1988–1989:
- Past day and past month use of cocaine powder and crack

Uniform Crime Reports (UCR) from Federal Bureau of Investigation (FBI), quarterly reports 1977–1989:
- Felony arrests for cocaine/opiate possession and sales/manufacture

Offender-Based Transaction Statistics (OBTS) from Bureau of Justice Statistics (BJS), annual reports 1983–1987:
- Incarceration rates for state drug-law felony arrests

National Narcotics Intelligence Consumers Committee (NNICC) Reports from Drug Enforcement Administration (DEA), annual reports 1977–1989:
- Federal seizures of cocaine

System to Retrieve Drug Evidence (STRIDE) from DEA, continuous seizure data base 1977–1990:
- Price, purity, and quantity per cocaine seizure

---

arrests for cocaine and opiates.[1] These time series all display a general pattern of little or no growth in the late 1970s, followed by rapid and accelerating growth during the 1980s. It is true that the DAWN and UCR trends do differ from each other somewhat at the very end of the 1980s: From 1987 to 1989 the growth in ER and ME mentions slows noticeably,[2] while the growth in arrests continues to accelerate. Despite this difference, these indicators clearly suggest enormous growth in the negative consequencs of cocaine use during just those years when self-reported past month use was increasing little or was on the decline. Similar growth during the 1980s is also seen in data on drug felony incarceration rates and cocaine seizures.

Growth in cocaine consequences and the cocaine trade during the 1980s coincided with the spread of crack cocaine. Crack, an easily processed, easily transported, smokable, and highly addictive form of cocaine, was first reported in Southern California and Texas in 1981, spread to New York City in 1984, and was found in urban areas all over the country by 1986 (NNICC 1987; Johnston

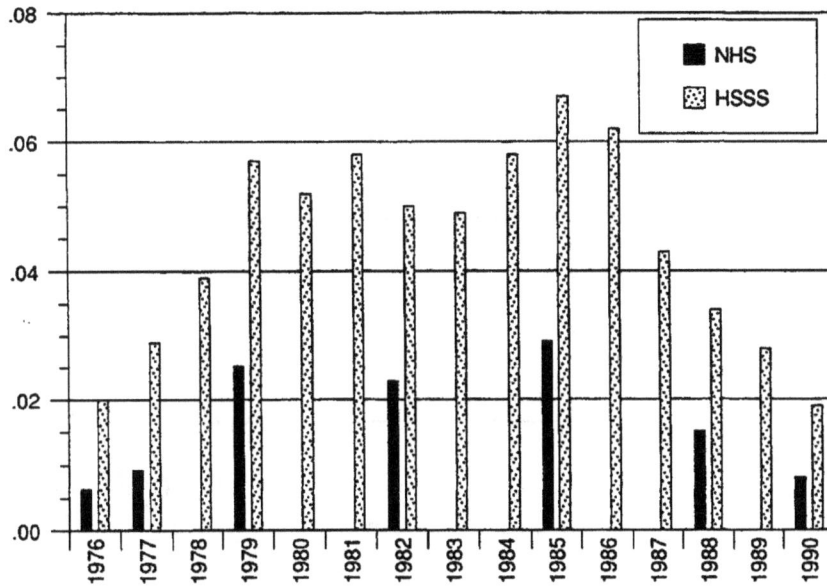

Fig. 1. Past month cocaine use self-report fractions from NHS and HSSS

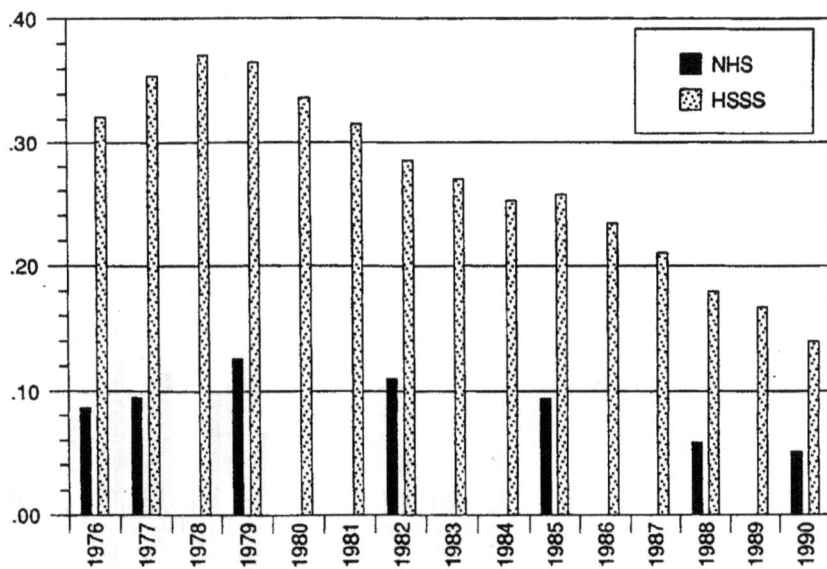

Fig. 2. Past month marijuana use self-report fractions from NHS and HSSS

Fig. 3. Emergency
room and medical
examiner cocaine
mentions from
DAWN

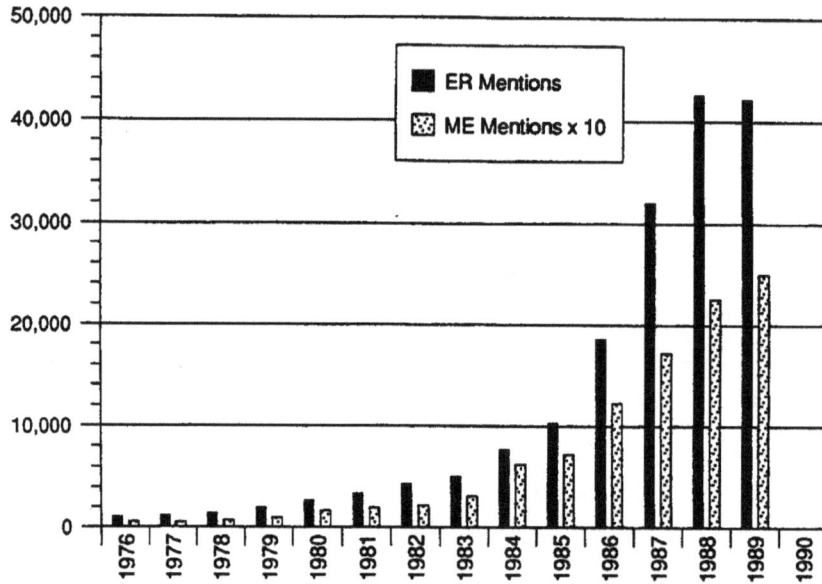

Fig. 4. Felony
arrests for cocaine/
opiate possession
and sales/manu-
facture from UCR

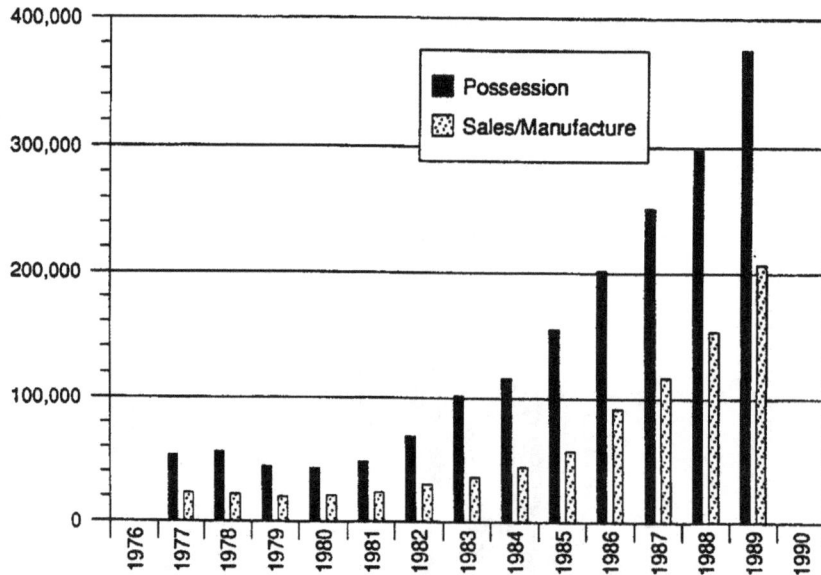

et al. 1989). Cocaine ER mentions attributed to smoking the drug grew from 2 percent of all cocaine ER mentions in 1983 to 37 percent in 1989, while ER mentions attributed to sniffing or injection declined from 86 percent to 35 percent of the total.[3] By 1987 and through 1990, crack accounted for 35–50 percent of self-reported past month use of cocaine, according to HSSS and DUF.

A factor that undoubtedly helped to fuel the growth in cocaine powder use during the late 1970s and crack use during the 1980s is the substantial decline in the price of cocaine throughout this period. Figure 5 presents estimates calculated from STRIDE seizure data on the average retail price (in 1990 dollars) per gram, expressed in terms of both adulterated (gross) and pure (net) quantities.[4] The decline in price is evident in both measures, but most dramatic for price per pure gram, the measure that more accurately describes the cost for a given amount of drug effect; price per pure gram fell from about $600 in the late 1970s, to about $300 in the early 1980s, to less than $200 in the late 1980s. This decline in price was concurrent with an increase in average retail purity, which grew from less than 40 percent in the late 1970s, to about 50 percent in the early 1980s, to over 70 percent in the late 1980s. These long-running trends continued until 1990, a year of relative shortage in which price increased and purity decreased moderately.

Other trends worthy of mention are those reflected in the HSSS data on cocaine-related attitudes, beliefs, and social milieu. In regard to attitudes and beliefs, these data show that the fraction of high school seniors and young

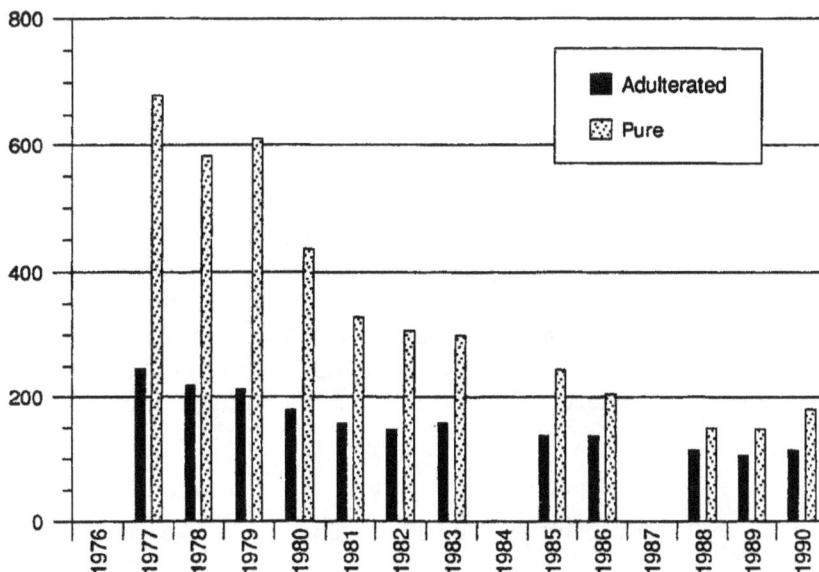

Fig. 5. Retail price of cocaine (in 1990 dollars) per adulterated gram and per pure gram from STRIDE

adults perceiving cocaine as a major health risk increased steadily throughout the 1980s, with the biggest changes in perception occurring prior to 1988. In regard to social milieu, the fraction of respondents who report having observed cocaine use during the past year or having friends who use cocaine has followed a rise-and-fall pattern over the years, looking similar to the HSSS data on use prevalence itself, as one might expect.

Model structure and parameters

The model presented here was developed for the purpose of generating national cocaine prevalence estimates and projections consistent with available indicator data and knowledge of underlying processes. During a development period of three years, various causal hypotheses were translated into equations, and these equations were tested to determine whether they were capable of reproducing historical data and of doing so with plausible parameter values.[5] Through this process of model testing and evaluation, a number of hypotheses were rejected and others were refined (Homer 1990). In line with system dynamics modeling principles, emphasis was placed on explaining trends as largely the result of endogenous feedback relations rather than exogenous factors, and on restricting the number of such relations so that the model would be manageable, understandable, and useful to decision makers.

An overview of the model's cause-and-effect structure is diagramed in Figure 6. All the indicator variables discussed in the preceding section are included in this structure, as well as those "hidden" variables that lie behind the indicators. For example, the model estimates reported cocaine use prevalence figures—intended to mimic those from the NHS—by starting from the underlying actual user population and applying reporting rates that are themselves subject to change. Similarly, price, purity, and seizure indicators change over time in response to such hidden market variables as imports and consumption. By following the arrows in Figure 6, one may see that most of the model's variables interact with one another, being elements of feedback loops rather than pure inputs or pure outputs. In particular, the consequences of cocaine use, such as morbidity, arrests, and effects on price, feed back to affect future use. However, the model does contain a small number of important exogenous variables, such as marijuana use prevalence, which also have a role in determining model behavior.

User population structure

Much of the model's structure is devoted to specifying the underlying user population. Any number of ways exist to subdivide this population, including

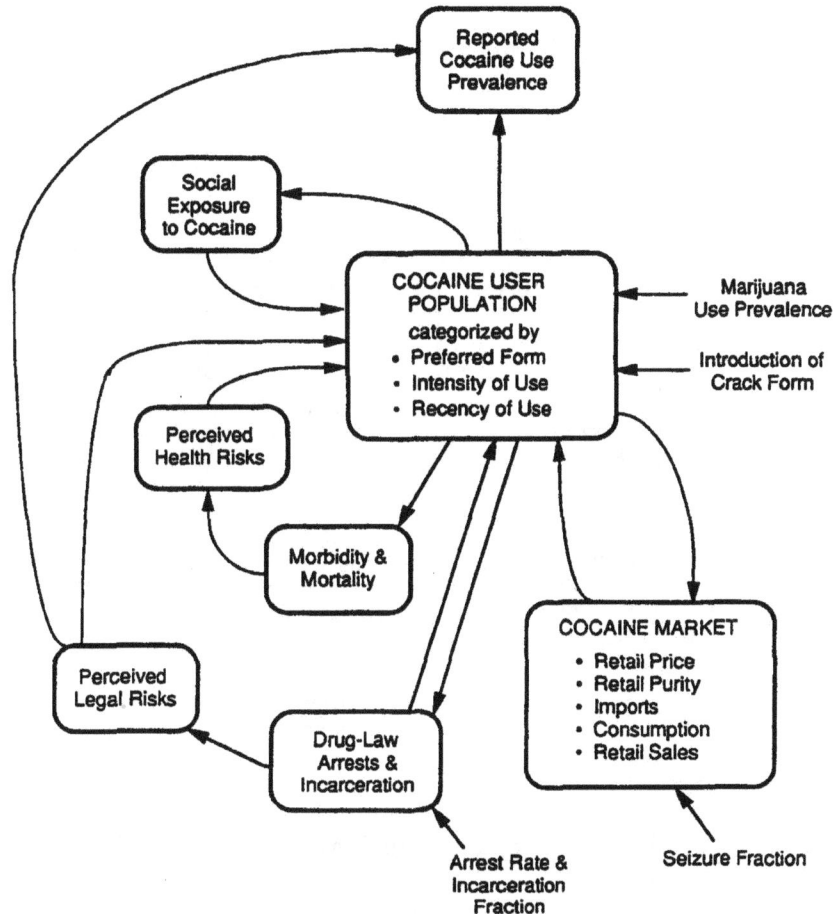

Fig. 6. Overview of
model structure

the use of such demographic variables as age, sex, race, region, and income. While acknowledging the importance of demographics, the model leaves them implicit and instead focuses on a categorization scheme more directly related to the amount and consequences of cocaine use. This scheme is shown in detail in Figures 7 and 8. Figure 7 presents the four major user categories defined by classifying users by intensity of use, either casual (light) or compulsive (heavy),[6] and by preferred drug form, either powder or crack. Within each of these four categories, there is a further breakdown by recency of use, according to the generic structure in Figure 8, where four population stocks are specified that combine to identify the number of past month, past year, and lifetime users.[7] For estimation and projection purposes, the model computes not only the individual stocks, such as past month compulsive crack users, but also

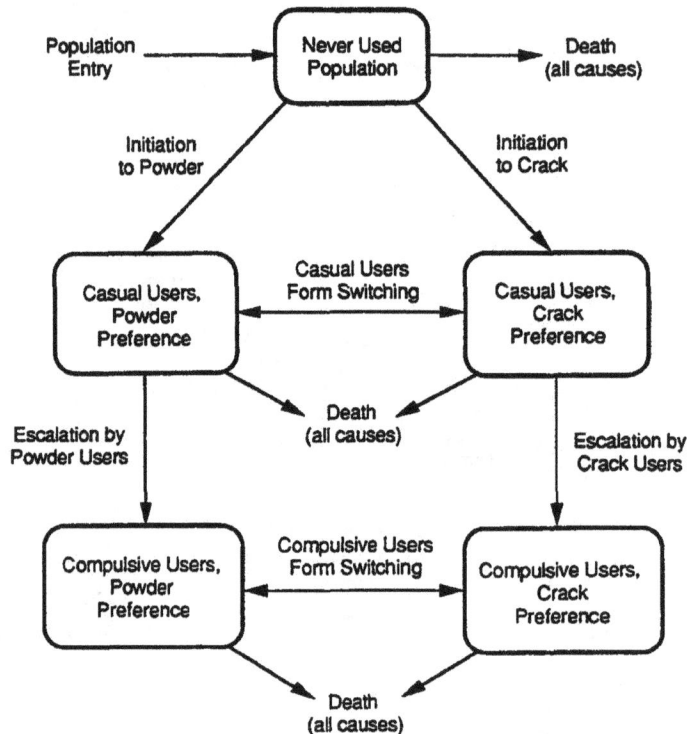

their key aggregates, such as total past month users and total lifetime crack users.

Changes in the model's user population are determined by the flows represented in Figures 7 and 8. The model tracks only that portion of the U.S. population aged twelve and over, in line with the National Household Survey, and follows members of this target population until their death. The process is as follows: Those who have never used cocaine may initiate into low-level, casual use of either cocaine powder or crack, from which they may escalate to compulsive use. Active (past month) users, both casual and compulsive, may switch their form preference from powder to crack or from crack to powder. They may abstain from active use, either voluntarily (including through treatment) or as the result of incarceration. Abstainers may then relapse to active use, some doing so within the first year of abstention and others doing so after the first year.

Each of the model's user flows (36 in all) is expressed as an annual fraction, or rate, multiplied by the size of the originating stock.[8] For example, the

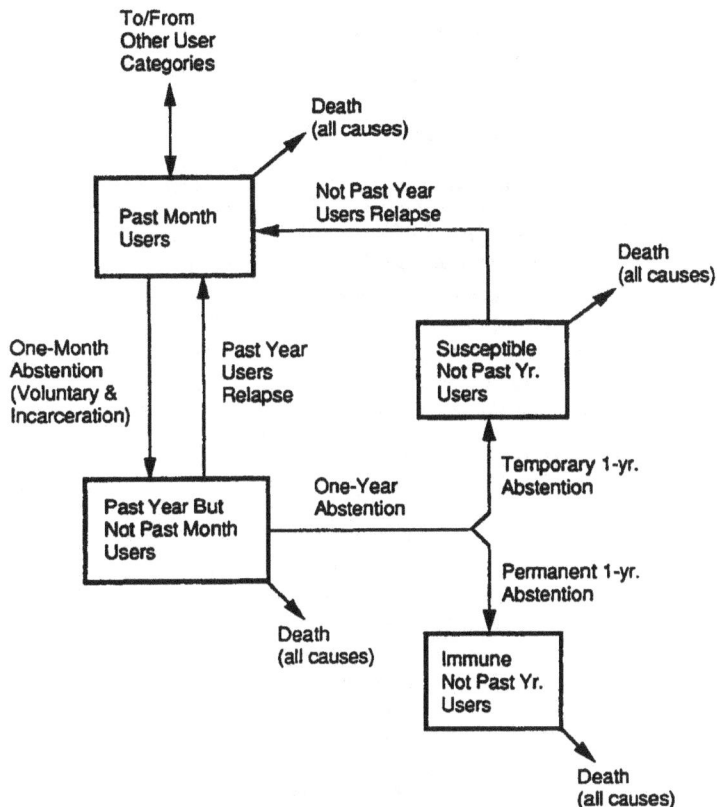

Fig. 8. Recency-of-use stocks and flows. This structure is replicated for each user category in Fig. 7.

number of new initiates to cocaine powder is calculated as the never used population multiplied by the initiation-to-powder rate. Each flow rate, in turn, is given a standard value for the year 1976 and, as time goes along, may or may not vary from this value in response to changes in other model factors. In particular, flow rates for initiation, casual user voluntary abstention, and casual user first-year relapse are assumed to be sensitive to an entire cluster of factors seen in Figure 6; these factors include marijuana prevalence (as an indicator of illicit drug use in general), social exposure to cocaine, retail price, perceived health risks, and perceived legal risks. Another important factor is the introduction of crack, which affects the flow rates for crack initiation and switching to crack from powder.

User population parameters

The various categories of users in the model are assumed to differ parametrically along a number of the dimensions listed in Table 2. First of all, they differ in

Table 2. Model parameters

Cocaine user categories and flows
- Target population (gradual growth over time)
- Initial (1976) past month, past year, and lifetime prevalence fractions
- Initial compulsive user fraction
- Crack introduction year (1981)
- Standard value for each of 36 user flow rates
- Initiation, casual user abstention, and casual user relapse rate sensitivities to marijuana prevalence, social exposure, price, perceived health risks, and perceived legal risks
- Initiate-to-crack and switch-to-crack rate sensitivities to crack prevalence

Cocaine user prevalence reporting
- Standard reporting rates (by intensity of use and form preference)
- Reporting rate sensitivities to perceived legal risks (by recency and intensity of use)

Cocaine market
- Consumption demand per user (by intensity of use)
- Standard imports expansion rate, sensitivity of imports to supply-demand ratio, and limit on importers' ability to offset seizures
- Seized fraction of imports (minor growth 1980–1982, major growth 1983–1990)
- Parameters relating retail price to retail supply
- Initial (1976) retail purity

Marijuana prevalence and social exposure to cocaine
- Marijuana past month prevalence fraction (growth 1976–1979, decline 1980–1990)
- Parameter relating social exposure to cocaine past month prevalence

Perceived health risks, morbidity, and mortality
- Parameters relating perceived health risks to past morbidity
- Morbidity and mortality rates per past month user (by intensity of use and form preference)

Perceived legal risks, drug-law arrests, and incarceration
- Parameters relating perceived legal risks to past arrests
- Arrest rates per past month user (by intensity of use and form preference, and with moderate growth 1988–1990)
- Incarceration fraction of arrests (minor growth 1977–1983, major growth 1984–1990)

---

terms of flow rate parameters. Casual users of crack escalate to compulsive use at a much higher rate (estimated at 30 percent per year) than casual users of powder (3 percent per year), abstain from use at a lower rate, and are more sensitive to price when considering whether to remain users. Compulsive users of both powder and crack abstain voluntarily at a much lower rate than casual users (even allowing for treatment—which probably explains over half of the compulsive user abstentions)[9] but are more frequently incarcerated. Compulsive users are also assumed to be insensitive to changes in the cluster of factors described above that influence casual use. Also, for both casual and compulsive users, the switch-to-crack rates are higher than the switch-to-powder rates.

A second important parametric distinction among user categories in the model is in the area of prevalence reporting, where intensity of use, form preference, and recency of use may all have a bearing on the rate at which actual use gets reflected in the National Household Survey. It appears that as perceived legal risks increase all reporting rates are suppressed, but much more so for individuals who have not used in the past year than for those who have used more recently. Also, regardless of legal risks, the reporting rate for compulsive crack users is considerably lower (with a standard value of 40 percent) than that for casual powder users (80 percent), with reporting rates for compulsive powder users and casual crack users falling between the two.

The third broad area in which parameter values differ significantly by user category is the consequences of use, specifically for past month users. Because compulsive users consume much more cocaine per capita than casual users, their impact on the cocaine market may be great even if their numbers are relatively small. Per-capita rates for cocaine-related ER mentions, ME mentions, and felony arrests (and, consequently, incarcerations) are much higher for compulsive crack users than for casual powder users, with the rates for compulsive powder users and casual crack users again falling between these extremes. For example, the standard annual drug-law arrest rate for compulsive crack users is estimated at 30 percent, compared with 8 percent for compulsive powder users, 4 percent for casual crack users, and less than 1 percent for casual powder users.

Cocaine market structure and parameters

The model's representation of the cocaine market is diagramed in Figure 9. This structure produces estimates for retail price, retail purity, and domestic seizures, which may be compared with national indicator data, and which, in the case of price, feed back to influence the user population. The market structure has undergone extensive revision over the course of model development, as alternative economic hypotheses (see Reuter et al. 1988 and Caulkins 1990) have been tested and modified in order to accommodate the historical data. The parameters required to fill out this structure appear in the third item of Table 2.

In the model, the retail price per pure gram is formulated simply as a downward sloping function of retail supply. The apparent success of this formulation supports the theory that the marginal cost of distribution may be the dominant component of an illicit drug's retail price and that this cost declines as the market expands (Caulkins 1990). In other words, cocaine's retail price has historically been driven primarily by economies of scale rather than by imbalances of supply and demand—imbalances that have been relatively small and short-lived.

Figure 9 shows how the balance of supply and demand is maintained. It is assumed that cocaine is imported largely in response to recent consumption demand. In particular, importers collectively tend to push somewhat beyond recent demand in an effort to expand the market, and they continue to push for

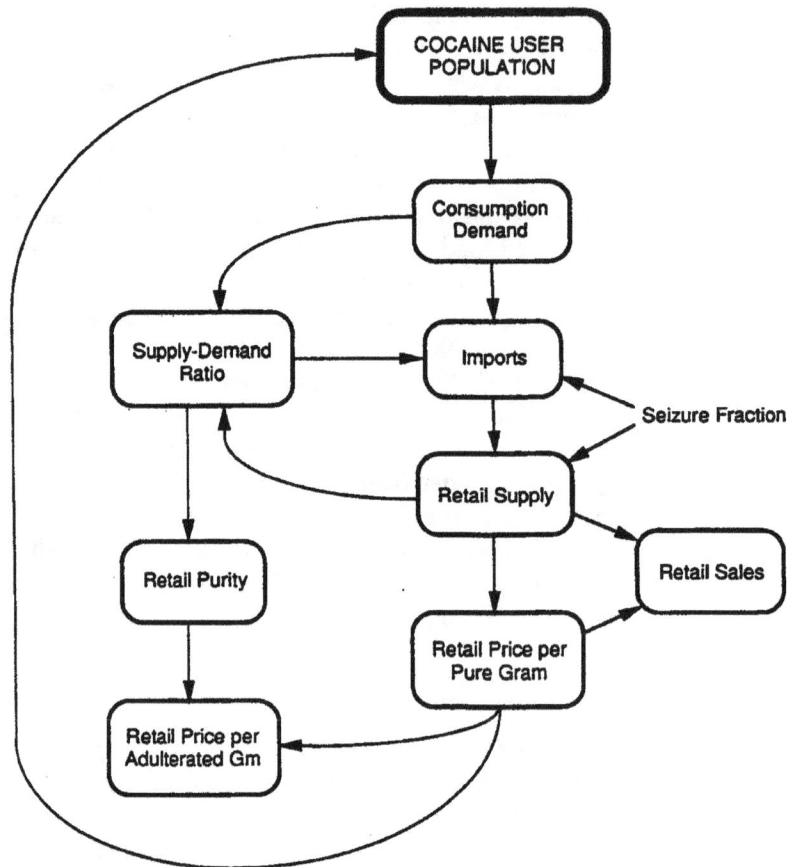

Fig. 9. Cocaine market causal structure

expansion as long as retail supply is not running too far ahead of demand. Also, importers boost their volumes in an effort to offset seizures, though there may be an upper limit to their ability or willingness to do so.[10] All imports not seized are assumed to be consumed,[11] causing excesses of supply over consumption demand to lead to increases in retail purity, while relative shortages lead to decreases in purity.

Understanding the past

Baseline estimates and trends

The model was simulated over the period 1976–1990 with a set of parameter values giving the closest possible overall fit to the historical indicator data. Some of the key results of this baseline simulation are shown in Table 3 and in

	1976	1980	1984	1988	1990
Table 3. Model-based estimates of cocaine prevalence and distributions among four user categories					
Used past month	1.4 mill.	7.8 mill.	6.6 mill.	4.2 mill.	3.6 mill.
Compulsive, prefer crack	0%	0%	1%	23%	36%
Compulsive, prefer powder	20%	8%	15%	18%	16%
Casual, prefer crack	0%	0%	5%	29%	28%
Casual, prefer powder	80%	92%	79%	30%	20%
Used past year	4.0 mill.	13.3 mill.	16.1 mill.	14.8 mill.	13.7 mill.
Compulsive, prefer crack	0%	0%	1%	8%	12%
Compulsive, prefer powder	8%	5%	7%	6%	5%
Casual, prefer crack	0%	0%	3%	19%	24%
Casual, prefer powder	92%	95%	90%	67%	59%
Used in lifetime	9.2 mill.	25.1 mill.	39.9 mill.	50.9 mill.	54.9 mill.
Compulsive, prefer crack	0%	0%	0%	3%	4%
Compulsive, prefer powder	11%	9%	6%	5%	4%
Casual, prefer crack	0%	0%	1%	9%	12%
Casual, prefer powder	89%	91%	93%	84%	80%
Consumption (pure)	33 MTons	104 MTons	145 MTons	183 MTons	177 MTons
Retail sales (1990 $)	$29.2 bill.	$35.4 bill.	$33.2 bill.	$34.2 bill.	$33.1 bill.
Compulsive, prefer crack	0%	0%	7%	51%	65%
Compulsive, prefer powder	80%	57%	69%	41%	30%
Casual, prefer crack	0%	0%	1%	4%	3%
Casual, prefer powder	20%	43%	23%	4%	2%

Note: Column entries may not add to 100% because of rounding.

Figures 10 through 17. In reviewing these numbers and graphs and assessing their validity, it is important to recognize that greater confidence may be placed in the simulated trends and relations than in the individual point estimates. Significant uncertainty exists regarding the values of certain key parameters, including the reporting rates and the consequence-of-use (consumption, morbidity, mortality, and arrest) rates that together determine the absolute numerical estimates of prevalence. Although this parameter uncertainty leads naturally to some uncertainty regarding the absolute estimates of users in Table 3, perhaps in the range of 10–20 percent, it has no significant effect on the detailed trends and relations among model variables necessary to produce a close fit to the historical data.

Figures 10–13 present baseline estimates of actual and reported prevalence fractions and show that the latter generally do a very good job of mimicking the NHS figures for past month, compulsive (weekly), past year, and lifetime use. One may see clearly the increasing gap over time between actual and reported prevalence, and the distortion this has caused in the reporting of trends for compulsive users and lifetime users in particular. Where compulsive use prevalence is reported to have grown a bit and then subsided from 1985 to 1990, the model suggests in reality quite substantial growth followed by leveling.

Fig. 10. Past month
user prevalence
fraction estimates

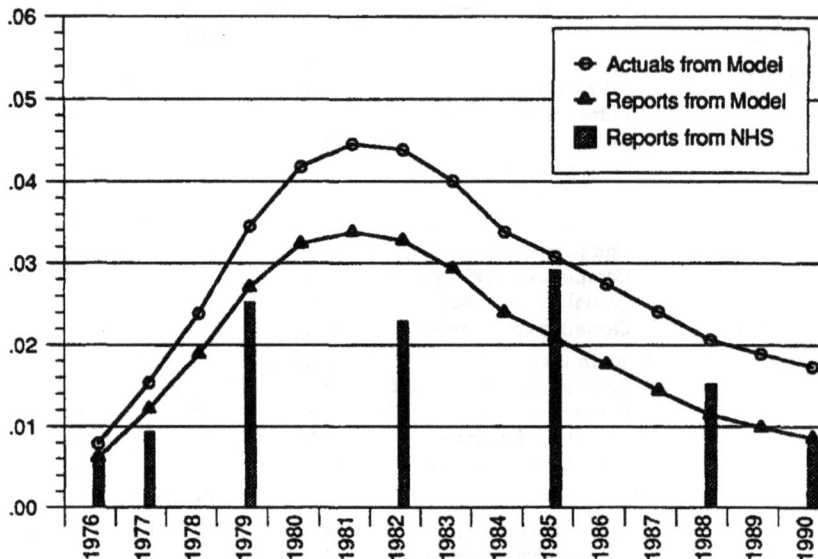

Fig. 10. Past month user prevalence fraction estimates

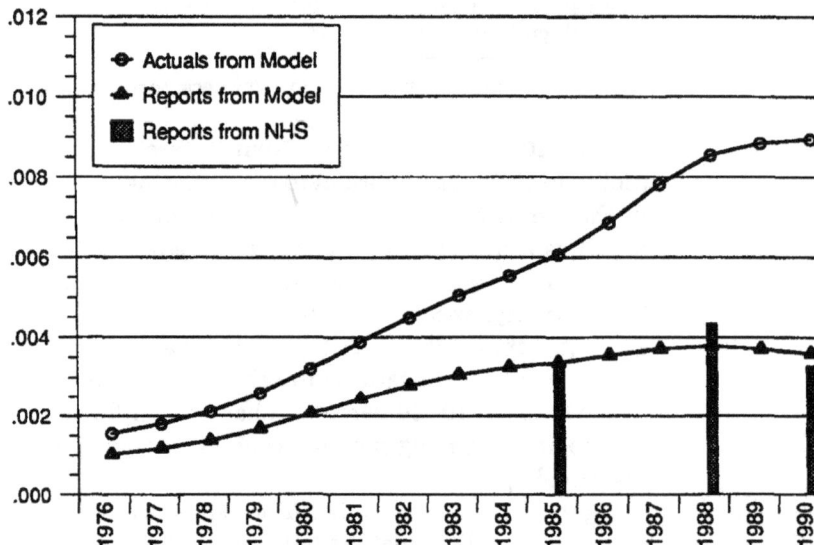

Fig. 11. Compulsive/weekly user prevalence fraction estimates

And where lifetime use prevalence is reported to have declined gradually from 1982 to 1990, the model suggests in reality an increase of over 50 percent. The model clearly suggests a sizable reduction in overall reporting rates during the 1980s, due in part to increased arrest rates and perceived legal risks for all users and in part to the spread of crack cocaine among a population less likely

Fig. 12. Past year
user prevalence
fraction estimates

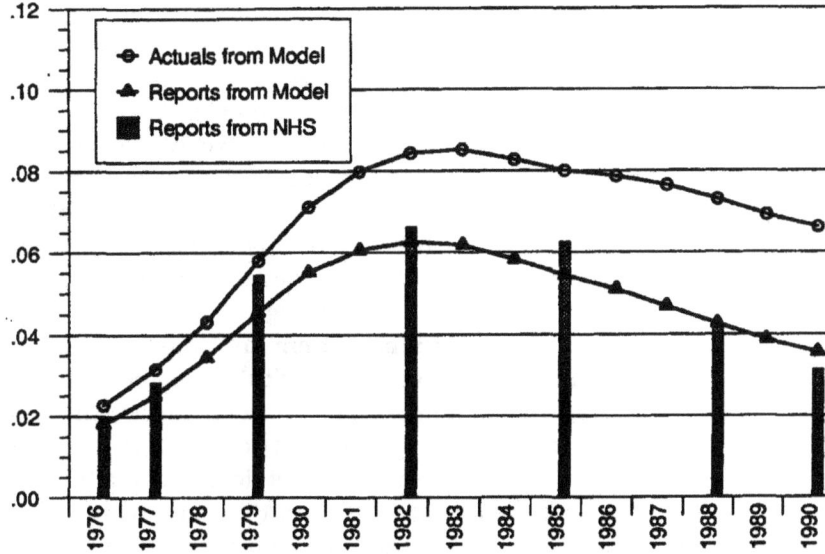

Fig. 13. Lifetime
user prevalence
fraction estimates

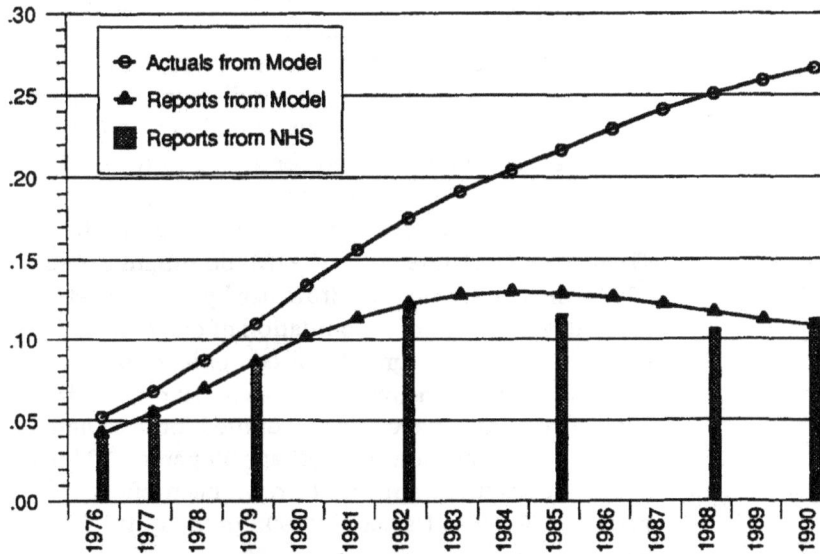

than powder users to be sampled by the NHS survey (and more likely to use
compulsively and to be arrested).

Trends in actual user prevalence are a direct consequence of changes in the
various population flows pictured in Figures 7 and 8. Specifically, one may
learn much about these trends by looking at the time paths of initiation and

escalation, as seen in Figures 14 and 15, respectively;[12] indeed, these key flows are fairly reliable as leading indicators of prevalence. The declines of past month use (Figure 10) and past year use (Figure 12) are foreshadowed, by two years or more, by a similar decline in initiation in Figure 14 following its peak in 1979. The decline in initiation also explains the gradually slowing growth of lifetime use (Figure 13) throughout the 1980s. Similarly, the uneven pattern of growth in compulsive use (Figure 11), which finally levels at the end of the 1980s, is foreshadowed by the escalation curve in Figure 15; the escalation curve levels in the early 1980s and then climbs again from 1984 to 1987, driven by crack, before finally subsiding.

Additional key estimates from the model are graphed in Figures 16 and 17, and again a good fit to the historical indicator data may be seen. The rapid rise in cocaine-related incarcerations in the late 1980s (Figure 16) helps to explain why compulsive use (Figure 11) did not grow as much as one might have expected during those years of rapid escalation to crack. And the rapid rise in seizures in the late 1980s (Figure 17) finally stymied importers enough to cause, at the end of the decade, an actual decrease in purity and pure consumption, despite the fact that compulsive users—who consume most of the cocaine— continued to grow in numbers.

Table 3 presents, for five selected years, numerical estimates of actual users, consumption, and retail sales, and their precentage breakdowns among the major user categories. It is estimated that by 1990, about half of all past month users were compulsives, and about two-thirds of all past month users preferred crack over powder. At the same time, only one-twelfth of lifetime users were current or past compulsives and only one-sixth current or past crack users. These numbers underscore the two dominant cocaine trends of the mid-to-late 1980s: the massive exodus from use by casual users of cocaine powder, and the rapid development of a population of crack users who use compulsively more often than not. The growth of compulsive use is directly responsible for an increase in pure consumption of some 70 percent from 1980 to 1990; and where compulsives may have once accounted for less than 60 percent of all consumption, they now account for perhaps 95 percent. This increase in the quantity of cocaine consumed appears to have been effectively offset by its decrease in price, so that retail sales (in 1990 dollars) did not increase and may actually have subsided a bit during the 1980s.

Feedback-loop explanation of historical trends

In the model, a small number of key feedback loops and exogenous factors, shown in Figure 18, are responsible for the observed trends for 1976–1990. The large increase in initiation and cocaine use from 1976 to 1979 is primarily

Fig. 14. Initiation to
casual use:
incidence fraction
estimates from
model

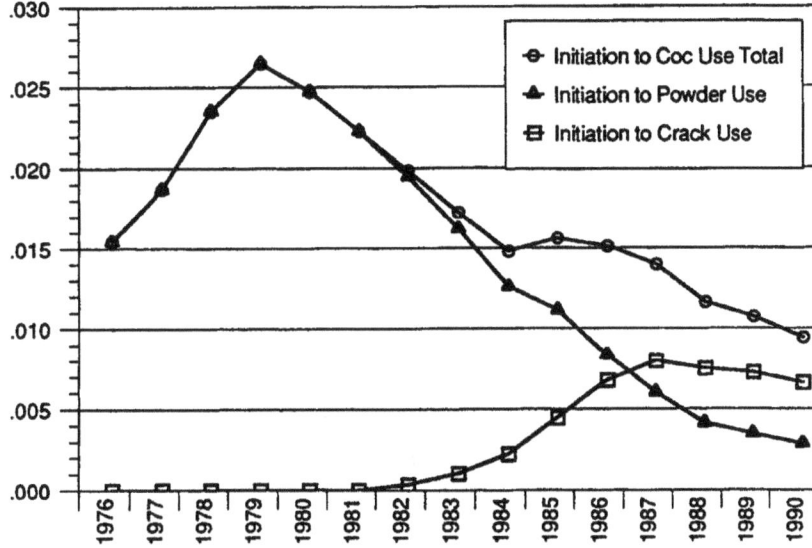

Fig. 15. Escalation
to compulsive use:
incidence fraction
estimates from
model

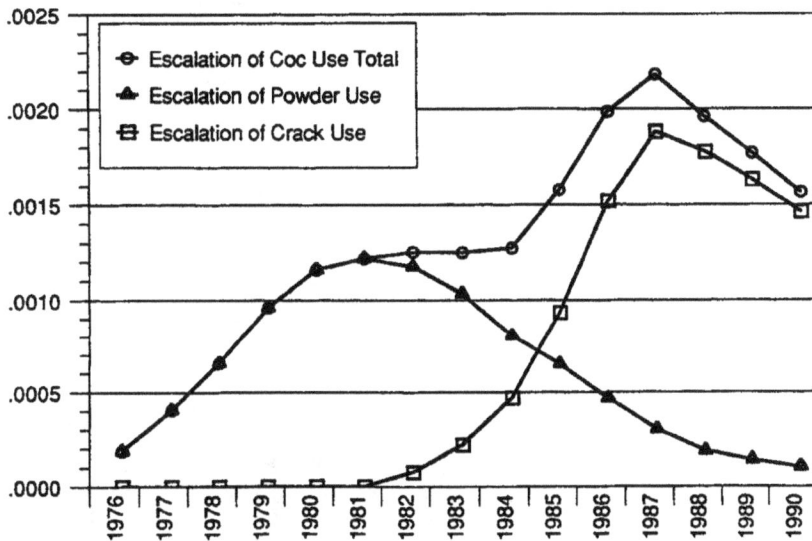

driven by two positive loops, numbered 1 and 2, and is also helped along by a
moderate increase in marijuana use during those years. Loop 1 says that an
increase in users leads to more social exposure to cocaine, which, in turn,
further increases initiation and the number of users. Loop 2 says that increasing
consumption drives up supply, which, in turn, drives down distribution costs

Fig. 16. Drug felony
arrests and
incarcerations
estimates
(thousands)

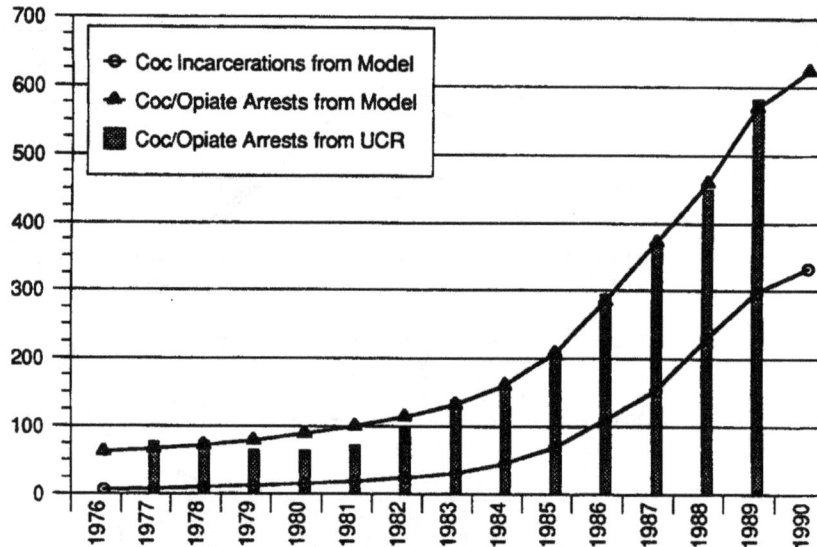

Fig. 17. Imports,
consumption, and
seizures estimates
(metric tons)

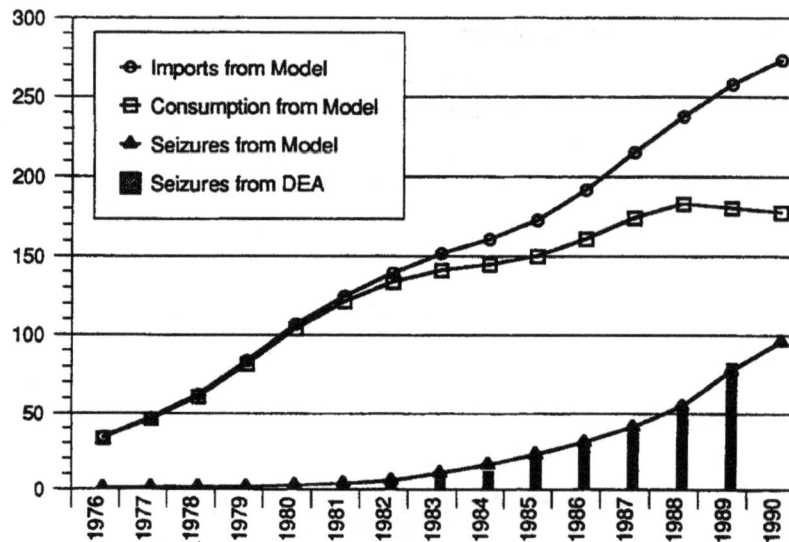

and retail price; declining price is a further spur to initiation and continued use.

Initiation starts to decline in 1980, followed by a decline in total past month use of cocaine from 1982 onward. The decline in total use is primarily driven by two negative loops, numbered 3 and 4, and is also helped along by a continuous decline in marijuana use during the 1980s, and reinforced by

Fig. 18. Key feedback loops driving cocaine prevalence

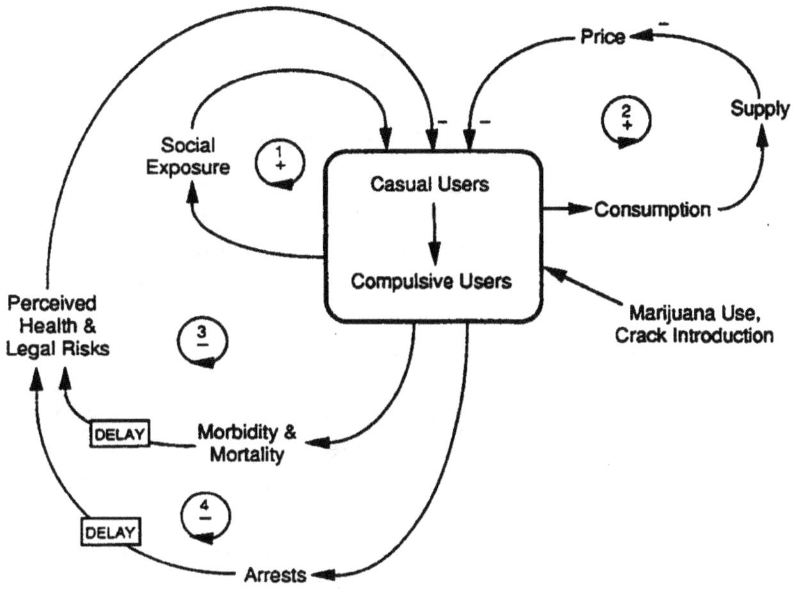

declining social exposure. In loops 3 and 4, initiation and use are suppressed by a growing perception, over time, of cocaine's health and legal risks, respectively. In loop 3, the growing perception of health risks follows (with a perception delay of three years) an increasing number of instances of cocaine-related morbidity and mortality. In loop 4, the growing perception of legal risks follows (again, with a delay of three years) an increasing number of drug-law arrests for cocaine. The perception of risks continues to grow throughout the 1980s, apparently in spite of the fact that total use is in decline.

To explain this phenomenon of growing risk perception, it is necessary to distinguish between casual users and compulsive users, because compulsive users are much more likely than casual users to suffer cocaine's more serious health and legal consequences. In particular, while casual use declines steadily during the 1980s, compulsive use increases continuously. The growth in compulsive use in the early 1980s may be viewed as a legacy of the growth in casual use in the 1970s. In contrast, the growth in compulsive use in the mid-to-late 1980s is due to the spread of crack, with its much greater rate of escalation relative to that of powder cocaine.[13]

In addition to their greater exposure to drug-related risks, compulsive users also consume much more cocaine per capita than do casual users, which explains why consumption and supply increase until the end of the 1980s. The increase in supply explains the continued decline of retail price during the

1980s (see Figure 5 and loop 2), which, in turn, helps to fuel the growth of crack and compulsive use.

Sensitivity testing: alternative past scenarios

Sensitivity testing is an essential part of system dynamics modeling methodology (Tank-Nielsen 1980) and may be used to better understand the roles and relative importance of key factors, such as those in Figure 6, in determining historical trends. Sensitivity testing involves altering baseline assumptions regarding input parameters or functional relations, effectively creating alternative scenarios for the past.

Figures 19 and 20 present prevalence results, describing total past month use and past month compulsive use, respectively, for four alternative past scenarios in addition to the baseline results discussed above. In the first alternative scenario, marijuana prevalence remains at its 1980 value throughout the 1980–1990 period, instead of declining as it did historically (Figure 2). In the second scenario, price per pure gram is fixed from 1980 onward, instead of being allowed to decline as it did historically (and as explained by loop 2 in Figure 18). In the third scenario, crack cocaine is simply never introduced. In the fourth scenario, crack is introduced as in the baseline, but the per-capita drug-law arrest rates for crack users are set equal to those of powder users, instead of being much higher as they are in the baseline.

PAST SCENARIO 1. In the absence of a decline in marijuana prevalence, the decline in casual use of powder is more gradual than in the baseline, and casual crack use takes off more rapidly, leading in the late 1980s to solid growth (rather than decline) in total use. Because crack has a high rate of escalation, the increase in casual crack use leads, in turn, to accelerating (rather than decelerating) growth in compulsive use, and results by 1990 in compulsive use prevalence more than double that of the baseline. The historical decline of marijuana (illicit drug use) prevalence has had a great influence on both casual and compulsive use of cocaine, indirectly cutting the latter to less than half of what it would have been had illicit drugs not lost much of their general popularity during the 1980s.

PAST SCENARIO 2. In the absence of a decline in price, both total and compulsive use grow less than in the baseline. This impact begins gradually in the early 1980s but picks up greater force during the era of crack, because high price creates more of a barrier to continued use for casual crack users than it does for casual powder users. As a result, the spread of crack is curtailed sooner and at a lower level than in the baseline, and compulsive use is stabilized by the mid-1980s. The historical decline in cocaine price per pure gram has had a

Fig. 19. Past month
user prevalence frac-
tion simulated under
alternative past
scenarios

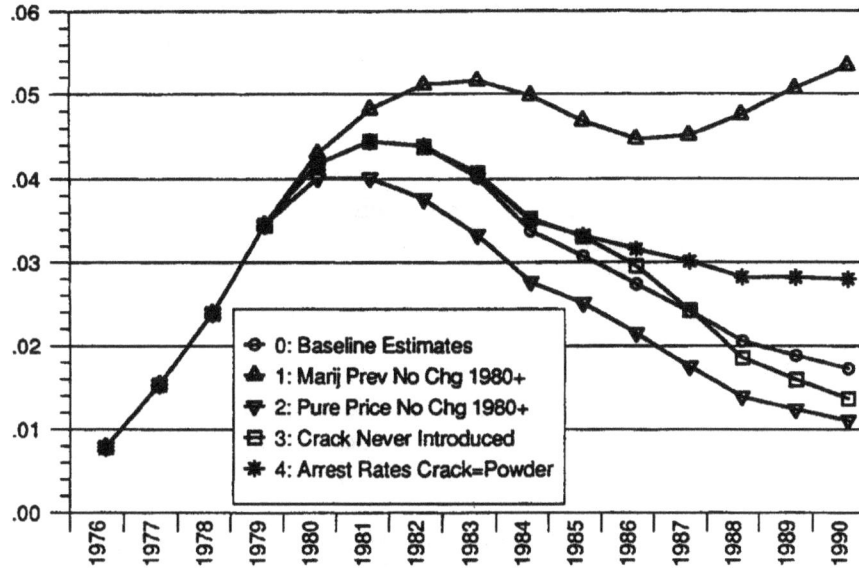

0: Baseline Estimates
1: Marij Prev No Chg 1980+
2: Pure Price No Chg 1980+
3: Crack Never Introduced
4: Arrest Rates Crack=Powder

Fig. 20. Compulsive
user prevalence frac-
tion simulated under
alternative past
scenarios

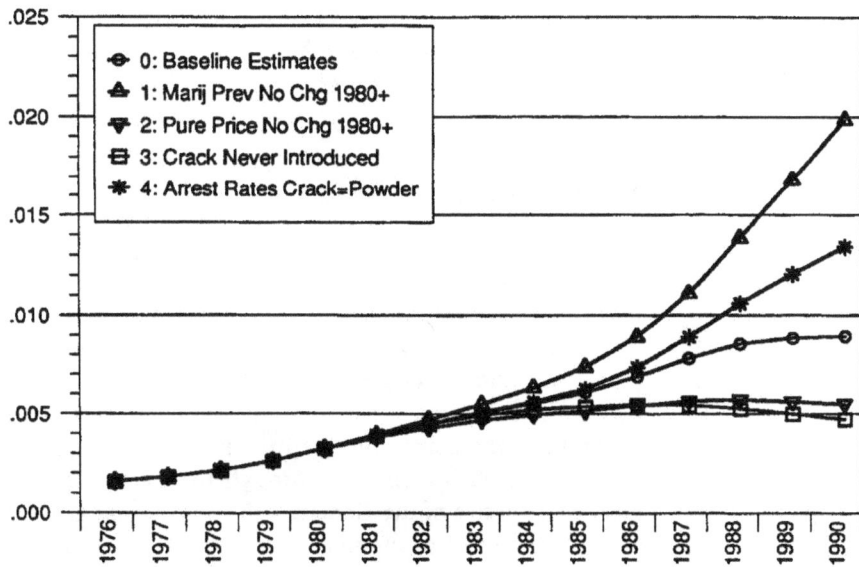

0: Baseline Estimates
1: Marij Prev No Chg 1980+
2: Pure Price No Chg 1980+
3: Crack Never Introduced
4: Arrest Rates Crack=Powder

significant influence on both casual and compulsive use, allowing both to grow about 60 percent more by 1990 than they would have had cocaine remained at its 1980 price.

PAST SCENARIO 3. In the absence of crack, compulsive use is held in check, even more so than in the fixed-price scenario. But the effect on total users is decidedly mixed, because an absence of crack actually leads to an increase in casual use relative to the baseline. Absent crack, cocaine-related morbidity, mortality, and arrests grow more slowly, causing perceived health and legal risks of cocaine to be less than in the baseline, and leading more individuals to experiment casually with the drug. (In effect, the absence of crack actually weakens loops 3 and 4, which restrain casual use.) The historical introduction of crack has led to a doubling of compulsive use over what it would have been otherwise but has discouraged the casual use of cocaine by making its risks more evident.

PAST SCENARIO 4. If crack users are no more likely than powder users to be arrested on drug-law charges, both total and compulsive users grow significantly more than in the baseline. This additional growth is primarily due to a relative reduction in perceived legal risks (thereby weakening loop 4), hence less deterrence effect; in addition, a relative reduction in arrests means fewer incarcerations, leaving more users, particularly compulsive users, on the street. The historical stepping-up of drug-law arrests to deal with crack has discouraged casual cocaine use and has cut compulsive use back to about two-thirds of what it would have been otherwise.

Anticipating the future

Baseline projections

To generate prevalence projections, the model is simulated into the future based on assumptions regarding possible changes in input parameters. A number of future scenarios have been tested that differ according to such assumptions, including parameter changes that may be interpreted as being the result of policy intervention. A baseline scenario was selected in which the only change of inputs is the continued gradual increase in the target (aged twelve and over) population, with estimates through the year 2002 based on U.S. Census projections (Bureau of the Census 1988). This scenario is perhaps conservative, in that it assumes no changes in other input parameters and, in particular, no further decline in marijuana prevalence and no further increases in per-capita rates of drug-law arrest and incarceration beyond 1990.

	1990	1992	1996	2000	2002
Used past month	3.6 mill.	3.4 mill.	3.5 mill.	3.6 mill.	3.6 mill.
Compulsive, prefer crack	36%	42%	49%	53%	54%
Compulsive, prefer powder	16%	13%	8%	6%	6%
Casual, prefer crack	28%	29%	30%	32%	32%
Casual, prefer powder	20%	17%	12%	9%	8%
Used past year	13.7 mill.	12.8 mill.	11.8 mill.	11.3 mill.	11.1 mill.
Compulsive, prefer crack	12%	14%	18%	21%	22%
Compulsive, prefer powder	5%	4%	3%	2%	2%
Casual, prefer crack	24%	28%	36%	41%	44%
Casual, prefer powder	59%	54%	43%	35%	32%
Used in lifetime	54.9 mill.	58.3 mill.	65.0 mill.	71.6 mill.	74.9 mill.
Compulsive, prefer crack	4%	5%	6%	7%	8%
Compulsive, prefer powder	4%	4%	3%	3%	3%
Casual, prefer crack	12%	15%	20%	24%	26%
Casual, prefer powder	80%	76%	71%	66%	64%
Consumption (pure)	177 MTons	181 MTons	191 MTons	201 MTons	206 MTons
Retail sales (1990 $)	$33.1 bill.	$33.6 bill.	$33.8 bill.	$34.0 bill.	$34.0 bill.
Compulsive, prefer crack	65%	72%	81%	86%	87%
Compulsive, prefer powder	30%	23%	14%	10%	9%
Casual, prefer crack	3%	3%	3%	3%	3%
Casual, prefer powder	2%	2%	1%	1%	1%

Table 4. Baseline projections of cocaine prevalence and distributions among four user categories

Note: Column entries may not add to 100% because of rounding.

Baseline projections of cocaine prevalence are presented, for selected years, in Table 4 and may be viewed as an extension of the historical estimates in Table 3. Corresponding graphs for total past month use and past month compulsive use prevalence fractions may be seen in Figures 21 and 22, respectively, where projections for the baseline scenario are shown along with results from other scenarios discussed later. The baseline projection for past month use indicates rather little change in the overall total but strong continuation through the end of the 1990s of the trend toward compulsive crack use and away from casual powder use. Consequently, the number of compulsive users is projected to increase by 17 percent from 1990 to 2002, and total cocaine consumption increases by 16 percent. (The continued growth of compulsive use in the 1990s occurs in the model as a downstream effect, a kind of legacy, of the enormous growth of crack use in the 1980s.) In contrast, the past year and lifetime user figures continue to be dominated by the ranks of casual users, the majority of whom abstain after only brief periods of experimentation with the drug. Past month casual use is projected to decline by 15 percent from 1990 to 2002, which explains the continued decline in past year use and decelerating growth in lifetime use, trends that began in the 1980s.

Fig. 21. Past month user prevalence fraction simulated under alternative future scenarios

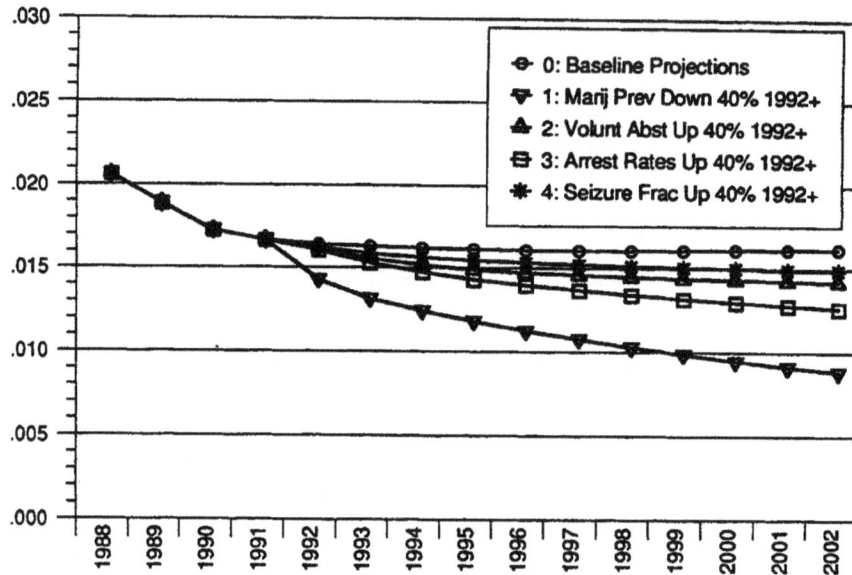

Fig. 22. Compulsive user prevalence fraction simulated under alternative future scenarios

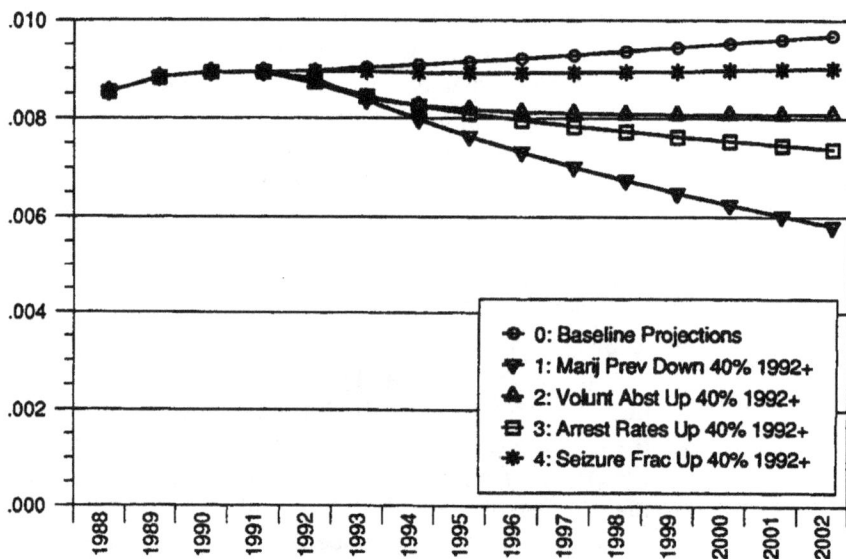

Alternative future scenarios

Prevalence projections corresponding to four alternative future scenarios, in addition to the baseline projections, are graphed in Figures 21 and 22. In each of these scenarios, a one-time change of 40 percent is made in a single key parameter, a change that (1) is made during 1992 and remains in place for the remainder of the simulation, and (2) has the effect of reducing the past month prevalence fractions. The testing of these scenarios is intended to aid understanding of the relative importance and over-time impacts of possible future events or policy interventions. Such testing is definitely not intended to establish confidence limits or lower bounds for future prevalence, and the 40 percent one-time change was selected for ease of analysis rather than as a statement of probable speed or magnitude of impact.

FUTURE SCENARIO 1. Marijuana prevalence drops by 40 percent in 1992. From a policy standpoint, such a drop might occur as the result of renewed efforts at illicit drug use prevention in the schools. Casual use is initially reduced, which cascades to reduction in compulsive use, and then feeds back for additional reduction of casual use because of reduced social exposure and increased retail price (see loops 1 and 2 in Figure 18). Relative to the baseline, casual use declines 55 percent during the subsequent ten years, compulsive use declines 40 percent, and both will continue to decline in the years beyond 2002. This simulation again points to the importance of illicit drug popularity in influencing cocaine user trends over the short and long term.

FUTURE SCENARIO 2. Voluntary abstention rates for compulsive users increase by 40 percent in 1992. Such an increase might occur as the result of a major expansion of treatment programs. Relative to the baseline, compulsive use declines by over 20 percent as a result, reaching this level within about five years. Because of this reduction in prevalence, there is some beneficial spillover effect on casual users (via loops 1 and 2), but this amounts to only 6 percent reduction relative to the baseline by 2002. Treatment programs have the primary effect of reducing compulsive use, leading to only minor reduction in casual use.

FUTURE SCENARIO 3. Drug-law arrest rates increase by 40 percent in 1992. Such a policy has the twin effects of increasing both incarcerations and deterrence. The increase in incarcerations primarily removes compulsive users from the population, having an essentially targeted effect similar to the expansion of treatment programs. In contrast, the deterrence effect acts like a reduction in marijuana prevalence, initially reducing casual use (via loop 4), then cascading to further reductions in both casual and compulsive use (via loops 1 and 2), which will continue beyond 2002. Relative to the baseline, these effects combine for a 25 percent reduction in compulsive use and a 20 percent reduction in casual use by 2002. Because of its direct impact on both compulsive and casual users, an increase in drug-law felony arrests may be an effective (though costly) way to reduce cocaine prevalence.

FUTURE SCENARIO 4. The domestic seizure fraction increases by 40 percent in 1992. This policy is less effective at reducing prevalence than the others because of the well-documented ability of importers to offset a significant portion of the seizures with additional imports and thereby minimize the effects on retail supply and price (Reuter et al. 1988). Relative to the baseline, casual use is reduced 10 percent and compulsive use only 7 percent by 2002. Some decision makers may hope that through increased seizures retail price will be increased enough to set a vicious cycle in motion—reduced demand leading to reduced supply, increased distribution costs, and yet further increases in price (see loop 2). The model suggests strongly that increased seizures will not "burst the balloon" of the national cocaine market in this way.[14]

Conclusion

Government decision makers and drug abuse researchers alike may benefit from the development of methods and models that will improve their understanding of drug prevalence trends and compensate for the shortcomings of self-report surveys. The system dynamics model presented here represents an attempt to integrate a variety of national indicators in a way that is (1) consistent with existing data and knowledge of cause and effect, (2) produces estimates for a variety of prevalence measures over an extended period of time, (3) clarifies the dynamic mechanisms underlying observed trends, and (4) can project those trends into the future under a variety of alternative scenarios with possible implications for public policy.

The model was revised until it was able to reproduce a wide variety of indicator data using plausible parameter values. This requirement has led to rather firm conclusions about past trends and relations but less certainty regarding absolute prevalence figures. Greater confidence in the absolute numbers could be obtained if better data were available for calculating parameters describing per-capita rates of drug use consequences, including consumption, morbidity, mortality, and arrest.

The model is able to produce detailed and relatively sophisticated projections,

projections that are largely the consequence of endogenous feedback structure and therefore require a minimum of assumptions about changing parameter values. However, one must recognize that even this complex structure represents a simplification of reality, and while it was adequate to explain the past, may fail to capture relations that will be important in the future. As a result, one must be careful not to view this model, or any particular model of a phenomenon as complex as illicit drug use, as a tool that promises forecasts having the accuracy of, say, a demographer's predictions of population growth. Instead, one should ask whether that model is able to shed light on future possibilities in a way that other models, including mental models, do not offer.

The model presented here appears to have value as a tool for improving understanding of national cocaine prevalence trends and policy impacts. It may also be useful as a starting point for further studies of illicit drug use and drug market dynamics. For example, the model may be extended to explore the dynamics of supply beyond the domestic borders and to look at such policies as crop substitution and eradication in coca-producing countries, as one preliminary system dynamics model has recently done (Wuestman 1990). The model may also be extended to explore the dynamics of complementary and supplementary use of illicit drugs, such as the relations between cocaine and amphetamine use, or cocaine and heroin use. Consideration should also be given to endogenous modeling of marijuana use, because of its apparent importance as a factor affecting cocaine use.

Notes

1. Cocaine and opiate arrests appear together in the UCR as a single aggregate measure.
2. Indeed, emergency room mentions for cocaine were slightly lower overall in 1989 than they were in 1988. Looking at the quarterly data, one finds that this decline is entirely due to a sudden drop-off of more than 20 percent in the fourth quarter of 1989, from a level which had changed little throughout the preceding five quarters. Reports for the first two quarters of 1990 do indicate some further decline in ER mentions, but at a considerably slower rate. The ER data for heroin are similar, with several quarters of relative stability being followed by a rapid drop-off in the fourth quarter of 1989 and the first quarter of 1990. For both cocaine and heroin, medical examiner mentions increased from 1988 to 1989. Although some may interpret the recent decline in ER mentions as an indication of less abusive usage habits (Sobel 1990), the unprecedented magnitude and suddenness of the drop-off, along with its apparent lack of corroboration by the ME data, seems curious and worthy of closer scrutiny.
3. The remaining percentages consist of other responses or no responses.
4. A retail seizure is defined here as one of six grams (gross) or less. Price

estimates for 1984 and 1987 are not presented because of small sample sizes and large variances for those years.

5. Parameter values were estimated through a combination of expert knowledge, which established plausible ranges, and model tuning to achieve the closest possible fit to the indicator data. The tuning process consisted of some statistical regression but more often involved running multiple simulations and evaluating their fit to the data.

6. An active compulsive user is defined as one who has used every week for the past year, in line with the NHS definition of weekly use. The average monthly per capita consumption for compulsive users is estimated to be eight grams (pure) versus one-half gram for casual users.

7. All users, past month or not, are identified by the intensity of use and form preference that applied when they last used. Past year users are the sum of past month users and past-year-but-not-past-month users. Lifetime users are the sum of all four user stocks in Figure 8.

8. In the case of one-month abstention, there are actually two separate flow rates for each past month user stock: one for voluntary abstention, the other for incarceration.

9. It has been estimated that intravenous drug users quit following treatment at the rate of 5–8 percent per year and quit without treatment at the rate of 4–5 percent per year (Homer and St. Clair 1991).

10. Some importers may be unprepared to replace large unexpected losses or reluctant to believe that their shipments might be seized. This limitation on the boosting of imports to offset seizures is used in the model to explain the possibility of relative shortage. Such relative shortage became a reality in 1990, by which time the seizure fraction may have grown to over 30 percent.

11. Domestic inventory adjustments and distribution losses not due to seizure are assumed insignificant. DEA experts estimate that domestic inventories comprise a few weeks' worth of consumption at most.

12. These graphs express initiation and escalation as annual fractions of the total target population.

13. The spread of crack is itself modeled as a positive feedback process (similar to loop 1) in which increasing prevalence (hence, exposure) to crack attracts more initiation to crack and more switching from powder to crack.

14. Caulkins (1990) presents a "balloon model" of illicit drug markets and explores its implications on both a national level and a local level.

References

Adams, E. H., J. C. Gfroerer, B. A. Rouse, and N. J. Kozel. 1986. Trends in Prevalence and Consequences of Cocaine Use. *Advances in Alcohol and Substance Abuse* 6: 49–71.

Bureau of the Census. U.S. Department of Commerce. 1988. *Statistical Abstracts of the United States 1988*. Washington, D.C.: Government Printing Office.

Caulkins, J. P. 1990. The Distribution and Consumption of Illicit Drugs: Some Mathematical Models and Their Policy Implications. Ph.D. Dissertation, MIT, Cambridge, MA 02139.

Gardiner, L. K., and R. C. Shreckengost. 1985. Estimating Heroin Imports into the United States. In *Self-Report Methods of Estimating Drug Use: Current Challenges to Validity*, ed. B. A. Rouse, N. J. Kozel, and L. G. Richards. NIDA Research Monograph 57. DHHS Publication (ADM)85–1402. Washington, D.C.: Government Printing Office.

———. 1987. A System Dynamics Model for Estimating Heroin Imports into the United States. *System Dynamics Review* 3 (1): 8–27.

Gawin, F. H., and E. H. Ellinwood, Jr. 1988. Cocaine and Other Stimulants: Actions, Abuse, and Treatment. *New England Journal of Medicine* 318 (18): 1173–1182.

Homer, J. B. 1990. *A System Dynamics Simulation Model of Cocaine Prevalence.* Final Report to the National Institute of Justice, Washington, D.C. Grant 87–IJ–CX–0042.

Homer, J. B., and C. L. St. Clair. 1991. A Model of HIV Transmission Through Needle Sharing. *Interfaces* 21 (3): 26–49.

Hser, Y. I., M. D. Anglin, T. D. Wickens, M. L. Brecht, and J. B. Homer. 1992. *Techniques for the Estimation of Illicit Drug Use Prevalence: An Overview of Relevant Issues.* Research Report NCJ 133786. National Institute of Justice, Washington, D.C.

JCC (Judicial Council of California. Administrative Office of the Courts). 1990. *Sentencing Practices Quarterly* (no. 51), March 31, 1990.

Johnston, L. D., P. M. O'Malley, and J. G. Bachman. 1989. *Drug Use, Drinking, and Smoking: National Survey Results from High School, College, and Young Adult Populations, 1975–1988.* DHHS Publication (ADM)89–1638. Washington, D.C.: Government Printing Office.

Levin, G., E. B. Roberts, and G. B. Hirsch. 1975. *The Persistent Poppy.* Cambridge, Mass.: Ballinger.

NNICC (National Narcotics Intelligence Consumers Committee). 1987. The NNICC Report 1985–1986: The Supply of Illicit Drugs to the United States from Foreign and Domestic Sources in 1985 and 1986. Unpublished manuscript (June).

Reuter, P., G. Crawford, and J. Cave. 1988. *Sealing the Borders: The Effects of Increased Military Participation in Drug Interdiction.* R-3594-USDP. Santa Monica, Calif.: The RAND Corporation.

Shreckengost, R. C. 1984. How Many Heroin Users Are There? Unpublished manuscript (June 4).

———. 1985. Estimating Cocaine Imports into the United States. Unpublished manuscript (October).

Sobel, K. H. 1990. Cocaine-Related Hospital Emergency Room Visits Drop 30 Percent. *NIDA Notes* (Fall), 6–7.

Tank-Nielsen, C. 1980. Sensitivity Analysis in System Dynamics. In *Elements of the System Dynamics Method*, ed. J. Randers. Cambridge, Mass.: Productivity Press.

Wuestman, E. A. 1990. Sowing Supply: Compensating Responses by Rural Coca Economies to Intervention in the Cocaine Market. In *Proceedings of the 1990 International System Dynamics Conference*, Chestnut Hill, Mass.

5

Why We Iterate: Scientific Modeling in Theory and Practice (1996)

Homer J. *System Dynamics Review*, 12(1): 1-19

Why we iterate: scientific modeling in theory and practice

Jack B. Homer

An approach to system dynamics modeling is advocated that adheres to the scientific method, and that may be applied regardless of model scope or size. Scientific modeling is distinguished from other approaches largely by the quality of evaluation and revision performed and by an insistence upon empirical evidence to support hypotheses and formulations. Three case studies drawn from the author's experience are presented. Practical lessons for scientific modeling are given to help guide expectations and maximize effectiveness of the approach. Modelers and clients should clearly understand the level of rigor they wish to pursue and what this means for the degree of confidence that may be placed in model results and insights.

Jack B. Homer is a system dynamics modeling consultant and president of Homer Consulting, with clients in business and government. He was formerly an assistant professor at the University of Southern California and received a PhD

iteration (n): a procedure in which repetition of a sequence of operations yields results successively closer to a desired result.

[Real] science is, quite simply, the scientific method, the relentless iteration of induction and deduction, of precise hypothesis-formation and careful experimentation—Meadows and Robinson 1985, p. 419

As a business consultant, I am often asked by prospective clients, "What's so great about system dynamics? What will my organization get that we can't get from other, more familiar, methods of strategic planning and analysis?" I tell them that system dynamics is the one method that will allow them to make all of their assumptions explicit and integrate them in a logical and testable way. A system dynamics model will also challenge the assumptions, provide a broader perspective, and put clients in a better position to make decisions that will stand the test of time. I also say that system dynamics models don't come ready-made off the shelf, but must be customized, tailored to the specific issues and peculiarities at hand, and tested to insure their semblance to reality and usefulness for "what-if" testing. And I describe model development as a process that is iterative, involving a certain amount of trial and error, and often requiring significant time and effort to come to fruition.

Clearly, I equate good system dynamics practice with care and rigor and I believe these are qualities that clients should expect from the modeling process. The characterization of system dynamics modeling as a rigorous process should seem familiar and natural to those who have worked long in the field or have read its classical works. But in practice today one also finds a brand of system dynamics that is less painstaking and, in particular, less concerned about the validation of assumptions through the use of numerical data and written documents. Clients as well as modelers are understandably intrigued by the idea of simulation models that do not require large investments of time and money. I myself have built such exploratory or impressionistic models on occasion (see, for example, Homer 1985) and believe they have a proper role to play in the spectrum of modeling practice. Indeed, one might say that every system dynamics model goes through an early exploratory stage in which some hypotheses are formulated with little or no empirical foundation. But models that go no further than the exploratory stage should not be confused with those that are subjected to the rigors of scientific evaluation.

I think it is vitally important that clients understand the difference between modeling approaches that adhere to the scientific method and those that do not. I am concerned that organizations may become enamored of "insights"

System Dynamics Review Vol. 12, no. 1, (Spring 1996): 1–19
© 1996 by John Wiley & Sons, Ltd. CCC 0833–7066/96/010001–19

Received April 1995
Accepted September 1995

from the Sloan School of Management at MIT for a thesis modeling the diffusion of new medical technologies. *Address*: 36 Covington Lane, Voorhees, NJ 08043, U.S.A.

and "learnings" without stopping to evaluate their validity. To put it plainly, feedback models are not exempt from the law of "garbage in, garbage out". It may be exciting to discover a feedback loop, but the fact that a loop exists does not mean it is dominant or even important. Instead, the loop should be considered carefully in the light of many other loops and pieces of evidence before it becomes the basis for real-world decisions. A client who expects solid results from a model should also demand a rigorous process of evaluation that digs below the surface of things to determine the relative importance of apparent insights.

Theoretical foundations

The important thing is not to stop questioning—Albert Einstein

Facts do not cease to exist because they are ignored—Aldous Huxley

If error is corrected whenever it is recognized as such, the path of error is the path of truth—Hans Reichenbach

The theoretical roots of scientific modeling may be found largely in the classical system dynamics literature on model evaluation and validation. Forrester (1961, Chapter 13; Forrester and Senge 1980) describes a multi-faceted approach to building confidence in a model's structure and behavior and thereby enhancing its effectiveness. Randers refers to the process as "generalized evaluation" (1973, p. 41) or "generalized testing" (1980). Such a demanding, wide-ranging approach is required largely because a model capable of reproducing past history may still be far from adequate for investigating possible futures. To be an effective tool for studying the future, a model must also have: equations that are consistent with reality and robust in the face of plausible extreme conditions; plausible input values; a boundary adequate for testing all relevant "what-if" questions; plausible outputs under all test conditions; and qualities of clarity, explanatory power, and novelty or surprise (Mass 1991).

Given the challenging criteria of generalized testing, it is natural that a well developed system dynamics model should go through multiple rounds of revision and evaluation. The initial focus of evaluation is typically the replication of reference behaviors, with the emphasis later shifting to questions of robustness, flexibility, and clarity. Model revision may consist of changes in system boundary, level of aggregation, or detailed formulation. Revisions often result in greater model complexity, but realistic detail should not be introduced at the expense of comprehensibility (Randers 1973, p. 282). The iterative process may, in theory, continue as long as the model fails to satisfy some evaluative criterion (Randers 1980). However, there is always some further refinement that may be made. Ideally, that is, given enough time and resources, one would like to weigh the benefits of further refinement against their costs as an indication of when to call a halt to iteration (Richardson and Pugh 1981, p. 305).

Scientific modeling may be distinguished from other approaches not by the presence or absence of evaluation and revision—indeed, all modeling is iterative to some extent—but by the manner in which evaluation and revision are performed. In scientific (or "refutationist") modeling the evaluation is done carefully and in depth, with a critical eye and an insistence on empirical evidence (Bell and Bell 1980). A concerted effort is made to gather historical data on as many relevant variables as possible for use as reference behaviors. In regard to the gathering of evidence in general, the modeler does not stop at the "easy pickings", but delves deeply into all relevant storehouses of data and experience, be they numerical, written, or mental (Forrester 1980). The initial model is based on accepted concepts and relationships, rather than speculations that run far afield from what is known. All discrepancies between model and evidence are investigated and their causes isolated to determine whether the model can not only reproduce history, but also do so for the right reasons. Revised hypotheses and formulations are supported with the same depth of empirical evidence used to support the original ones. The result is not necessarily a large model, but a model which takes into account a wide range of known details and which is therefore capable of making predictions with levels of confidence and insight greater than those of an exploratory model.

Modeling is largely a process of trial and error. Revisions are inevitable when a model is not taken from the shelf, but built from scratch to address a client's specific issues and conditions. Opportunities for model improvement are not always apparent or obvious and, indeed, often need to be drawn out through careful data investigation and close questioning of the client. Scientific modeling is not about minimizing the need for model revision through leaps of intuition (although they can help speed the process along), but rather about recognizing model shortcomings and following through with solid improvements.

Three case studies

It is important to understand not only the theory of scientific modeling, but also what one may expect to encounter in real-world applications. I believe there are some common themes or lessons that may help guide the expectations of both modelers and clients, and allow them to get the most out of the process while avoiding frustrations. Before discussing these lessons, however, I will first set the stage with a review of three modeling projects.[1] Although all three projects deal with product markets, they vary widely both in terms of scope and size and in terms of client type, and each illustrates some different aspects of my experience with scientific modeling. They will be presented in chronological order.

Cocaine prevalence model

In 1987, I began work on a system dynamics model of national cocaine prevalence funded by the National Institute of Justice (NIJ), the research arm of the U.S. Department of Justice. This model was developed in collaboration with a university-based team of drug abuse researchers and statisticians, with a team leader who is highly regarded in the field. In addition, I had several opportunities to present the work to the nation's leading drug abuse researchers and get their feedback. The work was funded for three years and took about 100 days of my time. Previous articles have described in detail the model's evolution (Homer 1990) and its final form (Homer 1993), so only the highlights will be discussed here.

The work began with a review of the existing literature and models and a limited amount of cocaine indicator data. Much of the literature suggested that price and availability of supply should be central to any analysis of illicit drug markets. Supporting this economic approach was an existing system dynamics model of the national heroin market (Gardiner and Shreckengost 1987), which was already familiar to the NIJ and other government agencies, was small and easy to understand, and had proved able to reproduce some key heroin indicator time series data.

From the early information, a small initial model was built with a basic causal structure as shown in Figure 1. The central hypothesis was that initiation to cocaine use was driven largely by price, with some fraction of casual users escalating to compulsive use. Price was modeled as a function of supply versus consumption, which also determines drug purity. Supply was modeled as a delayed response to price, with smuggling stimulated when domestic price is high relative to the foreign import price. Other factors, such as social acceptance and the development of crack cocaine, were considered secondary to price and perhaps superfluous from a modeling standpoint. The initial model was calibrated in a rough way and exhibited an interesting oscillation with a long period of several years. My team leader felt the model had enough face validity to be written up as a report and sent to several of his colleagues around the country for informal review.

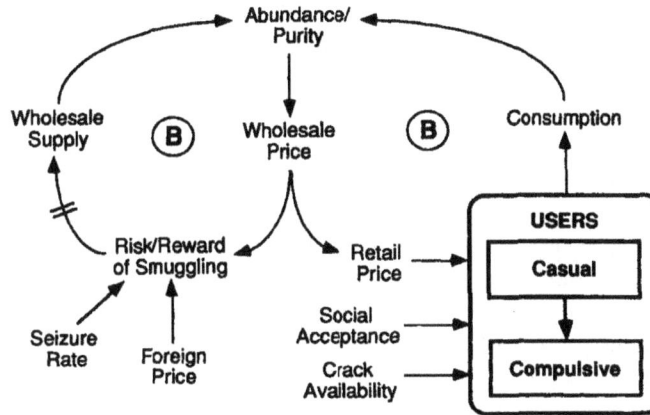

Fig. 1. Initial structure of cocaine prevalence model

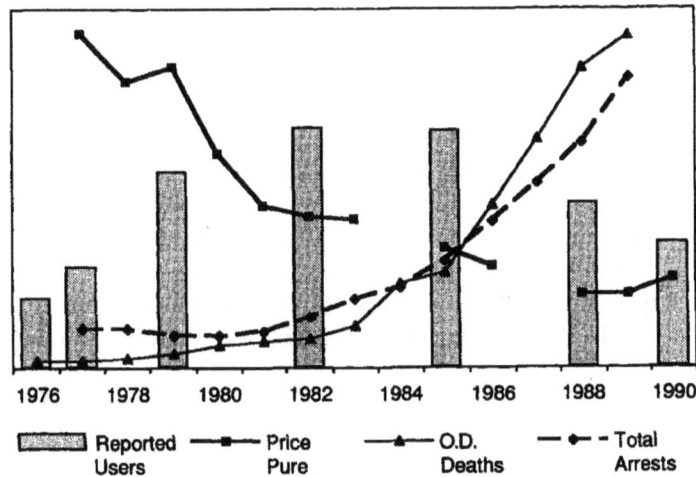

Fig. 2. Some of the historical data for cocaine prevalence model

Over the next $2^1/_2$ years, the model was revised repeatedly, in part because of experts' critiques, but more so because of the desire to create a model that could reproduce all relevant historical data for the right reasons. Compilation and updating of useful data became a major effort and continued nearly through the end of the project. We ultimately developed a rich and well documented picture of cocaine dynamics for the period 1976–1990, a small fraction of which is presented graphically in Figure 2. The indicator data came from nine different independent sources covering user categories, economics, morbidity and mortality, criminal justice, and social attitudes and beliefs.

Some of the data arrived in a raw or rough form, and required extensive numerical analysis to deduce the essential quantities. For example, the Drug Enforcement Administration (DEA) provided a raw data tape documenting, without summary, thousands of undercover cocaine purchases and seizures from 1977 to 1990. These data were our most reliable and complete source of information on price and purity at the retail and wholesale levels, but required extensive statistical analysis, resulting in large piles of computer print-out requiring further manual review. Though a time-consuming effort, this analysis turned out to be a turning point in the modeling project. In particular, the DEA data dispelled some key assumptions of the initial model, including the notion that there is a fixed relationship between wholesale and retail price and that supply is driven primarily by domestic price. As a result, we were forced to abandon temporarily that portion of the model explaining supply and pricing, subjects to which we returned only much later in the modeling project when a new theory of illicit drug pricing (Caulkins 1990) came to our attention.

An overview of the model's final causal structure is shown in Figure 3. The full model contains about 20 levels and 400 equations in all. The rise and fall in the total number of users is explained by four feedback loops, two reinforcing and two balancing. Sociological variables are central to the story. Supply and price still play a role, but they primarily reflect and amplify trends in demand rather than reshape them. The spread of highly addictive crack cocaine is depicted separately from that of powder cocaine and is the main reason for an explosion of consumption and its health and legal consequences in the late 1980s, at a time when powder cocaine use was on the decline. Also, the final model makes a distinction between apparent or

Fig. 3. Final structure of cocaine prevalence model

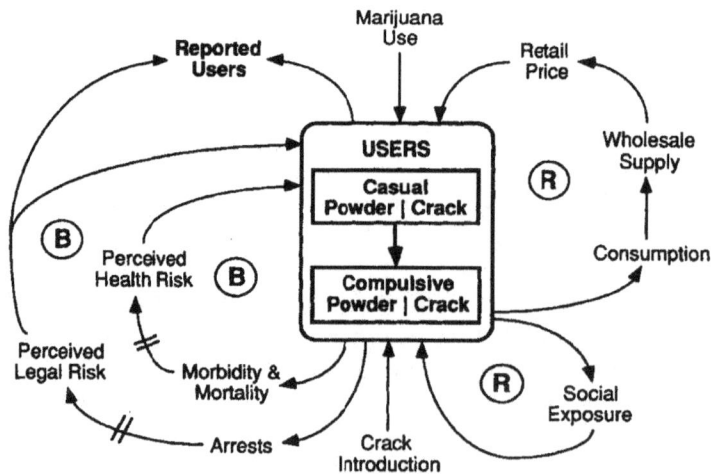

reported numbers of users and the actual underlying numbers. A rising rate of under-reporting in the 1980s is explained as partly the result of an increasing fear of legal sanctions and partly the result of a changing distribution of users.

The theory depicted in Figure 3 is an elaborate one by most standards and certainly in comparison with the initial model, but every element of this theory is supported by empirical evidence and helps to explain some aspect of the historical record. The model has been used to make future projections and for policy analysis, and has been generally well received by drug abuse researchers. One government economist, openly sceptical of system dynamics at the start of the project, stated publicly at a large interagency conference in 1991 that this was the best work on illicit drug markets he had seen in the past five years.

Cholesterol-lowering pharmaceuticals model

In 1991, I began work with Sandoz Pharmaceuticals on the first of what was to become a series of models dealing with the positioning of Sandoz products in the U.S.A. The first modeling project dealt with the market for cholesterol-lowering agents, of which Sandoz had one in development slated for launch in 1994. This project was broad in scope and very time-consuming, but was influential and became the foundation for the other models that followed. Carried out in two phases, the project took a total of 100 days of my time over an elapsed period of nine months. I worked intensively with the manager responsible for developing business plans for new drugs in the cardiovascular therapeutic area. We also received support from a group of internal experts at Sandoz, available as needed to assist with market research and issues including marketing, pricing, medical economics, and managed care.

Sandoz knew from the outset that its cholesterol-lowering drug, Lescol, was entering a market that would be crowded with strong competitors by 1994. Moreover, it was a "me-too" drug, one of a class of drugs, the first of which (Mevacor) had been introduced in 1987, and two more of which (Pravachol and Zocor) were to be introduced in 1992. Although nobody thought Lescol would be a breakaway success, previous industry experience suggested that a me-too drug—even the third or fourth of its kind—could grab some market share and turn a profit. But rarely had a me-too drug been introduced so late in the game as Lescol would be, a fact that led to genuine uncertainty about whether it could be positioned for success. The purpose of the model, then, was to investigate a variety of positioning alternatives, mainly focused on pricing and marketing, to determining whether Lescol could become a winner and what it would take.

To satisfy this purpose, the model would have to contain at center a well grounded theory of how patients end up on one cholesterol-lowering agent or another. In terms of data, we started with Sandoz market research (specifically, conjoint analysis) findings on physician preferences and biases, as well as basic information on competing drugs. The data on competitors covered efficacy, safety, and pricing, as well as five years of quarterly time series on prescriptions and marketing. Figure 4 shows total quarterly prescriptions for the four types of drugs available during the 1987–1991 period. (Lorelco, Lopid, and Mevacor are each individual agents, while the Resins are a class of drugs sold under a variety of forms and names.) It is evident from these data that Mevacor grew strongly starting from its introduction mid-1987, becoming the most prescribed agent by 1988 (while expanding the size of the total market) and continually increasing its market share thereafter. Figure 5

Fig. 4. Historical prescriptions data for cholesterol-lowering pharmaceuticals model

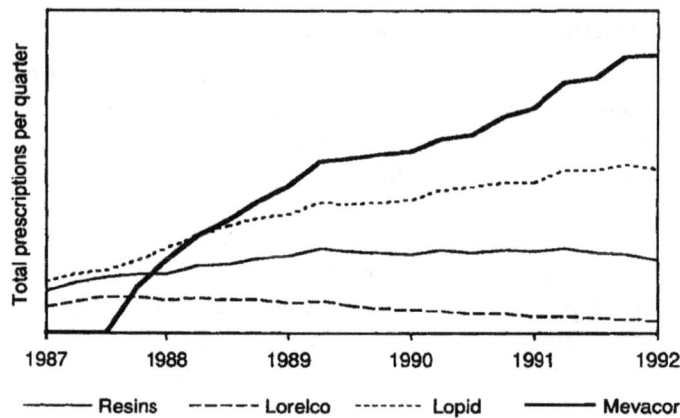

Fig. 5. Historical marketing data for pharmaceuticals model

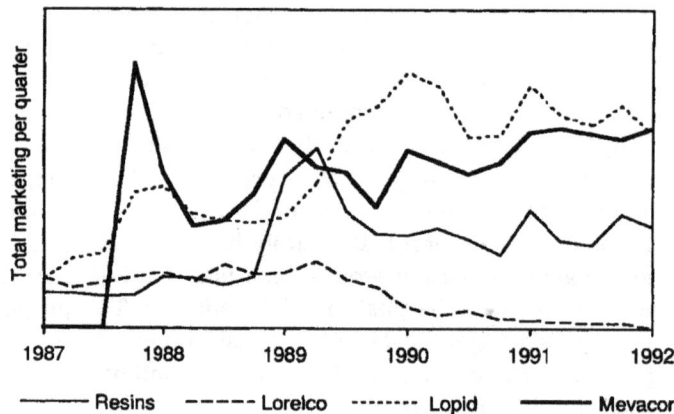

shows total quarterly marketing (including direct selling, advertising, and promotion) for these same drugs. Although a bit erratic, these data show a strong initial spike of post-launch marketing activity for Mevacor followed by gradual growth from 1988 onward, as well as marked increases in marketing of Lopid and the Resins starting in 1989.

Over a period of several months, the basic model structure seen in Figure 6 emerged. Although the model eventually became very detailed and complex, it started out relatively simply. The purpose of initial modeling was to develop a theory that could explain the historical prescriptions data in Figure 4, rather than to investigate the future. One basic hypothesis was that growth in the total number of patients starting drug use, and thus growth in the size of the market, was the result of physicians perceiving improvements in the utility of available drugs compared with diet and exercise regimens alone. Patients starting drug use would then be placed on a specific drug, on the basis of its perceived utility relative to other available drugs. Perceived utility was formulated as a function of intrinsic drug attributes and price, with marketing and word of mouth among physicians viewed as sources of information that are effective but have diminishing returns to scale. A specific drug could gain market share not only by attracting new drug users, but also by convincing existing patients to switch from other drugs, again for reason of superior perceived utility. However, patient switching was not initially thought to be a widespread phenomenon and the willingness-to-switch parameter in the model was accordingly given a small value.

Fig. 6. Feedback structure of pharmaceuticals model near completion

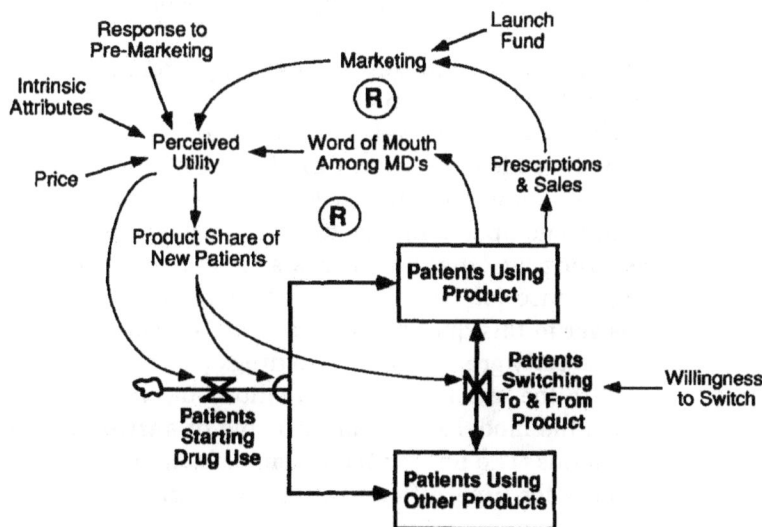

In our first attempts to reproduce the historical prescriptions data, we adjusted uncertain parameters related to perceived utility (including attribute importance weights and marketing and word-of-mouth impact parameters) within their plausible ranges, while treating marketing and price as exogenous. (Endogenous modeling of competitors' marketing and pricing decisions was not pursued until later in the project.) Although the model could reproduce the general trends, it consistently underestimated the overall growth of Mevacor, as well as the steepness of this growth immediately post-launch.

We next considered possible sources of the discrepancies, and found two important changes capable of significantly improving the model's fit. The first change consisted of raising the willingness-to-switch parameter value. With a substantial, but still plausible, increase in this parameter, the model was able to produce overall growth for Mevacor quite close to the historical level. We were thus led to the unexpected conclusion, later supported by anecdotal evidence, that many physicians were actually quite willing to switch their patients to a new cholesterol-lowering drug perceived as superior, even if the older drugs were causing no particular problems for these patients.

The second key change was the introduction of a new factor affecting perceived utility, seen in Figure 6 as the "response to pre-marketing". One of our internal experts had remarked on the fact that the initial model contained nothing about the pre-launch marketing activities that a pharmaceutical company may pursue in an attempt to build pent-up demand for a soon-to-be-released product. These activities often involve supporting the dissemination of favorable clinical trial results. They may also include "seeding trials", in which a large number of physicians are invited to try the product and perhaps complete some minimal evaluation. The response to pre-marketing was theorized to be strongest immediately after launch, dwindling away in importance after a year or so.

The response-to-pre-marketing factor proved capable of explaining the rapid growth of Mevacor within the first year after launch, thereby completing our goal of reproducing the historical prescriptions data.[2] However, we were left with some uncertainty as to whether the response to pre-marketing would be as strong for other new drugs as it was for Mevacor, a question of particular importance for Lescol and other me-too drugs in Mevacor's drug class. The answer to this question would have to wait until data were gathered in 1992 on the early growth of the two drugs launched in that year.

With the initial model evaluation objective satisfied, we moved forward to make the model richer and capable of answering a wide variety of "what-if" questions. The full model explains changing market shares and movements in pricing for nine different drugs and drug classes, including those available during the historical period and others (including Lescol) whose future introduction was likely or possible. The full model is also disaggregated into over 200 market segments defined by physician specialty, patient age and sex, severity of disease, and insurance type. (This disaggregation was supported by a detailed cross-sectional analysis of patient and prescription data for 1991.) Also, intrinsic drug attributes are defined along six different dimensions of efficacy, safety, and convenience. Altogether, the full model includes 12 basic levels for each drug or drug class, and over 1000 variables, many of them subscripted along multiple dimensions. It is a very large system dynamics model by any standard, its size determined mainly by the amount of detail the client deemed appropriate—given a very solid basis in data—to portray adequately the full array of positioning options and market scenarios under consideration.

The second-phase process of disaggregation and refinement to produce the full model was for the most part a smooth one, except for some frustrating delays in data collection. But as we entered the final month of the project, preparing to declare the model's structure "frozen" for final policy testing, some new information became available that led to a significant change in the model's basic patient flow structure.

The new data showed that patients on cholesterol-lowering drugs have an unusually high rate of non-compliance, about 50 per cent of all patients discontinuing their prescribed treatment within the first year. This fact by itself need not have led to a change in the model, if all drugs displayed equal rates of non-compliance.[3] But, in fact, the data showed a particularly high rate of non-compliance for the Resins, something like 80 per cent per year, no doubt due to their notable inconvenience of use. (Resins come in various non-tablet forms, all requiring large, multiple daily doses and having an unpleasant gritty consistency.) Also, the new compliance data were causing some Sandoz marketing people to think about a kind of packaging for Lescol that would make it easier to take than other drugs, that would actually encourage use in some way. Suddenly, the possibility of "compliance packaging" became a policy issue for the model.

The final change to the model resulted in the patient flow structure shown in Figure 7. In this structure, some patients are permanently compliant while others only temporarily so, with a non-compliance fraction specific to each drug product. Non-compliant patients usually return to their physicians within a year or so, at which point they generally receive another prescription for one or another cholesterol-lowering agent. The new structure improved the model's correspondence with documented reality and gave the client the desired flexibility to test all policy options under consideration.

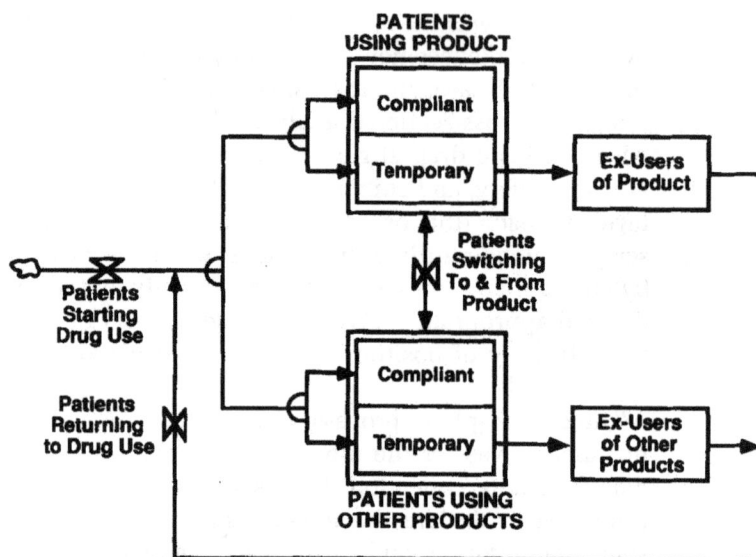

Commodity chemical model

I was in the process of completing a very large modeling project with one of
the plastics businesses at Dow Chemical in Europe when, at the end of 1994,
the director of marketing for chemicals had the notion that system dynamics
might be applicable in his area as well. Specifically, he was frustrated at the
inability of econometric models to forecast accurately industry prices and
demand for commodity chemicals and he spoke of the desire for a more realistic and useful tool, one that could support "grounded forecasting". As a
trained physicist, he intuitively understood the power of causal modeling and
had a natural affinity for system dynamics. It was decided to develop a small
model focusing on a particular commodity chemical product. The effort
ended up requiring only six days of my time over an elapsed period of three
months. The modeling team consisted of the marketing director, his product
manager, an internal modeling consultant, and myself.

The fact that we were able to pursue a path of scientific modeling in such a
short period of time is due, I think, to three factors: the availability of good
time-series data from the outset, the involvement of a client with decades of
experience in the business, and a decision to model for general understanding
rather than policy analysis. Three of the key historical time series, covering
the 1987–1994 period in monthly or quarterly intervals, are presented in
Figure 8. These data come from trade association figures which aggregate
across all producers in Western Europe.

Fig. 8. Some of the
historical data for
commodity chemical
model

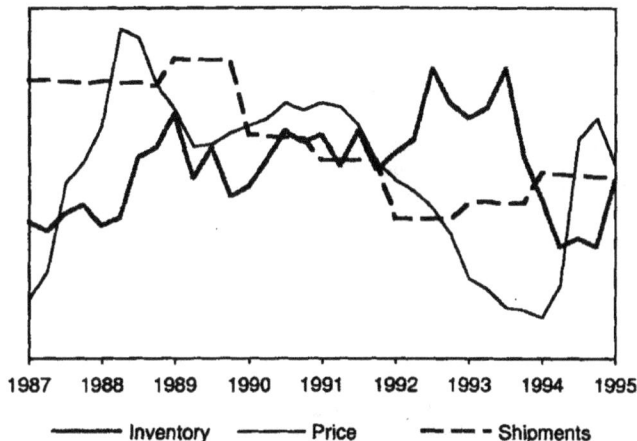

The curve for price in Figure 8 fluctuates widely, going through a bit more than one complete cycle during the eight-year period of interest.[4] The marketing director said that such large amplitude, multi-year price cycles had occurred repeatedly in the past, not only for the product in question but for other commodity chemicals as well. Production inventory grows more or less steadily through mid-1993, followed by rapid decline, and always stays within upper and lower economic limits of tank storage for the industry as a whole. Also seen in Figure 8 is a time series on shipments, actually the sum of separate data on domestic shipments and exports. Other available time series (not shown here) tracked production of the chemical and the price of a competing (substitute) product.

Our strategy for modeling was to start with many of the linear, econometric-type equations already familiar to the client and build from there. The first-version model contained plausible causal connections and equations, but could not generate output resembling the historical time series. This experience alone was an eye-opening one for the client and led quickly to the consideration of some alternative formulations that broke out of the econometric mold. Consequently, a second-version model was developed, with a feedback structure controlling shipments as shown in Figure 9.[5] This model contains a relatively slow loop from inventory to price to demand to shipments; and a fast "safety valve" loop that speeds shipments when inventory nears its upper storage limit and slows shipments when inventory nears its lower storage limit.

We found that the second version model could do a reasonably good job of reproducing history, but only when some implausible assumptions were made about parameter values. First, the model's formulation of price as a negatively sloped function of inventory (based on the original econometric

115

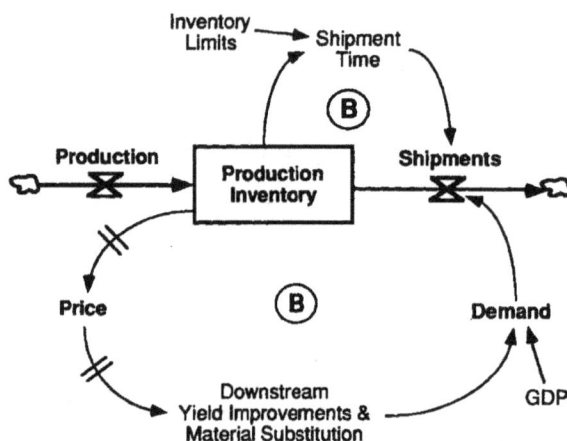

Fig. 9. Feedback structure of commodity chemical model, second version

model) gave a good fit only if inventory was first smoothed and an unrealistically long smoothing time used. Second, coefficients related to the effect of GDP (gross domestic product) on demand had to be unrealistically large to give a good fit. Third, a good fit could be obtained for the 1993–1994 period (when shipments rebounded after a decline) only if other demand coefficients were assumed to have change dramatically around 1991 or 1992. Clearly, there was something still wrong in the model's feedback structure.

We returned to the drawing board to revise our modeling of price and shipments, ending up with a third and final version model as represented in Figure 10. On the subject of price, a fresh look at the data in Figure 8 revealed periods, such as 1987 and 1991, when inventory was relatively flat but price changed quickly. Specifically, in 1987 inventory was low and price increased, whereas in 1991 inventory was higher and price fell. We hypothesized that, in general, low inventory levels cause price to increase while high inventory levels cause price to decline. The marketing director felt comfortable with this new theory and its implication that there was a desired level of inventory at which price would not change. Also, a partial-model test (Homer 1983), using historical inventory as the input, confirmed that the new pricing structure could give a close fit to the data on price.[6]

Thinking about shipments, we concluded that the problems of overly large coefficients and suddenly changing coefficients must both point to a lack of structure adequate to explain the rapid decline in shipments 1990–1992 as well as the rebound 1993–1994. We discussed industry trends of the 1990s, and the phrases "cost cutting" and "just-in-time" came up. The marketing director described how the chemical is stored not only by producers, but also in a long downstream chain of customer stocks that incorporate the chemical

Fig. 10. Feedback structure of commodity chemical model, third version

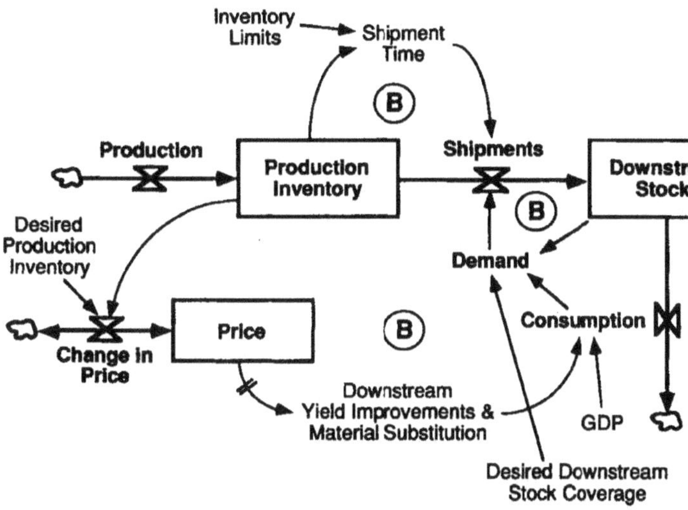

in some fashion. Successful efforts to reduce these downstream stocks had, in fact, been made in the early 1990s, and the extent of this "de-stocking" was roughly known. Incorporating the idea of de-stocking in the model led to the structure seen in Figure 10, in which the demand seen by producers reflects not only end-use consumption, but also the attempt to correct any discrepancy between actual and desired downstream stocks. Testing this structure, we realized that a process of de-stocking implies not only a period of rapid decline in shipments, when demand is suppressed below consumption, but also a period of rebound (as seen in the data) when the destocking nears completion and demand need no longer be suppressed.

The third-version model not only provided a close fit to the historical data, but did so with a set of parameter values and causal mechanisms that seemed solid and right to the client. The model is a small one, containing only about 60 equations in all. It has been used for making short-term forecasts and for scenario analysis. More importantly, it has received favorable reviews from Dow executives on both sides of the Atlantic and is being extended and adapted for use in global market analysis of various commodity chemicals.

Lessons

My experiences with real-world scientific modeling, including those described above, suggest that there are some practical lessons to be learned. These lessons do not contradict the theory of scientific modeling, but rather supplement it. The following list is undoubtedly incomplete and is based primarily on my own experience, but I hope it proves useful to others in sailing a steady course through what can sometimes be choppy waters.

1. Adherence to the scientific method is essential for developing models that will stand the test of time. Both client and modeler may be tempted at times to abandon the process, especially when needed evidence is not readily available, but should remind themselves that the devil is often in the details.
2. Model revision usually continues nearly to the end of the project. Do not expect the refinement process to come to a natural halt, nor for the modeling team to feel comfortable at any point in freezing the model. One usually faces a tradeoff toward the end of a project, in which additional model refinement will cut into the time available for policy and scenario testing. This tradeoff should be faced squarely by modeler and client and settled with the goal of maximizing the model's effectiveness while accepting certain limitations.
3. Clients often remain unaware of a model's flaws unless brought to their attention. As the one most familiar with the model, the modeler must repeatedly bring suspected shortcomings to the client's attention and not be satisfied with surface appearances. Clients should be involved in evaluating the model, of course, but deep evaluation is realistically more the modeler's responsibility than the client's.
4. Scientific modeling often encounters unexpected twists and turns, as it does in the physical sciences. Model structures that seemed settled may change dramatically, sometimes even late in the game. What seemed minor may become major and vice versa. Allow time for these surprises and do not steamroll over them or ignore them when they occur. Treat them as valuable discoveries, rather than as signs of stumbling or inefficiency.
5. Mental, written, and numerical data all have important roles to play in support of scientific system dynamics modeling. Mental and written data (the latter often in the form of company reports and memos) are rich and plentiful, but also imprecise and often ambiguous or even contradictory. They are much better at describing how things work and how people behave than they are at specifying the magnitude and timing of historical trends. Such measurements are the special domain of numerical data, but even in the case of supposedly solid numbers one must look closely at what exactly has been measured and the possibility of measurement error or bias.
6. Data collection requires persistence, patience, and immersion in messy details. Clients know more and have access to more information then they initially think they do. The modeler often ends up getting involved as a

data detective, compiler, and analyst. All pieces of evidence should be examined closely and their sources understood, so that suspect data may be given less weight than higher-quality data.

7. Some important data will be in error, irrelevant, missing, or slow to emerge. Such data problems can slow things down and may require a change in tactics. One may need to look to other sources for data, or consider using different variables or levels of aggregation to capture the same phenomena. Or, one may simply move forward with the existing model and data and accept a lower level of model confidence until such time as better data become available.

8. Insights gained from model evaluation and revision may equal or exceed those gained from initial conceptualization and final policy testing. Evaluation-based insights often help to resolve controversies about what is really important in a system and what is not. These insights should be documented and given as much weight as policy findings. Sponsors and decision-makers generally appreciate insights gained from model evaluation. They also find model results more persuasive if they see that the model's structure is not arbitrary but has been found superior to other plausible structures.

Conclusion

What would computer modeling be like if its practitioners . . . chose to identify themselves with the disciplines and responsibilities of science, rather than just its authority and privileges?—Meadows and Robinson 1985, p. 421

How many people today believe that empirical data are of relatively minor importance in system dynamics? I am concerned that some may have the impression that conceptualization is the main thing and takes priority over the careful gathering of evidence, or that all the important evidence will come naturally and easily to the surface. I wonder, in other words, if the lines between exploratory modeling and scientific modeling may have been blurred in some minds, including those of some consultants and practitioners.

We should resist the notion that system dynamics is mainly conceptual rather than empirical, that creativity is more important than thoroughness. System dynamics is much more than just another technique for brainstorming or critical thinking and will ultimately fade in significance and credibility to the extent that it backs away from its scientific roots. Scientific modeling can be pursued in any domain and regardless of whether the project is large or small in size. It is a method and a mindset, and its rationale lies in the distinction between apparent insights and validated insights.

In today's fast-paced and cost-conscious world, clients and consultants alike may feel the pressure to take shortcuts and reach conclusions quickly. The temptation may exist to make unsubstantiated claims for a model and to discount the need for further data collection or model refinement. Scientific modeling takes time and patience, just as most real learning does. But even when a tight deadline cannot be avoided, or when the best available data are still sketchy, the modeler's responsibility should be to tell the truth about the degree of certainty that can be claimed and to describe openly and fully the model's limitations.

Notes

1. I would like to acknowledge my sponsors at Sandoz Pharmaceuticals Corporation and Dow Europe SA for their contributions and for agreeing to have their projects described publicly. For the sake of simplicity, and to respect the wishes of my sponsors, y-axis values have been omitted from all data graphs in this article.
2. We found that the initial spike in post-launch marketing of Mevacor, as large as it was, was unable by itself to explain the rapid take-off in Mevacor prescriptions in 1987.
3. One could, in the case of identical non-compliance rates, safely interpret the model's flow of "patients starting drug use" as a net flow adjusted for non-compliance.
4. Price here refers to the spot price of the chemical, as opposed to its contract price. Contract price was also modeled, but its role in the model was less important than that of spot price.
5. In contrast with shipments, production was modeled without feedback loops from inventory. The chemical in question is actually the by-product of another chemical process, which means that its production is determined by a separate stream of demand and is not under the control of the by-product chemical's management.
6. Those familiar with feedback control theory will note that the new pricing theory used an integral control structure in place of a proportional control structure. Integral control structures with delays may lead to high-amplitude cycles, as seen in the chemical price data.

References

Bell, J. A. and J. F. Bell. 1980. System Dynamics and Scientific Method. In *Elements of the System Dynamics Method*, ed. J. Randers. Cambridge, Mass.: MIT Press. Reprinted by Productivity Press, Portland, OR.

Caulkins, J. P. 1990. The Distribution and Consumption of Illicit Drugs: Some Mathematical Models and Their Policy Implications. PhD. dissertation, MIT, Cambridge, Mass.

Forrester, J. W. 1961. *Industrial Dynamics*. Cambridge, Mass.: MIT Press. Reprinted by Productivity Press, Portland, OR.

———. 1980. Information Sources for Modeling the National Economy. *Journal of the American Statistical Association* 75(371): 555–574.

Forrester, J. W. and P. M. Senge. 1980. Tests for Building Confidence in System Dynamics Models. In *System Dynamics, TIMS Studies in the Management Sciences*, Vol. 14: 209–228. New York: North-Holland Publishing Co.

Gardiner, L. K. and R. C. Shreckengost. 1987. A System Dynamics Model for Estimating Heroin Imports into the United States. *System Dynamics Review* 3(1): 8–27.

Homer, J. B. 1983. Partial-Model Testing as a Validation Tool for System Dynamics. In *Proceedings of the 1983 International System Dynamics Conference*, Chestnut Hill, Mass.

———. 1985. Worker Burnout: A Dynamic Model with Implications for Prevention and Control. *System Dynamics Review 1(1): 42–62*.

———. 1990. *Cocaine Use in America: The Evolution of a Dynamic Model*. In Proceedings of the 1990 International System Dynamics Conference, Chestnut Hill, Mass.

———. 1993. A system Dynamics Model of National Cocaine Prevalence. *System Dynamics Review* 9(1): 49–78.

Mass, N. 1991 (written 1981). Diagnosing Surprise Model Behavior: A Tool for Evolving Behavioral and Policy Insights. *System Dynamics Review* 7(1): 68–86.

Meadows, D. H., and J. M. Robinson. 1985. *The Electronic Oracle: Computer Models and Social Decisions*. Chichester, UK: John Wiley & Sons.

Randers, J. 1973. Conceptualizing Dynamic Models of Social Systems: Lessons from a Study of Social Change. PhD dissertation, MIT, Cambridge, Mass.

———. 1980. Guidelines for Model Conceptualization. In *Elements of the System Dynamics Method*, ed. J. Randers. Cambridge, Mass.: MIT Press. Reprinted by Productivity Press, Portland, OR.

Richardson, G. P. and A. L. Pugh III. 1981. *Introduction to System Dynamics Modeling with DYNAMO*. Cambridge, Mass.: MIT Press. Reprinted by Productivity Press, Portland, OR.

6

Structure, Data, and Compelling Conclusions: Notes from the Field (1997)

Homer J. *System Dynamics Review*, 13(4): 293-309

Structure, Data, and Compelling Conclusions: Notes from the Field

Jack B. Homer[a]

Jay Wright Forrester Prize Lecture, 1997

Jack B. Homer is a system dynamics modeling consultant and has directed Homer Consulting since 1989. He was formerly an assistant professor at the University of Southern California and received a PhD in Management from MIT for a thesis modeling the diffusion of new medical technologies.

Abstract

Some system dynamics models are more effective than others in changing the thinking and actions of their audiences. In my experience, the models that prove most compelling to clients generally have two things in common: a potent stock and flow structure and a rich fabric of numerical data for calibrating that structure. Stock and flow structures focus attention on the intrinsic momentum of a situation and allow one to track movements of people and things in a clear and systematic way. Numerical data not only help to build a client's confidence in a model, but also can materially affect the final structure and key parameter values of a model. Three examples are presented that demonstrate the strong inferences one may draw when stock and flow structures are combined with sufficient numerical data. System dynamics models should be built on a foundation of straightforward core structures and the full range of available evidence. © 1997 John Wiley & Sons, Ltd.

Syst. Dyn. Rev. **13**, 293–309, (1997)

No. of Figures: 11.   No. of Tables: 0.   No. of References: 14.

This article is about the sorts of compelling conclusions that can come from the combination of model structure with empirical data. I will present some general thoughts on this subject and then give three examples from my own experience.

To set the stage, consider the examples of world dynamics and urban dynamics presented in Jay Forrester's classic paper, "Counterintuitive Behavior of Social Systems" (1971a). This paper made a great impression on me when I first read it as a high school student in 1972, in large part because it took future-oriented computer simulation from the realm of science fiction (as in Isaac Asimov's famous *Foundation* books) and turned it into a practical tool for dealing with modern complexities. The paper also impressed me, and still does, for the power and clarity of its arguments. These logical arguments start from premises about causal structure and empirical quantities to reach striking conclusions.

Figure 1 presents a simplified way of viewing *World Dynamics* (Forrester 1971b).

Fig. 1. Simplified view of *World Dynamics* model

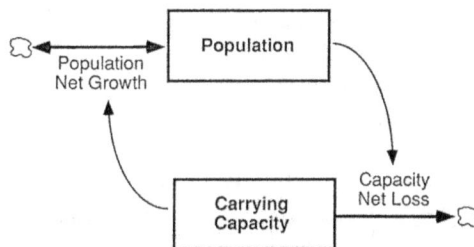

[a]36 Covington Lane, Voorhees, NJ 08043, U.S.A. E-mail: JackBHomer@compuserve.com

System Dynamics Review Vol. 13, No. 4, (Winter 1997): 293–309

© 1997 John Wiley & Sons, Ltd. CCC 0883-7066/97/040293–17 $17.50

Received September 1997

Accepted September 1997

The two stocks are *Population* and *Carrying Capacity*, the latter representing an upper limit on population based largely on natural resources. The model postulates that carrying capacity is not a constant, but will be reduced by a large enough population with a given pattern of resource consumption. Starting with this causal structure, one may infer:

> If carrying capacity is declining and the reason is population pressure, then population will overshoot and decline, unless its growth is halted short of capacity or its pattern of consumption radically altered.

This conclusion may be startling when first encountered, but is virtually inescapable once one accepts the empirical evidence that global carrying capacity is on the decline. The argument is a cornerstone of today's worldwide movement for sustainable development.

At the heart of the *Urban Dynamics* model (Forrester 1969) lie two progression chains of stocks and flows, one for housing and the other for businesses, as seen in Figure 2. When the usable land area becomes saturated with housing and businesses, construction slows and the ratio of old to new infrastructure increases. In other words, a process of urban aging or maturation occurs. Add to this process some general characteristics of older infrastructure, and one gets the following argument:

> If old housing supports more residents than new housing and old businesses support fewer jobs than new businesses, then urban maturation will increase unemployment, unless excess housing is removed and rezoned for new business.

Here, again, a powerful conclusion emerges from the combination of an easily understood dynamic structure with key pieces of empirical data. This conclusion has arguably led to the current U.S. policy of creating "empowerment zones" in the inner cities, with new businesses at their core.

It is not news that effective system dynamics models come from the integration of dynamic structures and empirical data. In "Information Sources for Modeling the National Economy", Forrester (1980) describes how modeling principles and

Fig. 2. Simplified view of *Urban Dynamics* model

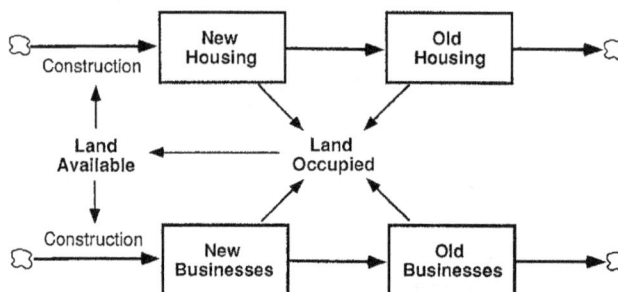

124

Fig. 3. The workings of a zipper (Macaulay 1988). Reprinted with permission. *Source*: Houghton Mifflin.

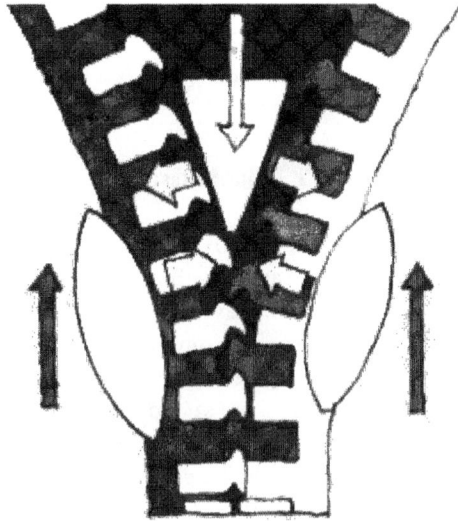

various types of information are brought together in the process of model construction. We all build models that consist of stocks and flows and feedback loops, and that use data sources both qualitative and quantitative.

But it is not every model that has the power to persuade and the power to generate valuable insights that *World Dynamics* and *Urban Dynamics* have. I know for certain that not all of my models have had this kind of power. The experience, when it does occur, makes a strong impression but one that is more subtle than, say, hitting a nail with a hammer or splitting wood with an axe.

In fact, I think the successful integration of structure and data is pretty well represented by a device we all use every day, the humble zipper. The illustrated zipper in Figure 3 is from David Macaulay's book, *The Way Things Work*. Here's what he has to say about it:

> As you open a zipper, the triangular wedge in the slide detaches the teeth and forces them apart. On closing, the two lower wedges force the teeth together so that they intermesh... Without the slide, it is practically impossible to free the teeth or make them mesh together. (Macaulay 1988, p 21).

I think of the modeling process working in both directions, both zipping open and zipping closed, using the combined force of structure and data. When one opens the zipper one is using a kind of analytical process, teasing apart truth from fiction, setting aside certain opinions in favor of others better supported by the evidence. When one closes the zipper one is engaged in a synthetic process, carefully intermeshing pieces of information in a way that allows one to reach new conclusions, in particular, conclusions about the future and conclusions about policy.

What kind of structure? What kind of data?

We know that models are not all equally effective, some lacking the zip that others have. It seems reasonable then to ask whether focusing on certain kinds of structure or certain kinds of data may improve the chances that a model will lead to conclusions that their audiences find compelling.

Although I have an opinion on this matter, I will admit that it comes from my own particular modeling experiences, and perhaps my own predilections. In any event, I find that of my models, the ones that prove most persuasive to clients generally have two things in common: a potent stock and flow structure and a rich fabric of numerical data for calibrating that structure.

On the importance of stock and flow structures, Barry Richmond has made the case convincingly in various writings (Richmond 1994a; 1994b). He uses terms like physics, plumbing, and infrastructure to refer to the stock and flow structures that lie at the heart of an issue, and he illustrates clearly the potential contained in certain types of these structures to produce valuable insights. For example, he notes that a progression chain, like those seen in *Urban Dynamics*, is a structure with a strong intrinsic momentum, or, as he says, "a mind of its own". He also describes the insights that can come from focusing attention on key flows and the stocks that undeniably drive those flows. *World Dynamics*, for example, focused attention on the fact that critical natural resources are being lost, a loss that is driven by the very stock of population that depends on those resources.

Stock and flow structures should also be recognized for their simple yet powerful ability to keep track of people and things as they move around. This tracking ability allows our models to make projections that adhere to basic laws of conservation that other methodologies ignore. Good stock and flow structures formally recognize that consumer products do not just disappear after purchase, that oil reserves are not infinite, and that a worker's time must be allocated among competing demands. This tracking ability also applies to co-flows, which, for example, can systematically calculate changes in the average employee skill level due to hiring, training, and turnover.

Not much has been written or spoken in our field about the insights that numerical data can provide. System dynamicists know that such data help us establish behavior reference modes, set parameter values, and perform model validation tests. But I have not sensed much enthusiasm for the collection and analysis of numerical data, something which is so central for other practitioners in the social and management sciences. Instead, system dynamicists tend to emphasize the importance of experiential data in model construction, and refer to the fact that feedback model behavior is often insensitive to changes in parameter values. These are valid points, of course, but taken to an extreme they may create a kind of tunnel vision, in which numerical data are viewed merely as a hurdle that must be cleared in order to gain the confidence of clients and other audiences.[1]

In my experience, there is much to be gained from not just clearing the perceived hurdle of numerical data but leaping well over it. I have previously written (Homer 1996) about the value of gathering extensive evidence for evaluating and refining a model. Hard data can materially affect the final structure and key parameter values of a model, and, consequently, its predictions and even its policy results.

Three case studies are presented in that previous paper, all showing just how much a model's structure may change as the result of a rigorous evaluation process. The examples presented below are perhaps less dramatic, but still just as significant to the models' sponsors. They are not about the modification of structure, but instead about the strong inferences one may draw when accepted stock and flow structures are combined with sufficient numerical data.

Three examples

Cocaine prevalence: the issue of reporting rates

I will start with the cocaine prevalence model which the Forrester Award committee has cited (Homer 1993). Probably the most striking conclusion of this model was that the U.S. National Household Survey of Drug Abuse has become more and more unreliable over time in its estimates of the number of cocaine users, and now underestimates by a large margin across all major categories of use. One key to reaching this conclusion was the development of a detailed stock and flow structure, a progression chain that produces explicit counterparts to the National Household Survey's own categories of weekly, past-month, past-year, and lifetime use.

Figure 4 presents that structure in simplified form. The model distinguishes between casual and compulsive users, between users who prefer cocaine's powder form as opposed to the more potent and addictive crack form, and between current and former users. Among former users, the full model distinguishes further between those who have used within the last year and those who have not. In all, there are 16 levels of users in the model.

The model also distinguishes between actual and reported users, with reported users intended to mimic the National Household Survey's own estimates. Reported users are calculated from actual users by applying reporting rates, fractions that

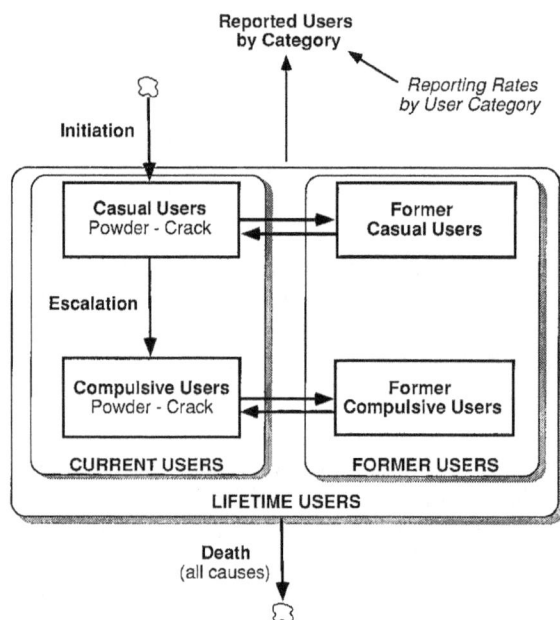

Fig. 4. Cocaine users stock and flow structure

reflect the availability of actual users or ex-users to be interviewed and their willingness to answer the survey questions truthfully.

These reporting rates vary by category of user and may change over time. The differences in reporting rates among categories of users have much to do with their socio-economic status and living conditions. Consequently, compulsive users have a lower reporting rate than casual users, and crack users have a lower reporting rate than powder users. Also, an increase over time in perceived legal risks and the social stigma of cocaine have caused reporting rates to decline for all user categories, but especially so for former users who have not used within the past year. There are many millions of such ex-users in the U.S.A., and as a group they tend to be older and more gainfully employed than current users, and have more to lose if the word about their former drug use were somehow to get out.

Figure 5 shows simulated reporting rates for the period 1976–1995 for three key aggregates measured by the National Household Survey: lifetime use, past-month use, and compulsive or weekly use. All three curves decline significantly over time, but for somewhat different reasons. The decline in lifetime use reporting is primarily due to the increasing legal risks and social stigma of cocaine. The decline in past-month use reporting is primarily due to a large increase in the number of compulsive users relative to casual users. And the decline in compulsive use

Fig. 5. Cocaine use
reporting rates
simulated 1976–1995

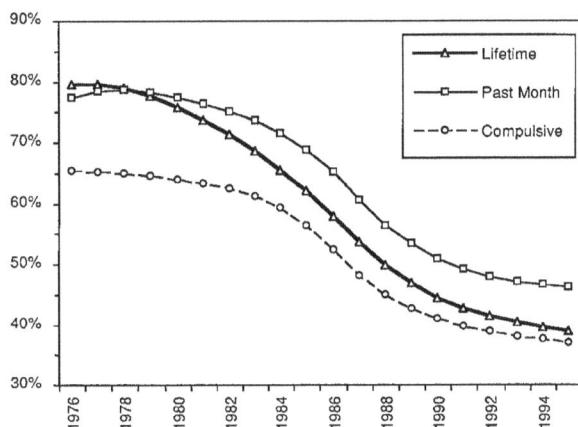

Fig. 5. Cocaine use reporting rates simulated 1976–1995

reporting itself is due to a large increase in the number of crack compulsives relative to powder compulsives.

The derivation of the reporting rates used in the model was not a simple process and involved integrating the model's structure with all of the other data that were available, data on morbidity and mortality, criminal arrests and incarcerations, cocaine seizures, and cocaine price. All of these other data, so-called indirect indicators of use, are much more reliable and more continuously measured than the National Household Survey, and together they tell an alarming story of increasing crack use and addiction which the Survey largely misses.

Though the reporting rates analysis became quite involved, it started with a simple observation. Figure 6 presents a graph of lifetime users, people who have ever tried cocaine, measured in millions for the period 1976–1995. The vertical bars are estimates from the National Household Survey, the line close to those bars is the simulation of those reported users, and the line that rises above those is the simulation of actual users.[2]

The Household Survey data show a pattern of robust growth from 1976 to 1982, followed by absolute flattening and even some periods of decline right through 1995. By definition, lifetime users represent the accumulation of new users, or initiates, minus deaths. With a death rate of less than 1% per year for this population, the data imply that initiation boomed along at two to three million new users per year right through 1982, and then quickly shrank to nothing.

The Survey data's implication that cocaine initiation was eliminated in the mid-1980s is simply not credible. In fact, that was precisely the time when crack started to spread and introduced a new generation and socio-economic class to cocaine. In contrast with the Household Survey data, the simulation of actual users suggests that initiation did decline overall in the 1980s and 1990s, but still sits at nearly 40%

Fig. 6. Lifetime users
of cocaine reported
and simulated
1976–1995

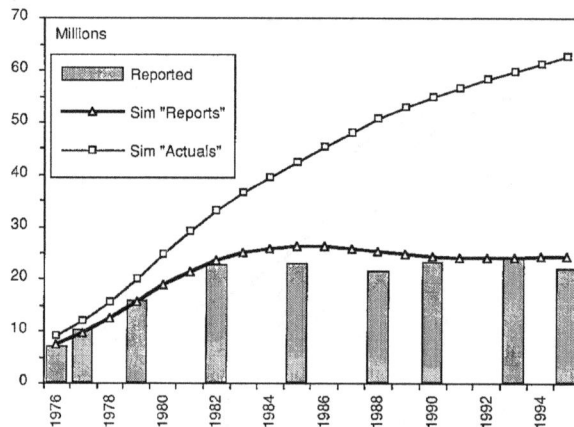

Fig. 6. Lifetime users of cocaine reported and simulated 1976–1995

of its 1979 peak, with three-quarters of current initiates using crack rather than powder. This is enough to add more than one million new users per year, so that rather than 20 to 25 million lifetime users of cocaine in the U.S.A. as the Household Survey estimates, there are now probably at least 50 to 60 million.

Motor vehicle parts sales: the issue of miles driven

The second example comes from a simpler model than the previous one and deals with a completely different subject, the sale of motor vehicle parts. But, like the cocaine example, this one illustrates how structure and data can be combined to draw logical inferences about key parameter values, even when some of the data are missing or imprecise. (See also Homer 1983 on the use of partial-model testing for parameter estimation.)

Figure 7 shows a portion of a model that was developed recently to analyze parts and accessories sales for a large U.S. manufacturer of motor vehicles. (The company must be left nameless here, but the other particulars described below are accurate and undisguised.) An owner of one of the company's vehicles may buy parts for routine maintenance, things like brakes and spark plugs, as well as replacement parts required in the event of breakdowns or collisions. These parts may be purchased from one of the company's own dealers, in which case they are supplied by the company itself, or from an unaffiliated mechanic who uses parts made by one of several competing parts manufacturers.

For planning and strategy purposes, the company needs to be able to project parts revenues several years into the future. Their primary concern is the flow shown in Figure 7 as parts shipments to dealers, which generates these revenues. The parts remain with the dealers as an inventory until they are sold to consumers. Until

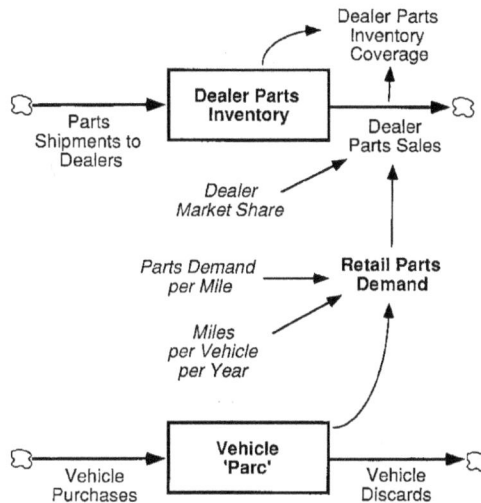

Fig. 7. Causal structure of motor vehicle parts business

recently, dealers have apparently made little attempt to control their parts inventories to maintain some desired level of coverage. In large part, this historical lack of inventory control is the fault of the company itself, which years ago created strong incentives for dealers to increase their parts and accessories orders each year at a rapid rate, regardless of consumer sales.

But now things are changing and these dealer incentives are weakening or being phased out. As a result, the company has recognized that, in order to anticipate future parts revenue growth, they must move their understanding from the wholesale level to the retail level, and analyze the factors affecting dealer parts sales to consumers. This analysis has boiled down to the handful of factors seen in Figure 7, all of which have been measured to some degree.

The starting point is the number of vehicles owned by consumers, a stock commonly known by the French term *parc*. Though it is not made explicit in Figure 7, the *parc* is modeled as an aging chain so that, for example, one may distinguish new vehicles under warranty from vehicles no longer under warranty.

The total retail demand for parts, it is believed, is best modeled simply by multiplying the total *parc* by two factors: the average number of miles driven per vehicle per year and the average demand for parts per mile driven. Dealer parts sales then equal some share of this total demand, a market share which exceeds 90% for vehicles under warranty but may be considerably lower for older vehicles.

Thus, in order to project dealer parts sales, the company must project four basic factors. Of these four factors, the size of the *parc* and dealer market share are modeled endogenously, and are the result of other assumptions for which future scenarios are well quantified; these include assumptions for vehicle sales and dealer service capacity. The other two factors, miles per vehicle and parts per mile, are modeled exogenously as time series. Parts per mile is a direct reflection of vehicle reliability and durability, which are closely monitored through study of warranty claims. These data suggest that parts per mile has been essentially constant since 1990.

The one factor that has not been adequately measured is the trend in average miles per vehicle. Three different periodic surveys done in the 1990s are unanimous in their conclusion that consumers are driving the company's vehicles fewer miles each year, and it is widely believed that this decline will continue into the future. Two of these surveys suggest the decline is at least 5% per year. But these surveys focus only on miles driven within the first year of vehicle purchase, a year in which driving habits are thought to be somewhat different than they are for older vehicles. The third survey makes it clear that there has been a decline in miles driven for older vehicles as well, but does not specify the rate of decline.

In order to get a better handle on this rate of decline, an analysis was done combining the structure in Figure 7 with available data from the 1990s. These data include annual figures on the vehicle *parc* and on parts shipments to dealers, annual estimates of dealer market share based on consumer surveys, and point-of-sale data on dealer part sales and inventory for 1996 only.

With these data, one first may derive the parameter values necessary to reproduce the 1996 dealer sales figure. Then, to generate a full historical curve for dealer sales, one must plug in an assumption for the rate of decline in average miles per vehicle during 1990–1996.

In fact, several different rates of decline in miles were tested, among them 0%, 2%, 4%, and 6% per year. Shown in Figure 8 are the four resulting dealer sales curves, superimposed on a bar graph of parts shipments to dealers. Note that the greater the assumed rate of decline in miles is, the higher the sales curve must be to end up at the same value in 1996.

Consider the implications for dealer inventory. When sales exceed shipments, dealer parts inventory will be decreasing, and when sales are less than shipments, inventory is increasing. So, although the four sales curves are similar in shape, they clearly have different implications in regard to changes in inventory. At the steepest decline rates of 4% and 6% per year, sales exceed shipments for the entire period 1990–1992, which is not the case at the lower decline rates of 0% and 2%.

Figure 9 shows the inventory results presented in terms of months' worth of sales coverage. The initial 1990 inventory levels have been adjusted so that by 1996, the four curves all converge at the same 1996 data point, an inventory coverage of about 11 months. If one assumes that the decline in miles has been as steep as 4% or 6% per year, then one must conclude that inventory coverage decreased for a few years

Fig. 8. Dealer parts sales under four scenarios super-imposed on shipments-to-dealers data

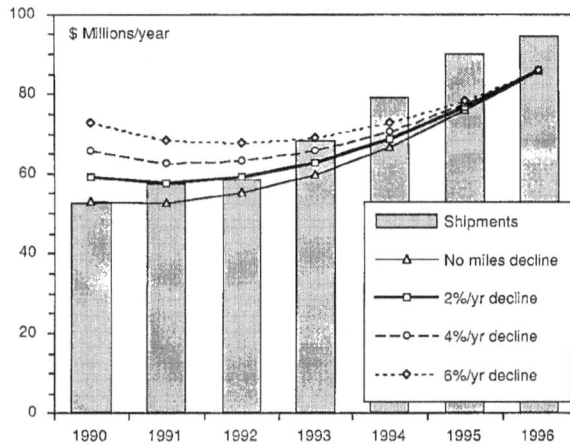

Fig. 9. Dealer parts inventory coverage under four scenarios

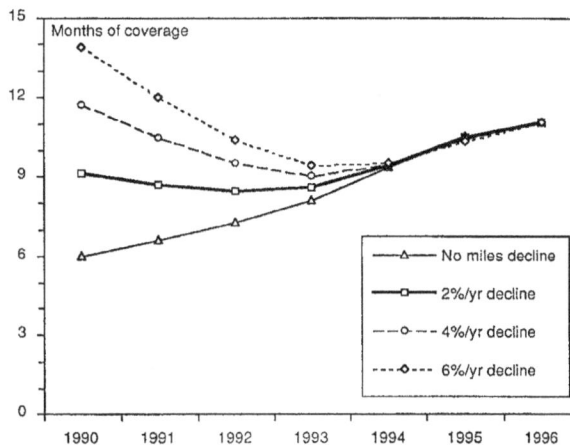

and, in fact, started in 1990 at a higher level than it ended up in 1996. But experts within the company agree that dealer inventory coverage levels did not decrease, but rather increased from 1990 to 1996. The conclusion is that the decline in average miles driven has probably not exceeded 2% per year in the recent past, and should not be expected to exceed 2% in the future. This conclusion is one of considerable significance to the company.

Computer-based educational testing: the issue of item inventory management

The last example deals with a major issue facing Educational Testing Service, or ETS, the world's leading developer and provider of standardized educational tests.

ETS tests are required for admission to most U.S. university programs, as well as many professional certification programs.

I started working with ETS in 1995 to help plan a transition from paper-and-pencil testing to computer-based testing. This transition, now in progress, is a complex undertaking that will require several years to complete and will affect the entire ETS organization (Homer and Nickols 1996). One major test, the Graduate Management Admission Test, or GMAT, has recently been transferred entirely to computer and is no longer available in its paper-and-pencil form.

A computer-based test is one in which all test items are given on-screen and all responses entered by mouse or keyboard. Unlike paper tests, which are administered *en masse* and offered only once every few months, computer-based tests are conveniently offered on an almost daily basis by individual appointment at specialized testing facilities. Also unlike paper tests, the computer-based testing software is "adaptive", which means that the difficulty of the next test item one receives is dependent on how well one has answered the questions up to that point. The test items are drawn from a pool many times larger than the test itself, so that different test takers generally receive different items. Moreover, the pool of test items is replaced each week with another pool drawn from an even larger inventory of available items.

The weekly refreshing of pool items is critical for minimizing the security risk that ETS today faces in the form of Internet discussion, direct word of mouth, and even commercial theft of test items. But this procedure requires that a rather large inventory of items be maintained at all times, something that represents a substantial investment to ETS. Before they may enter the inventory, items must be written and honed for precision, reviewed for cultural and gender bias, and pretested on large samples of test takers. Both item security and millions of dollars in item development costs are affected by the way in which the item inventory is managed.

Figure 10 is a stock-and-flow portrayal of the system that has been developed at ETS for management of a computer-based item inventory. Newly created items enter the inventory of available test items, from which they may be drawn anytime to assemble a weekly testing pool. A small number of items are permanently discarded, for example, if they fail to perform up to statistical expectations. In this system there are also outflows of items to the two way stations known as docks, namely the short-term dock and the long-term dock.

Docking is a procedure that affects certain of those items that have just served a week in a testing pool. The docking decision is based on exposures, or the number of times that an item is seen by test takers during its week in the pool. The number of exposures received by an entire pool is strictly determined by the number of items on the test and the number of test takers. But some items will receive a large number of exposures, while other items will receive a small number of exposures.

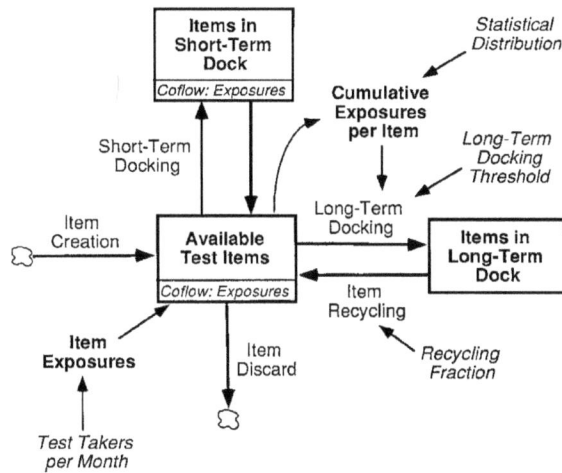

Fig. 10. Item inventory management system for computer-based educational testing

This distribution of exposures is the result of multiple characteristics of the items, such as difficulty, and of the test takers themselves, characteristics which make each individual item more or less likely than its poolmates to be selected by the adaptive software.

The short-term dock is where the vast majority of items go after they have served a week in a pool and received a minimum number of exposures. Items in this dock stay out of circulation for a few months, allowing the word of mouth and Internet chat to die down before they become available again. But very high-exposure items may be sent to the long-term dock, based on all exposures they have received both in the current pool and in the past. These high-exposure items are felt to pose the greatest security risk, and so are pulled more permanently out of circulation once their cumulative exposures exceed a specified threshold. In fact, until our simulation analysis was done, the long-term dock was generally conceived of as a life sentence for high-exposure items, with no opportunity for parole, that is, recycling.

The model analysis centered initially on simply projecting the number of available items over time, under different assumptions about security risk. Later analysis looked at policies for more cost-effective inventory management. The data came from one large testing program, the Graduate Record Examinations, or GRE, which has been making a gradual transition to computer-based testing for a few years now. These data included time series on item creation and discard, on available items and items in docks, and, perhaps most important, data on the statistical distribution of exposures across items.

Figure 11 shows how the inventory of available items evolves under five different

Fig. 11. Available test
items under five
scenarios

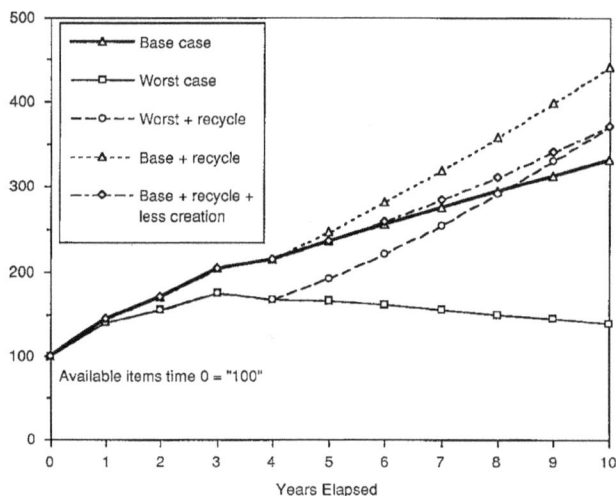

Fig. 11. Available test items under five scenarios

scenarios, based on the GRE data. Items are expressed in relative terms, with 100 representing the number of available items at time zero, when the test is first made available in computer-based form. The thicker line represents a base case in which available items grow continuously for the entire ten-year simulation period. This base-case trajectory is considered an acceptable one, in the sense that it builds an adequate buffer against catastrophic item losses; such catastrophic losses have occurred, though rarely, in the past.

A second scenario labeled "worst case" shows an unacceptable trajectory, in which available items end up at less than half of their base-case value. This scenario involves a change in just one model parameter, the long-term docking threshold. The current threshold in actual use by ETS, and assumed for the base case, was developed from early estimates of security risk for computer-based testing. But it is understood that this threshold may have to be lowered at some point if the damage done by Internet chat and item theft turns out to be greater than expected. The worst-case curve corresponds to a threshold level that is as low it might become, in the estimation of my client.

Once the dire consequences of the worst-case scenario had been made clear, my client wondered whether it was possible to alleviate the situation through a prudent policy of item recycling. Under this policy, items would be brought back from the long-term dock as if they were newly created (that is, with no cumulative exposures) but only after a period of at least a few years in the dock. The third curve in Figure 11 is the result of a scenario which adds to the worst case assumption a policy of recycling starting at year 4 and employing a recycling fraction of only 2% per month, corresponding to a time constant of over four years. Even such modest recycling adds a powerful balancing loop to the system, and the policy is clearly very effective in reversing the problem.

My client next wondered whether the same policy of recycling might be helpful with the base case, as a way of boosting the inflow of available items enough to allow for a reduction in the required rate of costly item creation. The fourth curve in Figure 11 is the result of adding recycling alone to the base case, while the fifth curve includes both recycling and a 20% reduction in item creation starting in year 4. The conclusion is that prudent item recycling can, in fact, return enough items under the baseline threshold assumption to allow for a significant reduction in item creation.

The model-based analysis described above has been presented in detail by my client to the ETS executive board, and is serving as a primary input to their decisions regarding test item inventory policy.

Conclusion

The purpose of this article has been to illustrate a certain potential for insight inherent in stock and flow structures when they are combined with a rich base of numerical data. Some of the conclusions may seem modest in a way, dealing as they do with the setting of parameter values and the playing out of dynamically simple structures. But I have found that these conclusions are meaningful to clients, even at the highest levels of an organization; conclusions that clients can easily understand, remember, and incorporate into their plans and discussions.

Of course, feedback loops are also essential, especially for helping clients to anticipate shifting trends and identify high and low leverage points. But feedback-based insights, especially those based on multiple loops of uncertain strength, can often be difficult things for people to understand and believe in. Before clients can engage effectively in complex feedback thinking, they often need to understand the more basic momentum of their situation and where that momentum is taking them. I think of it as first connecting the dots within the box of operational thinking, before one leaps outside the box to achieve more subtle insights.

Barry Richmond has written that the logic of stocks and flows is "the unique essence of system dynamics" (Richmond 1994a). These words undoubtedly have raised the eyebrows of some who feel that the identification of feedback loops should be a modeler's main concern. Similarly, when I suggest that a rich fabric of numerical data is an important asset for reaching compelling conclusions, I expect there will be some who disagree.

In fact, I am not dogmatic on this point. I do not think there is a single, one-size-fits-all answer to the question of what sorts of structure and data to emphasize in a modeling project, and I hope the field of system dynamics will continue to tolerate a wide spectrum of practice. This does not mean that "anything goes", but rather that it is perfectly acceptable for some models to emphasize feedback loops while others emphasize stocks and flows; and for some models to rely on relatively little numerical data, while others rely on a lot. Much may depend on subject matter and specific client needs, to which a modeler should be sensitive. But regardless of emphasis, I believe we should subject all of our models and findings to close scrutiny in the full light of available evidence. Certainly we should be known for our creativity and ability to look at the big picture, but we should also be known for our thoroughness and zipper-like logic.

Notes

1. The view of numerical data as being of limited value for modeling may perhaps be traced to a misinterpretation, or incomplete comprehension, of one of Jay Forrester's discussions in *Industrial Dynamics*, a section called "Sources of Information for Constructing Models" (Forrester 1961, 57–59). Forrester does indeed emphasize the importance of undocumented experiential data in model construction, and rejects an approach to modeling that relies exclusively and blindly on numerical data. He also states that structural assumptions should come before data collection, and that when a model's behavior is insensitive to changes in a particular parameter, that parameter simply need not be measured with precision. But he also says:

   > These comments are not to discourage the proper use of the data that are available nor the making of measurements that are shown to be justified; they are to challenge the common opinion that measurement comes first and foremost. Lord Kelvin's famed quotation, that we do not really understand until we can measure, still stands. But before we measure, we should name the quantity, select a scale of measurement, and in the interests of efficiency we should have a reason for wanting to know.

   Thus, Forrester clearly acknowledges the importance of gathering relevant numerical data as a modeling project proceeds, so as to gain the necessary precision about those parameters to which a model's behavior is sensitive.

2. The original study (Homer 1993) employed Household Survey data through 1990. I have now supplemented the cocaine user data with data from the 1993 and 1995 Household Surveys (SAMHSA 1994; SAMHSA 1996). The simulation results presented here also reflect small adjustments to the 1991–1995 portion of the model's input time series on past-month marijuana use, reflecting the 1993

and 1995 Survey data on this variable. No other adjustments were made to the model's input parameters in producing these new results.

References

Forrester, J. W. 1961. *Industrial Dynamics*. Cambridge, Massachusetts: MIT Press. (Reprinted by Productivity Press, Portland, OR.)

—. 1969. *Urban Dynamics*. Cambridge, Massachusetts: MIT Press. (Reprinted by Productivity Press, Portland, OR.)

—. 1971a. Counterintuitive Behavior of Social Systems. *Technology Review* 73(3). Also in *Collected Papers of Jay W. Forrester* (1975). Cambridge, MA: Wright-Allen Press. (Reprinted by Productivity Press, Portland, OR.)

—. 1971b. *World Dynamics*. Cambridge, MA: Wright-Allen Press. (Reprinted by Productivity Press, Portland, OR.)

—. 1980. Information Sources for Modeling the National Economy. *Journal of the American Statistical Association* 75(371): 555–574.

Homer, J. B. 1983. Partial-Model Testing as a Validation Tool for System Dynamics. *Proceedings of the 1983 International System Dynamics Conference* (*Chestnut Hill, MA*), pp. 919–932.

—. 1993. A System Dynamics Model of National Cocaine Prevalence. *System Dynamics Review* 9(1): 49–78.

—. 1996. Why We Iterate: Scientific Modeling in Theory and Practice. *System Dynamics Review* 12(1): 1–19.

Homer, J. B., and F. Nickols. 1996. Enterprise Modeling at ETS: Highways and Byways. *Proceedings of the 1996 International System Dynamics Conference* (Cambridge, MA), pp. 225–228.

Macaulay, D. 1988. *The Way Things Work*. Boston, MA: Houghton Mifflin.

Richmond, B. 1994a. Systems Thinking/System Dynamics: Let's Just Get On With It. *System Dynamics Review* 10(2–3): 135–157.

—. 1994b. Classic Core Business Infrastructures: Inadvertant Evolution and Fighting the Physics. In *ithink® Business Applications Guide*. Hanover, NH: High Performance Systems.

SAMHSA 1994. *National Household Survey on Drug Abuse: Population Estimates 1993*. DHHS Publication No. (SMA) 94-3017. Rockville, MD: Substance Abuse and Mental Health Services Administration.

SAMHSA 1996. *National Household Survey on Drug Abuse: Population Estimates 1995*. DHHS Publication No. (SMA) 96-3095. Rockville, MD: Substance Abuse and Mental Health Services Administration.

7

Macro- and Micro-Modeling of Field Service Dynamics (1999)

Homer J. *System Dynamics Review*, 15(2): 139-162

Macro- and micro-modeling of field service dynamics

Jack B. Homer[a]

Jack B. Homer is a system dynamics modeling consultant and has directed Homer Consulting since 1989. He was formerly an assistant professor at the University of Southern California and received a Ph.D. in Management from M.I.T. for a thesis modeling the diffusion of new medical technologies.

Abstract

A system dynamics model to investigate field service issues was developed for a major producer of equipment for semiconductor manufacturing. This strategic model has a broad scope and multi-year time horizon, and treats variables in an aggregate and deterministic way that is typical for such models. The high-level approach is adequate in most respects, but lacks the detail necessary to resolve a key issue regarding the impact of product cross-training on service readiness. As a result, it proved useful to supplement the strategic 'macro' model with a 'micro', OR-type model that portrays the daily queuing and assignment of service jobs. The micro model provides detailed what-if results that were used for calibrating the strategic model and may also be used for making tactical manpower decisions at the local level. Traditional OR tools may have a role to play in supporting strategic modeling efforts when important operations-level relationships are not adequately understood. Copyright ©1999 John Wiley & Sons, Ltd.

Syst. Dyn. Rev. **15**, 139−162, (1999)

A system dynamics modeling project was undertaken by a major producer of diagnostic equipment used in semiconductor wafer fabrication. The company's many products include both simple machines requiring little maintenance and complex systems requiring more frequent maintenance. All products come with an initial warranty on parts and service, after which the customer has the option of continuing with a service contract. Customers often select service contracts for the more complex and essential pieces of equipment, for which downtime must be minimized and do-it-yourself repairs are difficult at best.

The company's new product sales have grown rapidly over the last several years, though they do follow the ups and downs of the semiconductor industry's persistent two-year business cycle. This overall growth in sales has led to robust growth in the installed base of equipment and similar expansion of the workforce of field service engineers. Although field service does not generate much for the company in the way of profits, it is nearly as important as product performance and competitive price are for the company's continued success in the marketplace. As the field service workforce has grown, its planning, organization, and management have taken on increasing complexity and significance for the company. (Richmond 1994 describes the inevitable evolution of a high-tech manufacturer from a primary sales focus to an increasing field-service focus). For example, while the workforce has always been segmented by local territory or hub, it has only recently been segmented by customer account as well. This action was taken so that major customers

[a]J. B. Homer, 36 Covington Lane, Voorhees, NJ 08043, U.S.A.

System Dynamics Review Vol. 15, No. 2, (Summer 1999): 139−162   *Received August 1998*
Copyright ©1999 John Wiley & Sons, Ltd. CCC 1077-3495/99/020139−24 $17.50 *Accepted December 1998*

would generally have their service done by engineers with whom they are familiar and who have worked on their equipment before.

With the field service organization at a critical juncture, senior management turned to system dynamics as a tool for the analysis of strategies that seek to trim workforce costs without jeopardizing customer satisfaction. Some managers felt that the field workforce had grown too large relative to the installed base, as supported by anecdotal evidence of excess idle time, and managers advocated a hiring freeze or slowdown to raise workforce utilization. Other managers claimed that the idle time was primarily a reflection of workload imbalance due to inadequate product cross-training, and pointed to the fact that some field engineers worked long hours with no idle time while other engineers often had hours or days between service jobs.

Management also debated the value of policies intended to reduce workforce costs by reducing service demand or improving field productivity. One idea was to improve the products themselves, making them more reliable or easier to service. Another idea was to enhance the role of centralized technical support, either by lending more direct help to customers so that they could solve simpler equipment problems themselves, or by lending more on-the-job assistance to less experienced field engineers. Yet other proposals to improve service effectiveness included revised segmentation of the workforce and improved service parts distribution.

The modeling work was done with a team that included both senior staff and field managers. As often happens, our discussions tended to move back and forth from aggregate concepts, such as customer satisfaction or workforce utilization, to more detailed ones, such as distinctions among various types of products, customers, and service contracts. The team agreed that, for purposes of strategic analysis, the appropriate model would be a 'macro' one with a time horizon of five to ten years that aggregated across most of the detailed distinctions. However, the team also wanted to look closely at how a service job backlog can develop, along with idle time, when workforce skills do not match up well with service requirements. We required a separate 'micro' model to investigate this question, a stochastic job-queuing model that focused at the local hub level and represented product types and field engineers individually, and that had the ability to assign engineers to product-specific jobs in a realistic way.

The remainder of this paper presents an overview of the macro and micro models. It will be shown how the micro model has played a key role in supporting the macro model, particularly in helping to specify the nonlinear function through which cross-training impacts the workforce's ability to handle service volume.

A strategic model of field service

The idea of modeling service delivery is hardly new to system dynamics, though the number of published works is still modest. Much of the early work was in the area of health care and education (Levin *et al.* 1976). Later works of note include models of People Express Airlines (Graham *et al.* 1992), Hanover Insurance claims processing (Senge and Sterman 1992). NatWest Bank lending (Oliva 1996), and DuPont chemical plant equipment maintenance (Carroll, Sterman and Marcus 1998). Among other things, these models suggest that a service company should hire steadily rather than in spurts to avoid problems of inexperience, should hire enough workers to avoid overwork and a drift to low standards, and (in the case of equipment service) should give preventive maintenance high priority to avoid a spiral of equipment failures.

As shown in Figure 1, the field service strategic model (about 500 equations in all) contains sectors detailing with the installed base, service demand,

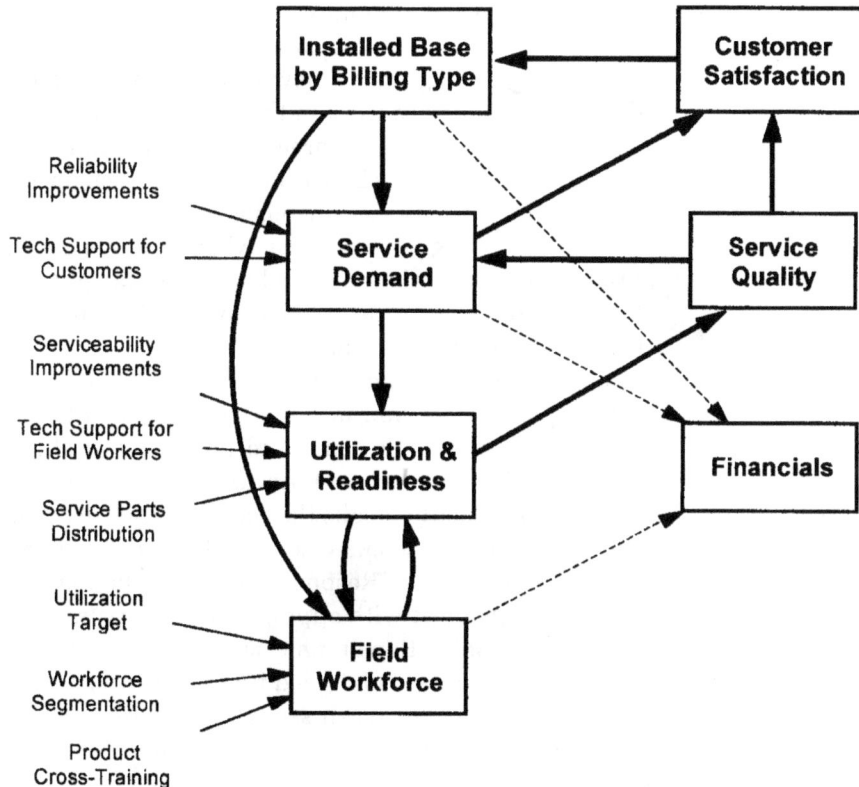

Fig. 1. Strategic model sectors and issues addressed

service quality, customer satisfaction, field workforce size and skill, workforce utilization and service readiness, and financial results. This figure also lists the key policy issues addressed by the model and shows where they fit into the model. It is worth touching upon some of the concepts in Figure 1 before moving on to discuss their dynamic interplay.

In the strategic model, the installed base of machines is disaggregated not by individual product type but by service billing category: machines under warranty, machines under service contract, and machines billed by time and material. These categories are important for financial reasons, but also because they differ in terms of typical service requirements. For example, machines billed by time and material typically generate fewer repair calls than do machines under contract or under warranty.

The volume of repair calls is driven not only by the size and billing-type breakdown of the installed base, but also by potential machine reliability, the effectiveness of customer technical support, the presence of bugs in new products, service quality, and the frequency of preventive maintenance (PM). PM improves equipment reliability, but it also requires time from the field workforce and so must be included when calculating total service demand. Similarly, the elimination of a bug from a new machine improves its reliability but requires additional field service work in the form of an engineering change order (ECO).

Service quality refers in large part to the ability of field engineers to 'fix it right the first time', which is a function of the engineers' skills and motivation, but also related to the phenomenon of shortcutting. Shortcutting (plus overwork and fatigue) can occur when workers are short of time and under pressure to get the job done quickly. Such time pressure, in turn, is a function of service readiness, the relative availability of appropriately trained field engineers to meet service demand.

Customer satisfaction with service is a function of both the required frequency of service and machine downtime. In the case of a repair call, the downtime represents the sum of response time and repair time, both of which may be considered aspects of service quality. Response time is measured from when the customer first calls the central response center until an assigned field engineer arrives at the customer site. Repair time is measured from arrival on site to job completion. Response time may be longer than normal when, for example, no engineer with appropriate product training is available during the current work shift. Repair time may be longer than it should be when, for example, the engineer is fatigued from overwork. Thus, both response time and repair time may suffer when service readiness is low.

Figure 2 is a causal-loop diagram showing the key feedback loops of the strategic model. At the center of the diagram is the concept of workforce

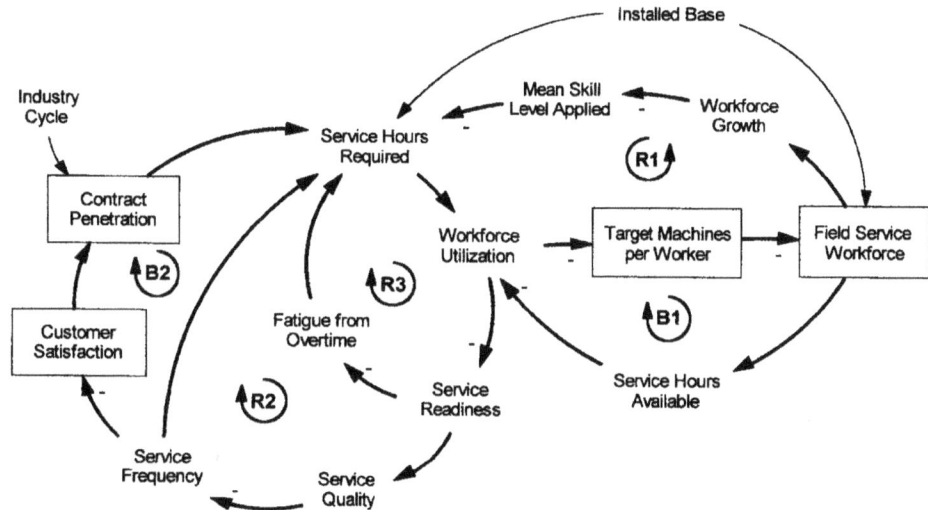

Fig. 2. Key feedback loops in the strategic model

utilization, the ratio of service hours required to service hours available. Utilization is one important determinant of service readiness, and management has traditionally considered it the best metric for gauging the relative balance of service demand and supply. When utilization (averaged, say, over the past year) has been lower than its target value, management will tend to increase the target number of machines per worker, and so will reduce hiring and slow down growth of the workforce relative to the installed base. Conversely, when recent utilization has been higher than its target, management will tend to reduce the target number of machines per worker, and so will speed up relative workforce growth. The result is a control mechanism shown as balancing loop B1.

One result of loop B1 is that the field service workforce may grow at a rate different from that of the installed base. This kind of behavior is illustrated in Figure 3, a simulation in which the workforce is initially flat or declining for two years before it starts to grow at about the same rate as the installed base. (Note that variables in Figure 3 are graphed in different scales; zero-based scales are used for Workforce, Installed base, and Machines per worker). The workforce growth pattern reflects the fact that workforce utilization starts below its target value. To correct this situation, hiring is reduced to less than replacement rate for the first two years, which succeeds in raising utilization and the number of machines per worker. Utilization fluctuates throughout the run, largely due to the industry business cycle that affects the fraction of the installed base getting regular service under contract. Despite this fluctuation,

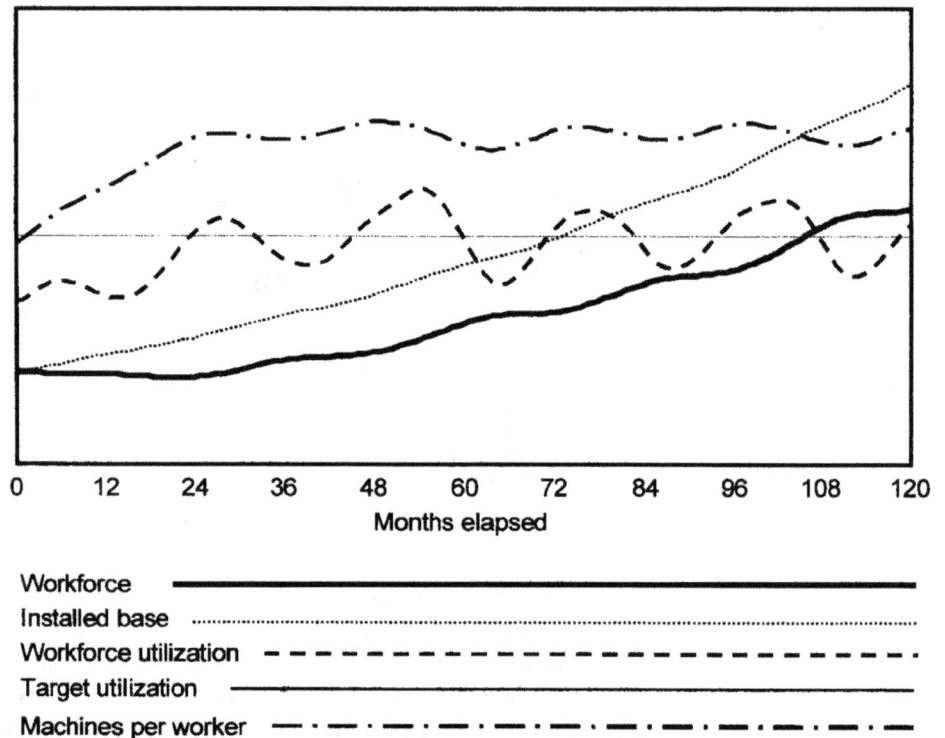

Fig. 3. Workforce growth as affected by utilization

Workforce	—————————
Installed base	··················
Workforce utilization	– – – – – – – – –
Target utilization	———————
Machines per worker	— · — · — · — · —

utilization remains close enough to its target value after the first two years that the number of machines per worker is allowed thereafter to remain essentially flat at its new higher level. Loop B1 does its job effectively in this run, and the field workforce is made more productive without triggering instabilities that might hurt customer satisfaction and company finances.

Under different assumptions, the strategic model may produce results that are not so rosy as those in Figure 3. Indeed, the reinforcing loops R1, R2 and R3 in Figure 2 are all capable of creating unintended consequences if the loop B1 control mechanism is not implemented properly. First consider loop R1. If utilization is initially below target, and the hiring slowdown response is too strong or sudden, the fraction of rookies in the workforce will drop quickly, and the average skill level will rise accordingly. Increased skill will reduce repair times and cause workforce utilization to remain below target. The hiring slowdown will remain in place until accumulated worker turnover forces an increase in hiring and starts the loop moving in the other direction. The result may be a hiring seesaw effect that lasts for several years. Even in the benign scenario of Figure 3, the observed fluctuations reflect some compounding of the industry cycle by loop R1. If managers were to react more strongly than they do in this scenario, their very attempt to maintain close control could compound the cycle to the point where it becomes debilitating.

Loops R2 and R3 are vicious cycles that become problematic when the workforce utilization target is set too high. If utilization is driven up to the point where it hurts service readiness, the result will be an increase in shortcutting by some field engineers and an increase in overtime hours by others. If shortcutting leads to errors and more frequent service calls, then utilization will be driven up still further. If long hours lead to fatigue, then the tired workers will require more time to complete a job and, again, utilization will be driven up still further. Another possible side effect of reduced service readiness (present in the model but not shown in Figure 3) is increased worker turnover, which can cause the situation to spin even more quickly out of control.

What may ultimately limit the effects of the reinforcing loops is the loss of customer satisfaction and contract penetration that results when service quality is low and service frequency is high. Shown in Figure 2 as balancing loop B2, this customer response represents a loss of business that the company feels strongly it must avoid. It is one more indication of how important service readiness is to the stability and success of field service operations.

Cross-training in the strategic model

In most field service operations, field engineers are responsible for a number of different product types. In this context, service readiness must be understood as a function of not only workforce utilization but also cross-training. If all engineers were trained on all product types, then service readiness could be defined purely based on utilization, the ratio of repair hours available to repair hours required. But if, practically speaking, each engineer can become trained and skilled on only a handful of products at most, service readiness will be reduced to some extent below what utilization alone would suggest. This means that there may be times when a service job cannot get immediate assignment, despite the availability of some engineers, because none of the available engineers is trained to work on the particular type of product requiring service.

Figure 4 presents the causal-loop diagram again, but now with the effects of cross-training included. The beneficial effect is shown as an arrow drawn directly from cross-training to service readiness. In the absence of any cross-training, each engineer is a narrow specialist unable to assist when, say, two different repair calls come in involving machines of the same product type for which there is only a single specialist. The greater the extent of cross-training

Fig. 4. Positive and negative effects of cross-training

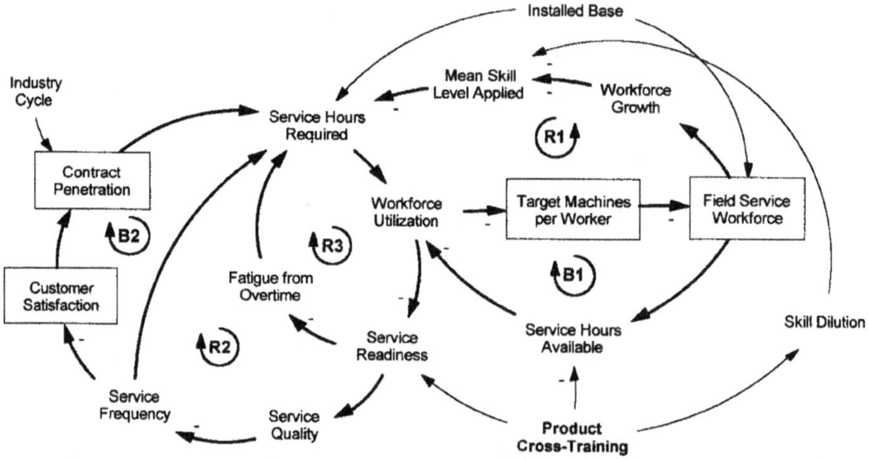

is, the more overlap of skills there will be in the workforce, and the greater the ability of engineers to act as back-ups when other engineers are already busy or otherwise unavailable.

Figure 4 also indicates two of the negative effects of cross-training. The first is a reduction in available service hours due to an increase in training time. This impact on training time (and associated training costs) can be considerable when the products are complex and require lengthy periods of classroom instruction. For example, the cross-training classes for the semiconductor equipment producer range from one month to three months full-time instruction, depending on the particular product type. Cross-training can thus represent a major investment of time and money for the company. This investment is not a one-time affair, but, like basic or primary training, must continue to be made as long as there is turnover or growth in the workforce and, thus, a cadre of newer field engineers who require secondary skills.

The second negative effect of cross-training is a phenomenon known as skill dilution, which tends to reduce the average level of skill applied to service jobs. In order to understand this phenomenon, it is useful to look in greater detail at how the building of skills is approached in the strategic model. As seen in Figure 5, the model uses two skill chains for this purpose, a primary skills chain and a secondary skills chain.

The primary skills chain follows engineers through their basic training on a single product and the building of skills through on-the-job experience with that product.[1] New hires receive their primary training and are considered rookies until they become certified to work on more challenging jobs. Over time,

Fig. 5. Field workforce
primary and
secondary skill chains

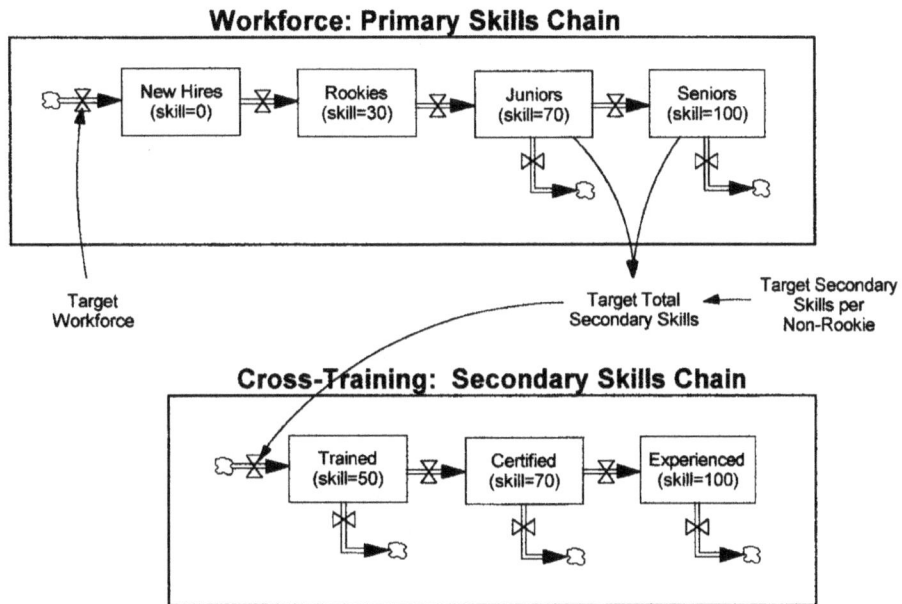

Workforce: Primary Skills Chain

New Hires (skill=0) → Rookies (skill=30) → Juniors (skill=70) → Seniors (skill=100)

Target Workforce

Target Total Secondary Skills ← Target Secondary Skills per Non-Rookie

Cross-Training: Secondary Skills Chain

Trained (skill=50) → Certified (skill=70) → Experienced (skill=100)

a certified engineer will gain the on-the-job experience that elevates him or her from junior status to senior status.

The secondary skills chain is a co-flow structure for following the building of skills on products for which engineers have been cross-trained. An engineer becomes eligible for cross-training once he or she has become certified on a primary product and has thereby demonstrated a fundamental grasp of the company's technology and procedures. In other words, only non-rookies (juniors and seniors) are eligible for cross-training. Secondary skills involve the same process of training, certification, and on-the-job learning as primary skills. But cross-trained workers must divide their time among the two or more products on which they are skilled. Consequently, it generally takes longer to climb the secondary skills chain than the primary skills chain. For example, the average time to become certified after training may be 12 months for a secondary skill where it is 6 months for a primary skill. Similarly, the average time to become fully experienced may be 24 months for a secondary skill where it is 18 months for a primary skill.[2]

The secondary skills chain contains one unit for every secondary skill in the workforce. The average number of secondary skills per non-rookie is thus simply the total number of secondary skills in the chain divided by the number of non-rookies: If there are 50 secondary skills spread among 50 non-rookies, then the average number of secondary skills per non-rookie is 1.0. Assume that the target is increased from 1.0 to 1.5 secondary skills per non-rookie on the average. This means that the target for the total number of secondary skills must increase from 50 to 75. The rate of cross-training will quickly increase above its baseline value and remain elevated until the new target is achieved.

The idea of skill dilution may now be understood. The mean skill level applied to service jobs is a weighted average of the skill level of rookies (e.g., 30/100) and the mean applied skill level of non-rookies. The mean applied skill of non-rookies, in turn, is a weighted average of their mean primary skill level (e.g., 90/100) and their mean secondary skill level (e.g., 80/100). An increased number of secondary skills per non-rookie will pull the mean applied skill level of non-rookies downward in the direction of the mean secondary skill level (e.g., from 90/100 toward 80/100).[3] In other words, the mean skill level applied by non-rookies is reduced, or diluted, to the extent that they are cross-trained and to the extent that non-rookies have lower skill levels for their secondary products than they do for their primary products.

Having identified the positive and negative effects of cross-training, management wanted to test the overall impact on service performance using the strategic model. They would then be able to evaluate alternative target levels for secondary skills, and reach consensus on cross-training policy. However, while it was straightforward to quantify the negative effects of cross-training, the same was not true of the positive effect on service readiness.

Logical considerations suggested, and the field managers on the modeling team agreed, that the service readiness gains from additional cross-training must diminish beyond a certain level. Indeed, the gut feel of the field managers was that there may be little gained from increasing the average number of secondary skills per engineer much beyond a value of one.[4] The effect of secondary skills on service readiness was accordingly formulated in the strategic model as a table function with decreasing slope. However, despite their intuitions, the field managers lacked any hard evidence on how steep that function should be or where it should start to flatten out; the research literature was moot on this point as well. This lack of data was unfortunate, as simulations showed that the behavior of the strategic model was quite sensitive to this table function.

In order to calibrate the table function with the accuracy necessary for reaching firm policy conclusions, the modeling team realized it would be necessary to do analysis below the level of aggregation of the strategic model. Our 'micro' job-queuing model, in fact, proved to be just the right tool for evaluating in detail how cross-training affects service readiness.

A model of service job queuing and assignment

Queuing models have been used for many years by operations researchers to address questions of how much and what configuration of service capacity will best meet randomly arriving service demand. The range of applications is very broad, including commercial services (e.g., bank tellers, grocery checkout, auto repair), transportation services (e.g., traffic lights, railroads, fire trucks), social services (e.g., health care, judicial, legislative), and industrial services (e.g., materials handling, equipment maintenance, quality control) (Hall 1991).

Queuing models are built to evaluate the statistical characteristics of a service system under various parametric assumptions over a period of hours, days, or weeks. These statistical characteristics include means and variances for such metrics as queue length, waiting time, and percentage of servers busy. Because they are not concerned with the evolution of the system over a longer period of months or years, basic parameters of service capacity and demand, such as the number of servers and the number of customers, are considered constant.

Figure 6 presents the basic structure of the field service job-queuing model. The focus is on a single regional hub with a complement of field engineers who work on service jobs of three types: repair, PM (preventive maintenance), and ECO (bug fixes). Time is measured in work shifts of eight hours. The hub's installed base is disaggregated by product type, and for each combination of product type and service job there is a job arrival probability per shift, based on a mean time between service that is assumed constant. The inflow of job

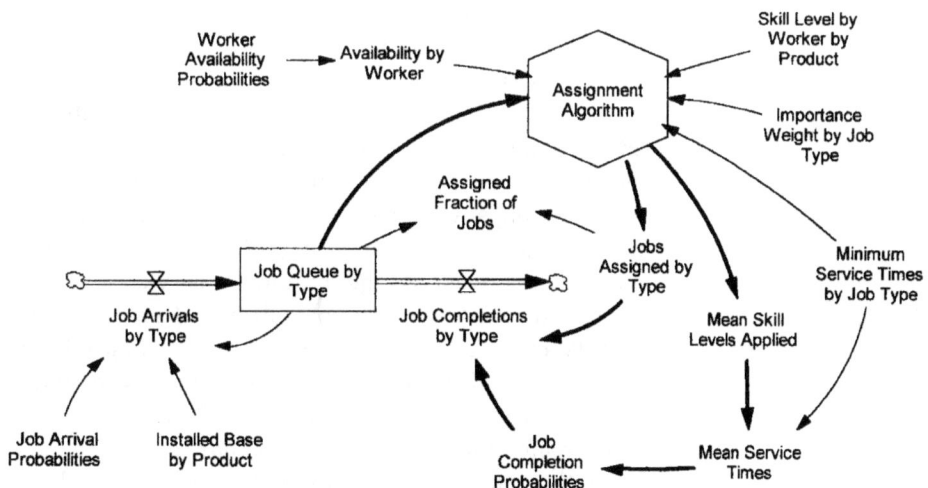

Fig. 6. Structure of the job-queuing model

arrivals by type in the model is a random variable, the mean of which is determined jointly by the installed base, the job arrival probability, and the number of jobs already in the queue. The nature of the service job types is such that, at any given time, a machine may be awaiting at most a single repair job or a single PM job, and any number of ECO jobs (corresponding to multiple bug fixes for that product type).

The remainder of the model is devoted to calculating the flow of job completions by type out of the queue. This outflow is a random variable, the mean of which is determined by the number of jobs successfully assigned to engineers and a job-completion probability per shift. The job-completion probability, in turn, is based on a mean service time reflecting normal preparation and travel time plus the time spent at the customer site. The on-site time itself is a function of two things, which may vary by product–job type: a specified minimum time based on the performance of fully skilled engineers, and the mean skill level of engineers assigned to the product–job type during the current work shift.

The effective assignment of engineers to service jobs is an intricate matter in real life, and the same is true in the model. On any given day, field service managers and dispatchers must attempt to achieve an effective match between the jobs in the queue and the skills of their available engineers, also taking into consideration any response-time agreements that exist with specific customers. A good approach generally is to match the more skilled engineers to the more difficult and urgent jobs in the queue, so as to minimize machine downtime and satisfy customer expectations. This is the approach taken in the model, where it has been implemented in the form of an optimization algorithm called upon each computation interval (twice per simulated work shift) to assign individual field engineers to product–job types for which they have the necessary skill.[5]

Figure 6 shows the ingredients that go into the assignment algorithm. Field engineers are identified individually, and a matrix specifies each engineer's skill level by product type. Engineers also differ in regard to their availability to perform service jobs, due to differences in number of personal days off, training days, and days assigned to other activities, such as machine installation and upgrade. The other required ingredients include size of the queue by product–job type, as well as the minimum service times and importance weights associated with each product–job type. Repair jobs are given a greater weight than PM or ECO jobs. This is because for repair jobs, every hour in the queue is another hour of machine downtime, whereas for PM and ECO jobs downtime occurs only when the job is being performed. Also, certain product types may have greater importance weights than others to reflect response-time agreements the company has with customers owning those products.

The queuing model was calibrated to represent two of the company's actual service hubs. In both cases, the model accurately reproduced the real-life workload situations in detail, at the level of individual product and job types. One of these hubs was troubled by a loss of engineers and was overwhelmed with workload. The model illustrated how the overload in this hub had led to virtual abandonment of PM and ECO jobs, and it correctly identified those product types that were hardest hit by the shortage of engineers. The second hub was in much better shape, with all service jobs getting attended to, although not always straight away. For example, the model confirmed that while 90% of repair jobs were assigned an engineer immediately upon arrival in the queue, the numbers were closer to 50% for ECO jobs and 20% for PM jobs. Figure 7 shows graphically how the repair jobs queue could be kept under rather good control in this hub, thanks to the high rate of job assignment.

Having validated the queuing model's behavior, we proceeded to do some what-if analysis for the two calibrated hubs. For example, for the troubled hub, one possible solution under discussion was to teach customers how to do their own preventive maintenance on certain products and thereby relieve the field workforce of the responsibility. This idea could be easily tested with the model by setting the PM job arrival probability to zero for those product types. This sort of tactical workload analysis proved quite illuminating and useful,

Fig. 7. Randomly arriving repair jobs kept under control through effective assignment (sample output from job-queuing model)

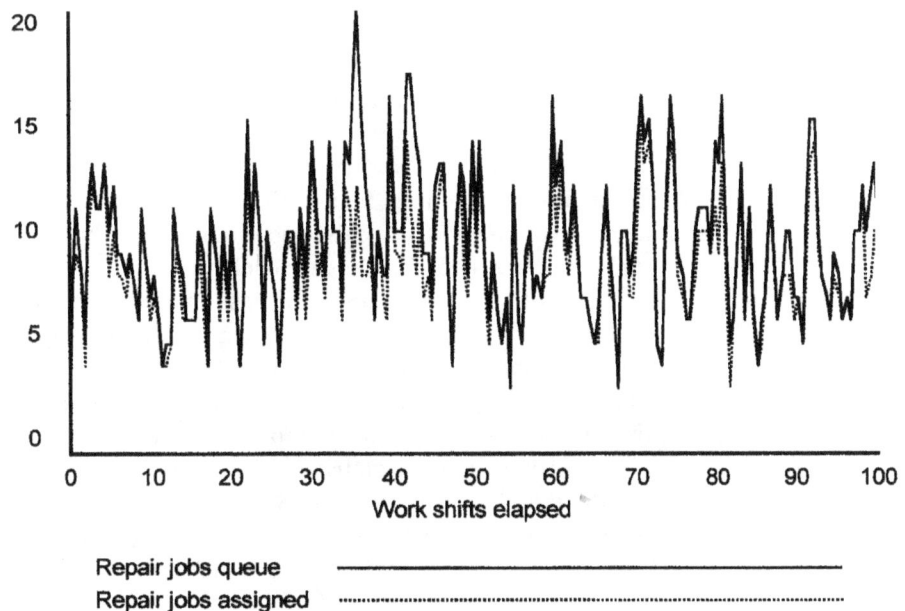

Repair jobs queue ————————

Repair jobs assigned ·······························

especially for the two members of the team who happened to be the service managers of the hubs in question.

We next turned to the task of using the queuing model to help calibrate the strategic model's table function relating secondary skills and service readiness. For this purpose, a battery of sensitivity tests was performed using the queuing model. First, however, we had to consider how one could draw inferences from a queuing model representing a local hub that would be applicable to a strategic model representing an entire country or world region. In the team's judgment, the more stable of the two hubs we had modeled was representative of most hubs across the U.S.A. On this basis, we decided to use the existing stable hub calibration of the queuing model as the basis for sensitivity testing.

The sensitivity testing was a bit tricky in both preparation and implementation. In the stable hub, non-rookie engineers had a current average of 0.9 secondary skills. The hub's service manager was asked to describe how he would go about changing the skill matrix if he were required to increase or reduce significantly the number of secondary skills in his workforce. (Presumably, a reduction in secondary skills would correspond to a decision to stop assigning some engineers to products for which they had been cross-trained). In order to assure as pure as possible a test of the secondary skills effect, each alternative skill matrix had to be constructed so that the secondary skills were distributed across the product types in a balanced way, as in the baseline matrix. Also, the mean secondary skill level (e.g., 80/100) had to be approximately the same in each alternative matrix as it was in the baseline matrix. Five alternative skill matrices were built in this way, with secondary skills per non-rookie varying from a low of 0.1 to a high of 2.5.

The goal of sensitivity testing was to find the impact on service readiness for each alternative skill matrix. This was not a trivial matter, because there is no metric in the queuing model that corresponds precisely to the concept of service readiness in the strategic model. Nonetheless, the task can be accomplished indirectly by finding, for each alternative skill matrix, a workload factor (affecting all job arrival probabilities equally) that offsets the impact of altered skills and causes the queuing system to achieve approximately the same level of performance seen in the base case.[6] The most reliable performance metric to use for this purpose turned out to be the assigned fraction of jobs (see Figure 6), averaged across all product–job types and over 100 simulated work shifts. When testing the skill matrix with only 0.1 secondary skills per engineer, the workload had to be reduced with a factor of 0.66 to bring the assigned fraction of jobs back up to its baseline value. At the other extreme, when the matrix was tested with 2.5 secondary skills per engineer, a workload factor of 1.65 had to be used to get the assigned fraction of jobs back down its baseline level.

The five workload factors found through sensitivity testing described how changes in the number of secondary skills would affect service readiness, relative to the base case. These relative values, combined with a baseline value for the service readiness effect (corresponding to 0.9 secondary skills per non-rookie), would provide six points for the table function in the strategy model. The baseline value for the service readiness effect was found adjusting this value until the strategic model's initial response time for repair jobs matched that actually experienced in the stable hub, our proxy for the country as a whole. This tuning process produced a baseline service readiness effect of 0.5, meaning that the level of service readiness with 0.9 secondary skills per non-rookie is half of what it would be if all non-rookies were trained on all products.

The complete table function is shown in Figure 8, which is derived from sensitivity testing of the job-queuing model and used as a table function in the strategic model. Note that the curve is roughly S-shaped rather than purely diminishing in slope. The slope is steepest where the average number of secondary skills per non-rookie is between 0.6 and 0.9, and is smaller both before and after that range. The slope becomes quite shallow beyond the point of two secondary skills per non-rookie, as the service readiness effect starts to approach its ultimate value of one. It seems that the field managers were correct in their intuition that the benefits of cross-training in their company start to diminish beyond the point of one secondary skill per engineer. However, the

Fig. 8. Multiplier on service readiness from cross-training

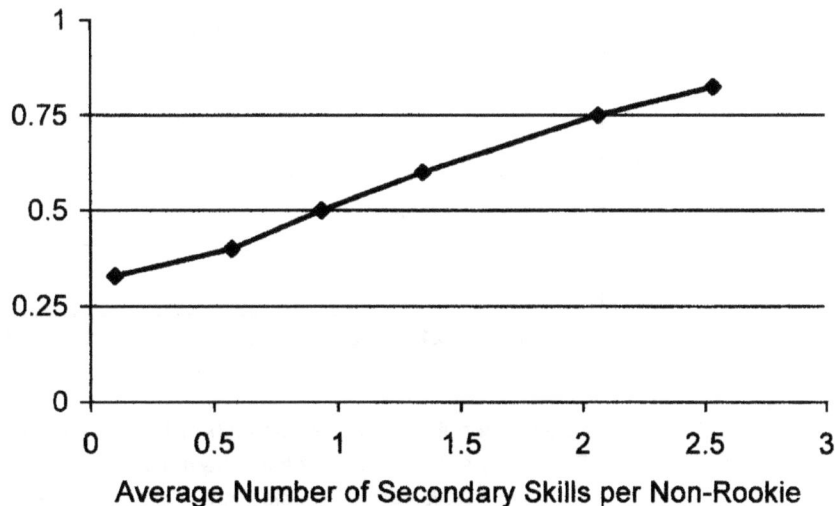

Average Number of Secondary Skills per Non-Rookie

decline in slope is gradual rather than sudden, which leaves open the question of what, given all the other assumptions in the strategic model, constitutes an optimal level of cross-training. To answer this question, one must go back to the strategic model and, with the table function now properly calibrated, test the impact of various levels of cross-training.

Cross-training results from the strategic model

Figures 9–14 illustrate how cross-training policy can affect field service in the strategic model. The three runs compared in each graph are identical for the first simulated year, with target and actual secondary skills per non-rookie steady at a value of 1. In the run named *xtrain 1*, this baseline condition continues through the end of the simulation. In *xtrain 0.5*, the target for secondary skills steps down to a value of 0.5 as of Month 12. In *xtrain 2*, the target steps up to a value of 2 as of Month 12. One may see both the transient and the steady-state impacts of a change in policy in the graphs, and so get a sense of the short-term and long-term benefits and disadvantages.

Figures 9 and 10 show how the secondary skill targets are achieved over time, and the amount of cross-training involved. In *xtrain 0.5*, the new lower

Fig. 9. Secondary skills per non-rookie under three scenarios

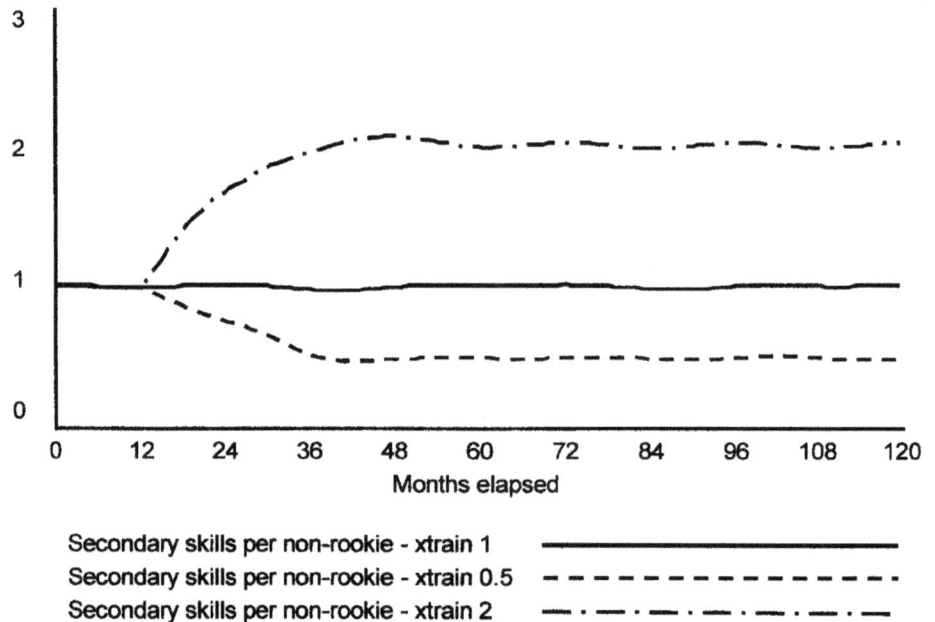

Secondary skills per non-rookie - xtrain 1 ——————————
Secondary skills per non-rookie - xtrain 0.5 – – – – – – – – – – –
Secondary skills per non-rookie - xtrain 2 —·——·——·——·——·——

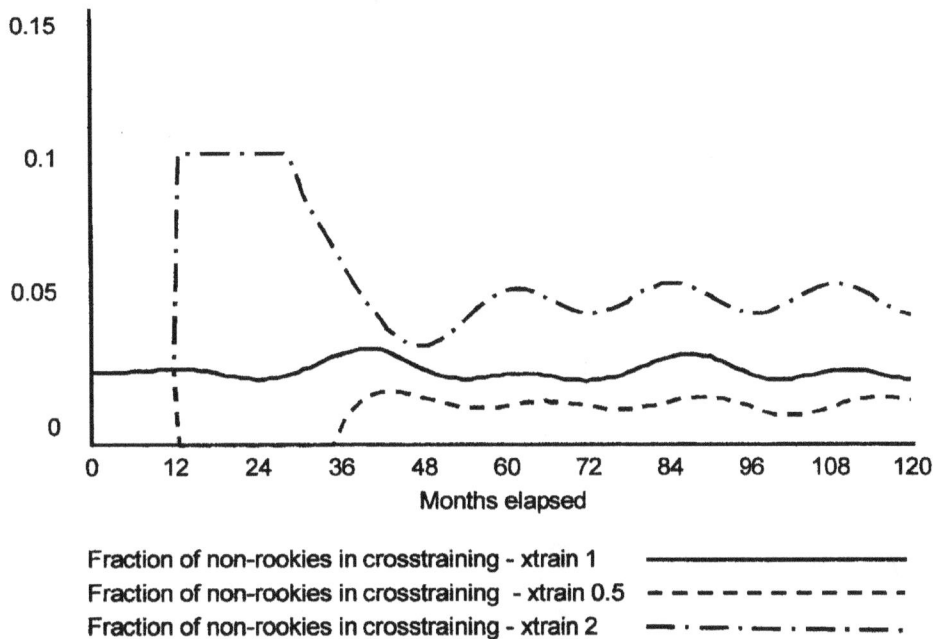

Fig. 10. Fraction of non-rookies in cross-training under three scenarios

Fraction of non-rookies in crosstraining - xtrain 1 ——————————
Fraction of non-rookies in crosstraining - xtrain 0.5 - - - - - - - - - -
Fraction of non-rookies in crosstraining - xtrain 2 — · — · — · — · — ·

level of secondary skills is achieved through two years of no new cross-training at all, after which the rate of cross-training remains permanently lower than in the base run. In *xtrain 2*, the new higher level of secondary skills is achieved through two years of quite intensive cross-training, after which the rate of cross-training must remain permanently greater than the base run in order to maintain the higher level of secondary skills.

It is possible to infer from Figure 10 just how much of a negative effect increased cross-training has on the service availability of field engineers. This effect is significant and is of particular concern in the short term after the new policy is implemented. Indeed, one may see in *xtrain 2* that a cap of 10% is placed on the fraction of non-rookies in cross-training, so as to avoid the short-term catastrophe that could occur if too many non-rookies were simultaneously unavailable as a result of cross-training.

The second downside of increased cross-training, after reduced availability, is skill dilution. This negative effect is seen in Figure 11 in terms of the mean non-rookie skill applied to service jobs. The short-term impact is again considerably greater than the long-term effect. It is not clear from Figures 10 and 11 which of the two negative effects of cross-training is the more significant, and it probably does not matter. The real concern is with their combined impacts on relative workload and service readiness.

Fig. 11. Mean non-
rookie skill applied
under three scenarios

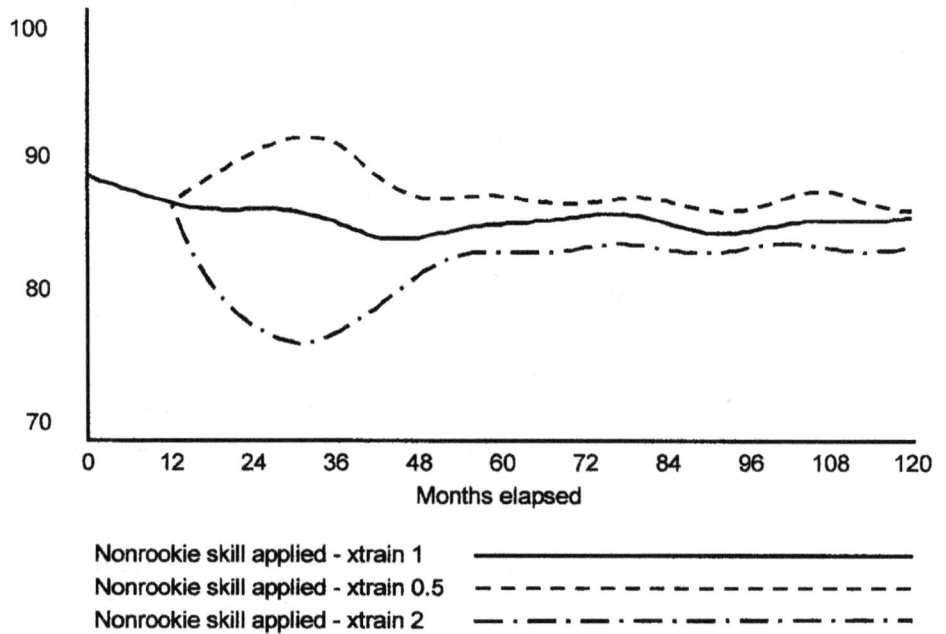

Fig. 11. Mean non-rookie skill applied under three scenarios

Figure 12 shows the net impact on service readiness (expressed relative to its initial baseline value) of the negative effects described above and the direct positive effect of increased secondary skills. Comparing the three runs, it is clear that the positive effect on service readiness strongly dominates the negative effects not only in the steady-state but for most of the simulation. For example, in moving from one secondary skill per non-rookie (*xtrain 1*) to two (*xtrain 2*), the positive effect becomes dominant after little more than a year. Referring back to Figure 9, one may say that the positive effect of cross-training overcomes the initial dominance of the negative effects once the number of secondary skills has moved a sufficient way toward its new steady-state value. These and other simulations indicate that increased cross-training unequivocally boosts mid-to-long-term service readiness throughout the range of secondary skills per non-rookie under consideration (that is, up to a value of 2.5).

Cross-training affects the profitability of field service operations from both revenue and cost standpoints. The operational variable most closely associated with service revenue is contract penetration, the dynamics of which are shown in Figure 13. The positive effect of service readiness (see Figure 12) on contract penetration is apparent, though it is neither immediate (due to fixed contract periods and inertia in customer satisfaction) nor linear (due to saturation in the effect of service performance on customer satisfaction). If increased

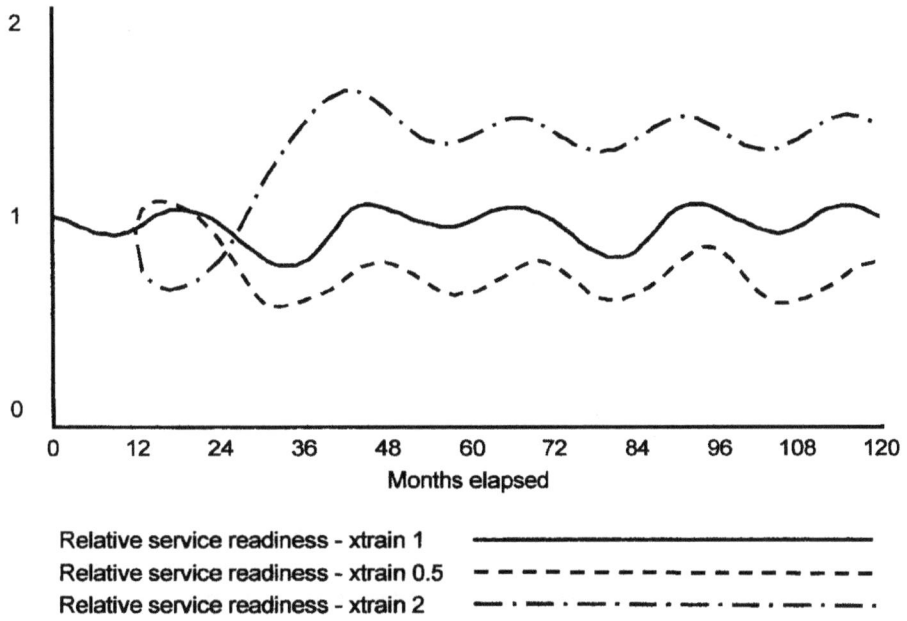

Fig. 12. Relative service readiness under three scenarios

Relative service readiness - xtrain 1 ————
Relative service readiness - xtrain 0.5 — — — —
Relative service readiness - xtrain 2 —·—·—·—

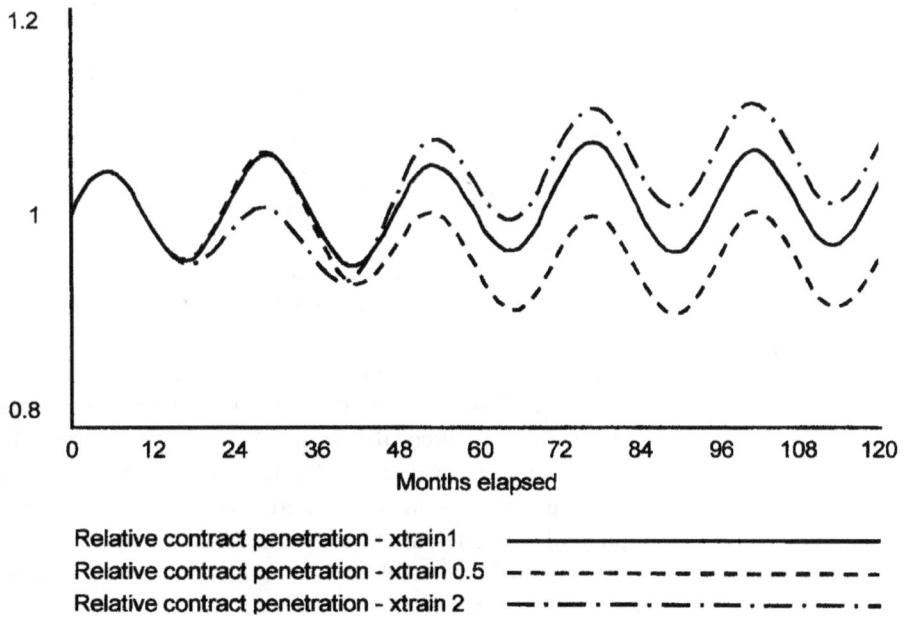

Fig. 13. Relative contract penetration under three scenarios

Relative contract penetration - xtrain1 ————
Relative contract penetration - xtrain 0.5 — — — —
Relative contract penetration - xtrain 2 —·—·—·—

159

Fig. 14. Relative
machines per worker
under three scenarios

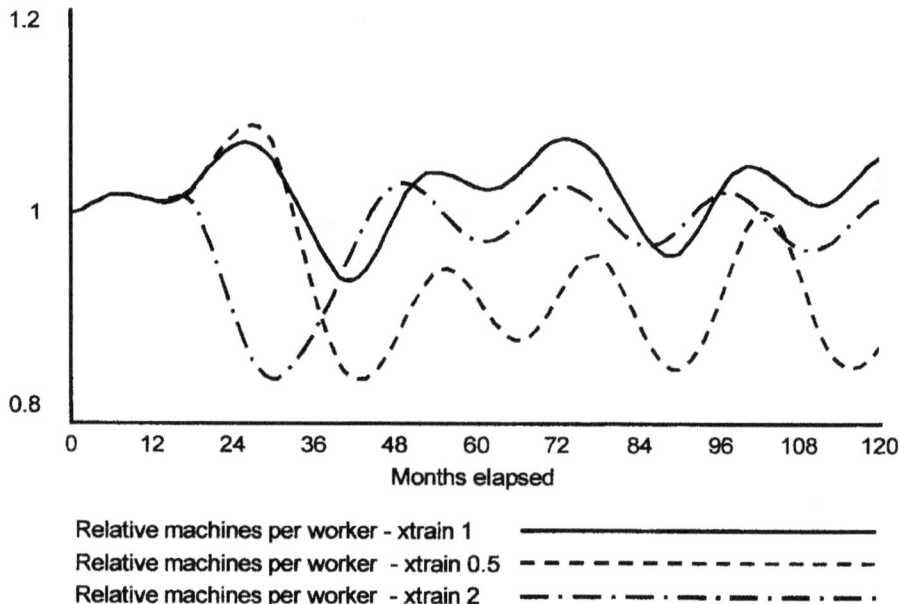

Fig. 14. Relative machines per worker under three scenarios

Relative machines per worker - xtrain 1 ————————

Relative machines per worker - xtrain 0.5 — — — — — — — —

Relative machines per worker - xtrain 2 — · — · — · — · — ·

cross-training always improves mid-to-long-term service readiness, then it has a similar positive impact on contract penetration and, thus, service revenue.

The impact of cross-training on field service costs is more complex and conditional than the impact on revenue. The operational variable most closely associated with costs is machines per worker, the dynamics of which are shown in Figure 14. As discussed previously, the number of machines per worker is adjusted based on workforce utilization; for example, low utilization leads to less hiring and more machines per worker (as in Figure 3), which tends to reduce service costs.

Figure 14 indicates that, relative to *xtrain 1*, there is a steady-state reduction in machines per worker (hence, higher costs) for *xtrain 0.5* as well as *xtrain 2*. Thus, it appears that increased cross-training may either increase or decrease service costs, depending on where one starts. To understand why, return to the causal-loop diagram of Figure 4 and consider the various paths from cross-training to workforce utilization. First, the negative effects of increased cross-training tend to increase utilization, by leaving fewer service hours available and more service hours required. However, at the same time, the positive net impact of cross-training on service readiness leads to less shortcutting and service rework, as well as less overtime-related fatigue, and consequently, a reduction in service hours required.

The net impact on utilization of a change in cross-training depends upon the extent to which shortcutting and fatigue are affected. In *xtrain 1*, shortcutting and fatigue are only minor problems to begin with, so that the increase from one to two secondary skills in *xtrain 2* does relatively little to reduce the extra service hours generated by these problems. Thus, the additional cross-training in this case has the net effect of increasing utilization, owing to reduced worker availability and skill dilution. However, when secondary skills are reduced from one to one-half in *xtrain 0.5*, shortcutting and fatigue become quite significant problems and, in fact, destabilize the system by triggering the vicious cycles R2 and R3 in Figure 4. So, despite the fact that availability is greater and skill dilution is less in *xtrain 0.5* than in *xtrain 1*, the number of service hours required increases so much due to shortcutting and fatigue that utilization and costs are driven up considerably more than they are in *xtrain 2*.

One may conclude that there exists an optimal level of secondary skills from the standpoint of field service costs. Simulations of the strategic model, given all of its assumptions regarding the company's field service operations in the U.S.A., indicate that this optimal level is between 1 and 1.5 secondary skills per non-rookie. At this level, service readiness is high enough so that the system suffers few problems of shortcutting and fatigue, while avoiding serious issues of worker unavailability and skill dilution.

Conclusion

It is a well-known maxim in system dynamics that model parameters should be estimated using data below the level of aggregation of model variables wherever possible. Such detailed data characterize the individual members of a level, or the individual events within a rate. For example, the best way to estimate the average lifetime of housing in an urban area may be to use official records detailing the dates of construction and demolition of individual houses (Graham 1980). However, it is not always easy or straightforward to make the leap from individual-level data to the aggregate concepts used in a system dynamics model, especially when that model explores policies and scenarios beyond the range of past experience.

This paper has presented a case in which a key table function affecting service readiness could only be properly estimated by analyzing a separate micro-level, OR-type model that mimics the daily queuing of service jobs and their assignment to individual field engineers. Although the development and testing of the micro model was a time-consuming affair, it allowed the modelling team to draw important conclusions about the appropriate level of cross-training, one of several issues considered by the higher-level strategic model. The modeling team and senior management told me directly that these conclusions would have been viewed with scepticism had the micro model not been developed to establish and validate the underlying numerical assumptions.

For forty years now, there has been a conscious divide between system dynamics and traditional operations research (Lane 1994). Where system dynamics looks for strategies to improve the vitality and stability of multi-faceted organizations, traditional OR focuses more narrowly on the performance of individual components of those organizations. Clearly, 'hard' OR methods such as linear programming, queuing theory, and network analysis are unable to deal with the dynamics that emerge from the interactions of subsystems, and therefore are inappropriate—as Jay Forrester (1994) has observed—for addressing truly strategic issues.

Nevertheless, traditional OR tools do have their proper place, and their practical value has been demonstrated time and time again in operational applications of many sorts. As system dynamicists, we should set aside the mistaken notion that hard OR is disengaged from the real world or of no real use to managers. Once we do this, I think we will see that there are numerous opportunities for system dynamics and hard OR to be used together. We are already learning from the 'soft OR' community about how to structure issues and design strategic options with clients (see Lane 1994, Forrester 1994, and other articles in *System Dynamics Review* 1994, 10(2–3)). When we similarly learn to adapt the language and tools of hard OR, we will gain the ability to put some of our models' key relationships under the microscope of individual-level analysis when they are not otherwise adequately understood. I believe our methodology will benefit from such adaptation, and in many cases clients will get added value out of projects that combine macro- and micro-modeling.

Notes

1. In the strategic model, engineers are represented in the aggregate, not as individuals, and are not mapped to specific product types. The skill-building processes, both primary and secondary, are assumed to be structurally identical and parametrically similar for all of the company's engineers and product types.
2. To avoid the longer skill-building time for secondary skills, management could conceivably dictate that only seniors and not juniors are eligible for cross-training. Seniors have already finished climbing one learning curve,

and so need climb only one more after cross-training, rather than the two learning curves cross-trained juniors must climb. For this reason, the stated policy of the client company is, in fact, to select seniors over juniors for cross-training whenever possible. In reality, however, there are never enough seniors in the workforce to comprise the kind of well-distributed back-up capability that management would like to have, so that juniors end up being cross-trained nearly as much as seniors.

3. It is assumed that a cross-trained engineer is assigned with roughly equal frequency to each of the products for which he or she has been trained. Under this assumption, the weight (W) on secondary skills for non-rookies is related to the average number of secondary skills per non-rookie (S) as follows: $W = S/(1 + S)$.

4. This gut feel was undoubtedly related to the field managers' actual range of experience. Fewer than half of the company's field engineers had more than one secondary skill, and only a fraction of these had more than two.

5. The optimization routine, called from Vensim as an external function, was designed and developed by Robert Brooks of RBA Consultants, Los Angeles, California. The routine employs integer programming to solve a linear model assigning jobs in the queue to available workers with appropriate skills. This model has an objective function that represents the expected value of total job completions for the current work shift, adjusted for product–job type importance weights. The linear model also includes constraints that ensure that an engineer is assigned to no more than one job at a time, and that a job is worked by no more than one engineer at a time.

6. Service readiness in the strategic model is represented by the aggregate ratio of repair hours available to workload (repair hours required), multiplied by the cross-training or secondary skills effect. Suppose that one can increase secondary skills to the point where service readiness is doubled. One could then double the workload and still have the same level of service readiness as one had originally. Thus, an appropriate way of inferring the impact on service readiness of a change in secondary skills is to find a workload multiplier that brings the system back to its original level of performance.

References

Carroll, J. S., J. D. Sterman and A. A. Marcus. 1998. Playing the maintenance game: How mental models drive organizational decisions. *Debating Rationality: Non-rational Elements of Organizational Decision Making*, eds. J. J. Halpern and R. N. Stern. Ithaca, NY: Cornell University Press.

Forrester, J. W. 1994. System dynamics, systems thinking, and soft OR. *System Dynamics Review* 10(2–3): 245–256.

Graham, A. K. 1980. Parameter estimation in system dynamics modeling. *Elements of the System Dynamics Method*, ed. J. Randers. Cambridge, MA: MIT Press. Reprinted by Pegasus Communications, Waltham, MA.

Graham, A. K., J. D. W. Morecroft, P. M. Senge and J. D. Sterman. 1992. Model-supported case studies for management education. *European Journal of Operational Research* 59(1): 151–166.

Hall, R. W. 1991. *Queuing Methods for Services and Manufacturing*. Englewood Cliffs, NJ: Prentice-Hall.

Lane, D. C. 1994. With a little help from our friends: How system dynamics and soft OR can learn from each other. *System Dynamics Review* 10(2–3): 101–134.

Levin, G., E. B. Roberts and G. B. Hirsch *et al.*. 1976. *The Dynamics of Human Service Delivery*. Cambridge, MA: Ballinger.

Oliva, R. 1996. A Dynamic Theory of Service Delivery: Implications for Managing Service Quality. Ph.D. Thesis, Sloan School of Management, Massachusetts Institute of Technology.

Richmond, B. 1994. Classic core business infrastructures: Inadvertent evolution and fighting the physics. In *ithink*® *Business Applications Guide*. Hanover, NH: High Performance Systems.

Senge, P. M. and J. D. Sterman. 1992. Systems thinking and organizational learning: Acting locally and thinking globally in the organization of the future. *European Journal of Operational Research* 59(1): 137–150.

8

Toward a Dynamic Theory of Antibiotic Resistance (2000)

Homer J, Ritchie-Dunham J, Rabbino H, Puente L, Jorgensen J, Hendricks K. *System Dynamics Review*, 16(4): 287-319

Related writing

Homer J, Jorgensen J, Hendricks K. Modeling the Emergence of Multi-Drug Antibiotic Resistance. *Proceedings of the 19ʰ International System Dynamics Conference*, Atlanta, GA, 2001.

Toward a dynamic theory of antibiotic resistance

Jack Homer,[a]*   James Ritchie-Dunham,[b]   Hal Rabbino,[b]   Luz Maria Puente,[b]
James Jorgensen,[c] and Kate Hendricks[d]

Jack B. Homer is a system dynamics modeling consultant and has directed Homer Consulting since 1989. He was formerly an assistant professor at the University of Southern California and received a Ph.D. in Management from M.I.T. for a thesis modeling the diffusion of new medical technologies.

James L. Ritchie-Dunham is President of the Strategic Decision Simulation Group (SDSG) and co-author with Hal Rabbino of *GRASP: The Art and Practice of Strategic Clarity* (Wiley, 2001). He is currently completing a Ph.D. in Decision Sciences at the University of Texas at Austin. He was formerly a professor at the ITAM in Mexico City and a petroleum engineer at Conoco.

Abstract

Many common bacterial pathogens have become increasingly resistant to the antibiotics used to treat them. The evidence suggests that the essential cause of the problem is the extensive and often inappropriate use of antibiotics, a practice that encourages the proliferation of resistant mutant strains of bacteria while suppressing the susceptible strains. However, it is not clear to what extent antibiotic use must be reduced to avoid or reverse an epidemic of antibiotic resistance, and how early the interventions must be made to be effective. To investigate these questions, we have developed a small system dynamics model that portrays changes over a period of years to three subsets of a bacterial population—antibiotic-susceptible, intermediately resistant, and highly resistant. The details and continuing refinement of this model are based on a case study of *Streptococcus pneumoniae*, a leading cause of illness and death worldwide. The paper presents the model's structure and behavior and identifies open questions for future work. Copyright © 2000 John Wiley & Sons, Ltd.

Syst. Dyn. Rev. **16**, 287–319, (2000)

Over the past few decades and around the world, a wide variety of common bacterial pathogens have become increasingly resistant to the antibiotics used to treat them. As a consequence, the illness and death rates for some formerly well-controlled diseases, such as tuberculosis, are now climbing in worrisome fashion. In the United States, it is estimated that drug-resistant bacteria are now responsible for some 70 percent of the 90 thousand fatal hospital infections that occur annually (Bright 1999). Antibiotic resistance was estimated in 1993 to add $200 million a year to U.S. medical bills in the form of more expensive antibiotics, and over $30 billion a year when the costs of extended hospital care are included (Garrett 1994).

A large body of evidence suggests that the essential cause of the problem is the extensive and often inappropriate use of antibiotics, a practice that encourages resistant mutant strains of bacteria to proliferate while suppressing the susceptible strains. In the U.S.A., it has been estimated that one third of the 150 million annual prescriptions for antibiotics are inappropriate, as defined by established clinical standards (Levy 1998).[1] Public policies that would help reduce the inappropriate use of antibiotics include educational campaigns and antibiotic use monitoring and control, as well as the development and dissemination of quick and accurate diagnostic tests. Of equal importance are

[a] Homer Consulting, Voorhees, New Jersey, U.S.A.
[b] Strategic Decision Simulation Group, Austin, Texas, U.S.A.
[c] University of Texas Health Science Center, San Antonio, Texas, U.S.A.
[d] Texas Department of Health, Austin, Texas, U.S.A.
* Correspondence to: Jack Homer, 36 Covington Lane, Voorhees, NJ 08043, U.S.A.

System Dynamics Review Vol. 16, No. 4, (Winter 2000): 287–319
Copyright © 2000 John Wiley & Sons, Ltd.

Received August 2000
Accepted September 2000

Hal Rabbino is a partner with SDSG and has advised clients throughout North and Latin America in the oil, financial services, and public health industries for more than ten years. He specializes in improving the ability of leadership teams to formulate, test and communicate strategy. He has an M.B.A. and a Masters in International Management.

Luz Maria Puente is a partner with SDSG and has spent her career working in industry and as a management consultant. She has assisted clients throughout the Americas in areas including urban planning, banking, and public health. She has an M.B.A. and a degree in actuarial mathematics.

policies that reduce the spread of infection, both because they limit the spread of already-resistant bacteria and because fewer infections implies less need for antibiotic use. Such policies include vaccination and traditional hygiene and infection control measures such as the isolation of patients with drug-resistant infections (Neu 1992; Baquero *et al.* 1996; Okie 1997, Ekdahl *et al.* 1998; Levy 1998).[2]

Although the policy options are known, their relative effectiveness is not. Every policy has its costs and risks, and agencies with limited budgets and limited authority must decide when and how to intervene.

A systemic approach to the problem

In the spring of 1998, the Infectious Disease Epidemiology and Surveillance Division of the Texas Department of Health (TDH), in partnership with the Strategic Decision Simulation Group (SDSG), assembled a group of experts to begin thinking systemically about the problem of antibiotic resistance and possible policy options at the state and national levels. The meeting produced a causal loop diagram reflecting a diversity of perspectives, a version of which is presented in Figure 1. This diagram identifies the basic dilemma associated with antibiotics, that is, the fact that they can reduce illness for individuals in the short term, but if overused may become ineffective—regionally, nationally, or even worldwide—as a result of bacterial resistance. Such a local-versus-global dilemma has been described as a "tragedy of the commons" (Hardin 1968). Figure 1 identifies how various responses, such as education and regulation, may eventually be tried as this problem becomes more apparent to policymakers and pressure grows to do something about it.

After development of the causal loop diagram, simulation modeling was employed to look more closely at specific policy options. The initial models considered the dynamics of both human and bacterial populations, with an approach to human population dynamics much in line with prior system dynamics models of epidemics (Homer and St. Clair 1991; Ritchie-Dunham and Méndez Galvan 1999; Sterman 2000). As this initial modeling progressed, we came to realize that all proposed policies rested on the presumption that an emerging problem of antibiotic resistance would be recognized in a timely fashion, its severity properly assessed, and that appropriate goals for behavioral change would be set. We came to learn that these conditions do not often hold; in particular, it is not known when or by how much antibiotic use must be reduced to avoid or reverse an epidemic of resistance.

To help develop such an understanding, our more recent efforts have focused on grounding the work more solidly in the scientific literature, collecting data not available in the literature, and revising our model of bacterial population

Fig. 1. Antibiotic
resistance policy loops

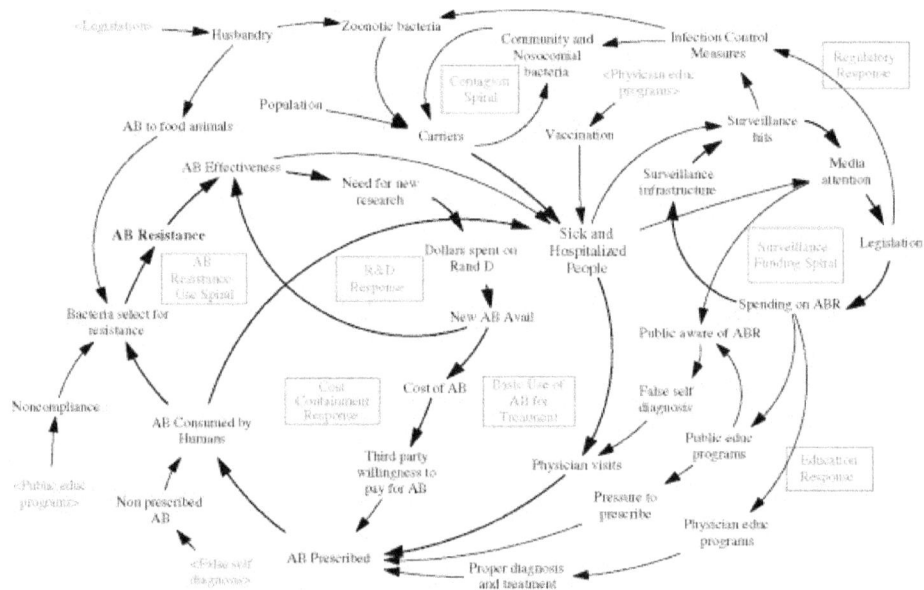

James H. Jorgensen is
Professor of Pathology
and Medicine at the
University of Texas
Health Science Center
in San Antonio and
Director of the
Microbiology
Laboratory of
University Hospital.
He is a clinical
microbiologist whose
research focuses on
emerging antibiotic
resistance. He
received a Ph.D. from
the University of
Texas Medical Branch
and has been a faculty
member at UTHSCSA
since 1973.

Kate Hendricks, MD,
MPH&TM, is the
Division Director for
Infectious Disease
Epidemiology and
Surveillance at Texas
Department of Health.
She is Board Certified
in Preventive
Medicine and serves
as the medical editor
for the state of Texas'
biweekly morbidity
report *Disease
Prevention News*.

dynamics so that it reflects this knowledge as much as possible. Our purpose
in this paper is to present what we have discovered so far, and to identify some
of the open questions that remain.

Fundamentals of antibiotic resistance

The literature on antibiotic resistance is vast; many books are entirely devoted
to the subject (e.g., Levy 1992; Chadwick and Goode 1997) and papers
on it appear monthly in infectious disease journals. Although the basic
concepts derive intuitively from population genetics, the mechanisms differ
considerably by bacterial species and antibiotic class, and the relationship
between antibiotic use and antibiotic resistance is complex and a subject of
some controversy.

Bacteria are ubiquitous in our environment and within our bodies; only a
fraction are pathogenic. Infectious disease occurs when the immune system
encounters a bacterial pathogen for which it is not prepared, perhaps due to a
lack of previous exposure. The bacterial flora with which we coexist, known
as commensals, are not only benign but often protect us from pathogenic
bacteria by competing with them for niche turf and thereby preventing their
proliferation (Levy 1998).

Antibiotics typically retard bacterial proliferation by interfering with the
production of components needed to form new cells; specifically, they may
inhibit bacterial cell wall synthesis, protein synthesis, folic acid synthesis, or

DNA replication. The number of antibiotics on the market is large and includes more than 50 variants of penicillin, 70 variants of cephalosporin, 12 tetracyclines, 9 macrolides, 8 aminoglycosides, and so forth. But bacteria evolve continuously and have developed responses, in the form of resistance genes, for every type of antibiotic in existence; indeed, at least three bacterial species have strains that are resistant to every antibiotic available (Neu 1992, Levy 1998).

A bacterium may acquire a resistance gene by direct inheritance, by spontaneous mutation (which may produce a new resistance trait or strengthen an existing one), or by accepting genes from other bacteria in a process known as gene exchange. Most gene exchange passes traits among members of a species (e.g., *Streptococcus mitis*), but in some cases may pass traits to members of a different species (e.g., *Streptococcus pneumoniae*).[3] As one leading researcher has stated (Levy 1998), "The exchange of genes is so pervasive that the entire bacterial world can be thought of as one huge multicellular organism."

In terms of population genetics, antibiotics are said to put selective pressure on susceptible strains of bacteria, diminishing their ability to proliferate and thereby putting them at a reproductive disadvantage relative to resistant strains. Although antibiotics are used to fight pathogenic bacteria, they also tend to put selective pressure on commensal bacteria and encourage the spread of resistance traits among these benign flora. Indeed, resistance often develops among commensals before being passed to pathogenic intruders (Levy 1998; Austin *et al.* 1997). Consequently, antibiotic resistance has been a particular problem with broad-spectrum antibiotics, such as third-generation cephalosporins, that are active against a wide variety of bacterial genera. Such antibiotics encourage the spread of resistance traits among multiple types of commensals and subsequently to their pathogenic cousins, and not only to the original pathogen for which the drug is given (Levy 1998; Ballow and Schentag 1992; Frimodt-Møller *et al.* 1997).

While researchers agree on a positive link between antibiotic use and resistance, the exact nature of this link is hotly debated. On an individual level, evidence suggests that resistance becomes more likely when patients fail to take the full recommended daily dosage of antibiotic (Guillernot *et al.* 1998) or to complete the full course of treatment (Okie 1997). On a population level, empirical evidence and epidemiological models suggest that resistance may emerge only when antibiotic use exceeds some threshold (Levy 1998; Austin *et al.* 1997). Such a threshold may vary by hospital or community, because of differences in contact rates (as affected, for example, by the prevalence of day-care programs for young children), hygiene, and other factors affecting the spread of pathogens. Unfortunately, no-one yet knows how to determine the threshold, and most hospitals and communities lack the data that would be required to do so (Levy 1998).

If resistance has already emerged, the key question concerns the extent to which antibiotic use must be reduced to reverse the problem to an acceptable level, or even whether such a reversal is possible. Laboratory studies of *Escherichia coli* have shown that resistance usually, but not always, imposes what is called a fitness cost, expressed, for example, as 5 percent per generation (Levin *et al.* 1997). This fitness cost has been explained as being the result of resistant strains having to divert some of their limited energy from reproduction to maintaining their antibiotic resistance traits (Levy 1998). The existence of a fitness cost suggests that, at a lower level of antibiotic use, susceptible bacteria will regain a natural reproductive edge over resistant strains, leading in theory to a reversal of resistance.

Outside the laboratory, it has been demonstrated that stringent restrictions on the use of cephalosporins in a hospital can, in a year's time, reverse even high levels of antibiotic resistance (Ballow and Schentag 1992; Rahal *et al.* 1998). Perhaps more instructive for society at large is the case of Finland, where macrolide (mainly erythromycin) consumption was reduced by more than 40 percent after outpatient usage guidelines were issued at the end of 1991 in response to the emergence of resistance in group A *Streptococcus*. Resistance continued to climb from a value of 13 percent in 1990 to a peak of 19 percent in 1993, before starting a decline that by 1996 left resistance at under 9 percent (Seppälä *et al.* 1997). A similar story is reported from Iceland, where a national campaign to reduce antibiotic use has been moderately successful at bringing down penicillin resistance in *Streptococcus pneumoniae* among children, which prior to the campaign had climbed from less than 1 percent in 1988 to 20 percent in 1993 (Kristinsson 1997).

But the news about the reversability of resistance is not uniformly good. Samples of *E. coli* taken from a day-care center in Atlanta were found to be resistant to streptomycin even though that antibiotic has been rarely used in the last thirty years. Laboratory studies confirmed that these streptomycin-resistant *E. coli* had virtually no fitness cost relative to susceptibles. Other laboratory studies involving the coevolution of resistant and susceptible species indicate that the fitness cost of resistance can decline over time through adaptation, specifically through the development of so-called compensatory mutations. In such cases, even the radical reduction of antibiotic use may do little or nothing to reverse resistance (Levin *et al.* 1997; Morell 1997).

Case study: beta-lactam resistance in *Streptococcus pneumoniae*

Our modeling work has, from the beginning, focused on a case of particular concern to health agencies throughout the United States: the growing resistance of pathogenic *S. pneumoniae*, commonly known as pneumococcus, to the broad class of antibiotics known as beta-lactams, which include the penicillins, the cephalosporins, and the carbapenems.

Pneumococci were first identified by Louis Pasteur in 1881. They are the leading cause of community-acquired pneumonia worldwide, as well as the leading bacterial pathogen in children and adolescents. In the U.S.A. alone, pneumococci cause each year an estimated three thousand cases of meningitis, 50 thousand cases of bacteremia, 500 thousand cases of pneumonia, and seven million cases of otitis media (middle ear infections) (Centers for Disease Control and Prevention (CDC) 1996). Deaths from invasive pneumococcal disease exceed 40 thousand per year in the U.S.A. (Long 1999). Worldwide, an estimated 1.2 million children die annually from pneumococcal disease (FitzGerald 2000). A pneumococcal vaccine for use in young children has only recently been developed.[4]

Some 90 different capsular serotypes of pneumococci have been identified, though only a small subset of those are prevalent pathogens, with some variation in the distribution of serotypes by country and continent. Pneumococci colonize the nasopharynxes of 15 to 60 percent of individuals at any given time; the carriage rate has been shown to depend on age of the individual, season, crowding, and the prevalence of influenza and other viral infections that may compromise immunity. Most people are colonized for weeks to months and remain asymptomatic.[5] Disease, when it occurs, usually follows rapidly upon the acquisition of a serotype new to the individual, rather than after prolonged carriage. For example, about 15 percent of young children who acquire a new serotype become ill, most commonly with otitis media (Long 1999).

The first case of antibiotic resistant pneumococci was reported in Australia in 1967, followed by reports from New Guinea in 1969 and South Africa in 1977 (CDC 1996). Evidence suggests that resistance often emerges first in oral viridans group streptococci commensals, such as *S. mitis* and *S. oralis*, after which it may be transferred to colonizing pneumococci (Koornhof *et al.* 1992; Reichmann *et al.* 1997).

Multidrug resistant strains of pneumococci emerged in the 1980s. By the end of the decade in Hungary, 70 percent of children with pneumococcal illness carried strains resistant to five different classes of antibiotics. By the 1990s, drug resistant pneumococci had turned up all over the world, some resistant to as many as six different classes of antibiotics (Neu 1992; Garrett 1994).

Data on pneumococcal resistance to penicillin

There have been many reports on the prevalence of antibiotic resistant pneumococcal infections, but such reporting is far from universal and the full extent of the problem is not known (CDC 1996). In the U.S.A., for example, it is a reportable condition in only some states, and national surveys by the Centers for Disease Control and Prevention (CDC) and by others are published only irregularly.[6] What is known is that there is great variation in prevalence by geographical region and by age of individual, and that prevalence is capable of changing quickly in any particular location (Cetron *et al.* 1996; CDC 1996; 1999a).

Despite the lack of universal reporting, most of what has been published is consistent in reporting on the prevalence of penicillin resistant pneumococci (PRP), and researchers all generally use the same accepted standards for determining whether a pneumococcal isolate should be considered susceptible, intermediately resistant, or highly resistant to penicillin.[7] Researchers often sample or survey multiple hospitals and medical centers, rather than just one. Also, the analysis often focuses on invasive pathogenic isolates drawn from normally sterile areas (e.g., blood, cerebrospinal fluid, pleural fluid, and middle-ear fluid), rather than on noninvasive isolates that may comprise a mix of those pneumococci causing the acute disease and those not responsible.

Time series on PRP for the U.S.A., Spain, South Africa, and Hungary are presented in Figures 2 and 3. Figure 2 shows "total PRP", the percentage of isolates that are either intermediately or highly resistant, for all four countries.[8] Figure 3 shows "high PRP", the percentage of isolates that are highly resistant, for the United States and Spain only. These are the most extensive time series we have found, each covering at least 11 years. Despite a few gaps and inconsistencies,[9] the data lead to some useful conclusions or working hypotheses about the dynamics of PRP on a national level:

Fig. 2. Total (intermediate plus high-level) pneumococcal resistance to penicillin in four countries
Sources: See Note 8

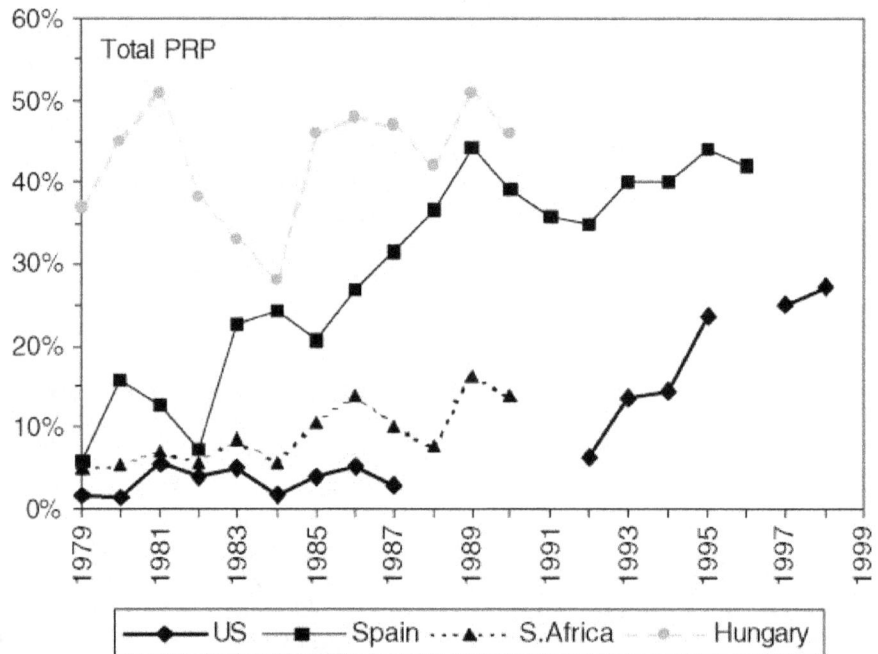

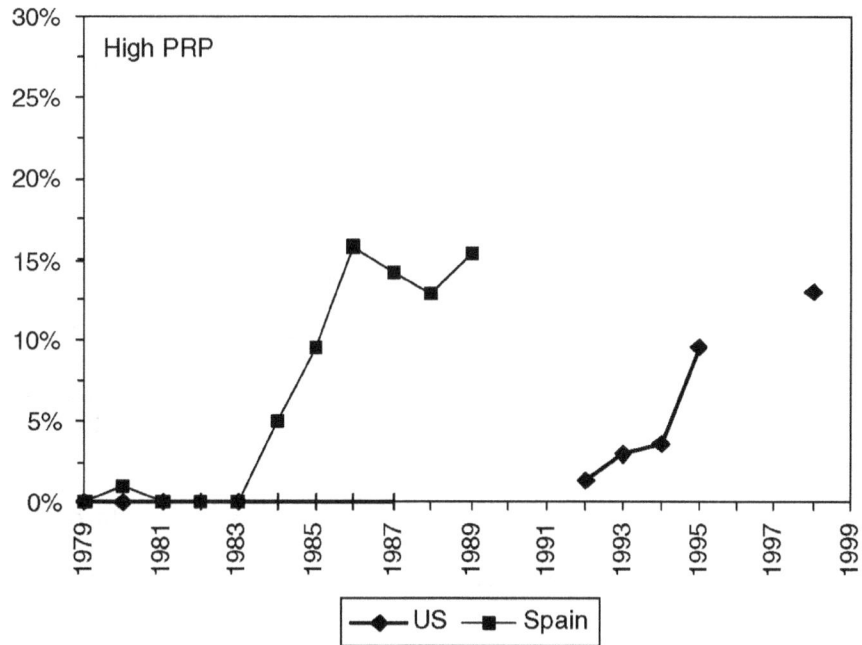

Fig. 3. High-level PRP in the United States and Spain *Sources*: See Note 8

- PRP may grow in a ten-year period, from a low level of 10 percent or less to a high level of 30 percent or more (Spain), but need not grow to that extent (U.S.A., South Africa).
- Even after rapid growth, PRP tends to plateau without climbing inexorably to 100 percent (Spain, Hungary).
- PRP tends to fluctuate (all four countries) and may at times reverse substantially and in a short period of time (Spain, Hungary).
- Growth in high resistance occurs only when intermediate resistance is already significant, e.g., greater than 10 percent (U.S.A., Spain).
- High resistance may plateau even while intermediate resistance continues to grow (Spain).

The fact that total PRP may plateau at levels well below 100 percent is of particular interest. One epidemiological model suggests that such stable coexistence of susceptible and resistant strains in a population should be possible under certain conditions involving the relative reproductive fitness and genetic convertibility of the different strains (Austin *et al.* 1997; 1999). This model is able to reproduce historical patterns of increasing, plateauing, and even decreasing levels of antibiotic resistance among children for pneumococci in Iceland and *Moraxella catarrhalis* in Finland (Austin *et al.* 1999). However, in order to achieve a reasonable fit to history, it is necessary in both cases to assume that resistant strains have a natural reproductive edge over susceptible strains, while at the same time assuming that resistant strains have a tendency to lose their resistance during gene exchange. These assumptions are not supported by direct evidence; indeed, the first assumption appears to contradict what is known about the fitness cost of resistance.

There may be a more compelling reason why, at least in the case of pneumococci, resistance can settle at levels below 100 percent. It seems that among pneumococci, some common serotypes develop resistance much more frequently than others, reflecting widely differing propensities to absorb resistance genes from the host environment. Figures 4 and 5 show the serotype distributions of PRP for the U.S.A. and Spain, respectively, based on aggregations of data covering 1979 and most of the 1980s.[10] In these figures, each line segment on the graph represents the incremental contribution of a single serotype (or, in a few cases, the combined contribution of multiple miscellaneous or unidentified serotypes), in terms of the total number of isolates (x axis) and the number of resistant isolates (y axis) in the sample. In the U.S.A. data, PRP averages only 5.0 percent overall, with a range of 0 to 38 percent for individual serotypes. In the Spanish data, PRP averages 27.8 percent overall, with a range of 0 to 89 percent for individual serotypes. The strongly bowed-out rather than linear curves in Figures 4 and 5 reflect

Fig. 4. Serotype distribution of PRP in the United States, 1979–1987. *Source*: Spika *et al.* 1991

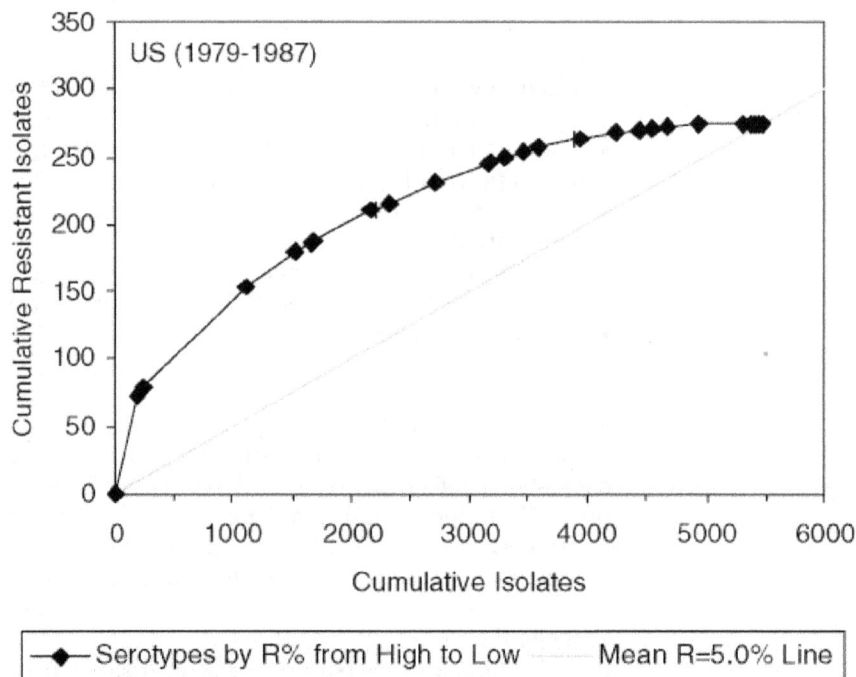

174

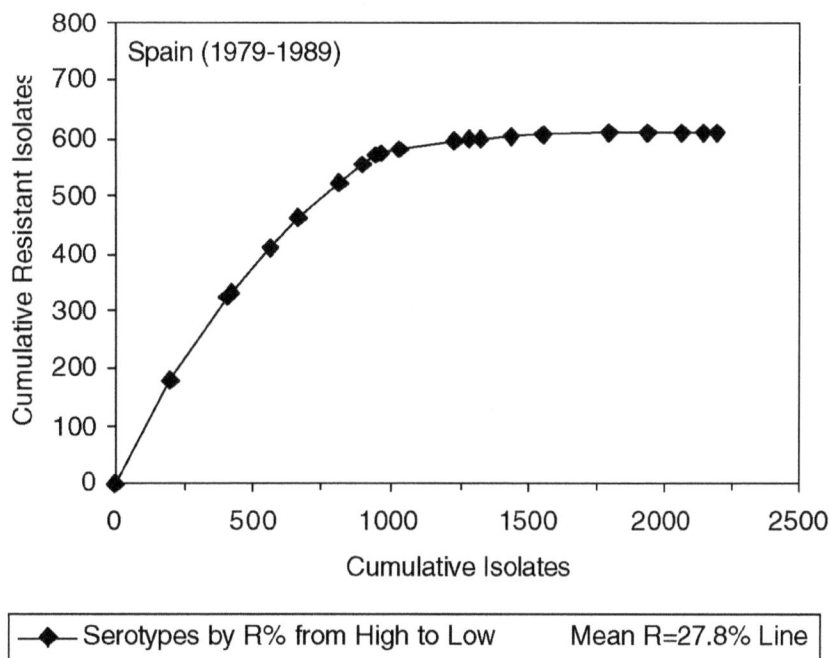

Fig. 5. Serotype distribution of PRP in Spain, 1979–1989. *Source*: Fenoll *et al.* 1991

the skewed distribution of resistance among the serotypes. It is evident that, in both cases, a small number of serotypes are responsible for the majority of resistant isolates. Moreover, the specific serotypes that account for most of the resistance in the U.S.A. data also account for most of the resistance in the Spanish data.[11] Taking the two data sets together, it appears that an increase in total PRP primarily reflects increasing resistance within a subset of serotypes, rather than increasing resistance among all serotypes. Even if the more resistance-prone serotypes eventually became 100 percent resistant, many of the less resistance-prone serotypes would still have much lower levels of resistance. Such heterogeneity probably explains why total PRP may settle at something considerably below 100 percent.

Next, one might ask why, in the Spanish time series data, the level of high PRP appears to plateau at around 15 percent in the late 1980s, even as intermediate PRP continues to climb. (High PRP data for the 1990s were not reported.) One hypothesis is that there are differing serotype propensities for high PRP just as there are for PRP generally. In particular, if an even smaller number of serotypes have the propensity to become highly resistant to penicillin (e.g., through the acquisition of even stronger resistance genes) than have the propensity to become intermediately resistant, this might explain a relatively low plateau for high PRP. Unfortunately, the Spanish data do not identify rates of high PRP by serotype, and we have not yet located any other data that would help clarify this matter.

The PRP time series data presented here all reflect relatively large samples taken from multiple locations throughout each country, and in that sense they are the best data we have found on national PRP patterns. But we cannot ignore other data that may indicate possibilities not apparent in these time series. In particular, smaller sample, non-longitudinal data from Hungary (Marton 1992) and Asia (Lee *et al.* 1995; Song *et al.* 1999) indicate the possibility of very high percentages of both total PRP and high PRP *per se*. The highest percentages are from Korea, where we have data from two periods of time. For 1991–1993, one hospital (131 isolates) experienced total PRP of 70 percent and high PRP of 33 percent, while for 1996–1997, two medical centers (177 isolates) experienced total PRP of 80 percent and high PRP of 55 percent.[12]

Data on beta-lactam antibiotic use

As stated previously, our modeling work has centered on pneumococcal resistance to the whole beta-lactam class of antibiotics, of which the penicillins and cephalosporins comprise the major portion. (The carbapenems are used much less frequently and almost exclusively for severe infections in hospital settings.) For modeling purposes, we assume that the PRP data presented above give an accurate indication of trends in resistance by pneumococci to beta-lactams generally. In fact, the available evidence suggests that pneumococcal resistance to cephalosporins tends to move in parallel with resistance to penicillin (Butler *et al.* 1996; Doern *et al.* 1999; Fenoll *et al.* 1998; Marton 1992; Lee *et al.* 1995; Song *et al.* 1999). Although the PRP data are specific to penicillin, it must be recognized that all beta-lactams, and not only penicillin, are capable of encouraging the spread of the sort of resistance genes that can inactivate penicillin.[13] In other words, if one is trying to find a link between antibiotic use and PRP, it is important to track the use of beta-lactams generally and not only penicillin.

Unfortunately, nationwide data on beta-lactam use are neither published nor perhaps even recorded for most countries. Thus far, we have managed only to construct a time series for the United States, as shown in Figure 6, and even this is incomplete and does not cover all categories of use. These data are drawn from the National Ambulatory Medical Care Survey (NAMCS), the only national survey in the U.S.A. to provide information on the prescribing of oral antibiotics by office-based physicians. The NAMCS has been conducted annually from 1973 through 1981, in 1985, and again annually since 1989 (McCaig and Hughes 1995). Figure 6 presents estimates of office-based beta-lactam prescriptions (Rxs) per thousand persons per year.[14] From 1990 to 1995 (but excluding 1992, a year of unusually high use), these rates translate

Fig. 6. Office-based
beta-lactam antibiotic
use in the United
States. *Sources*:
McCaig and Hughes
1995, NCHS 1997

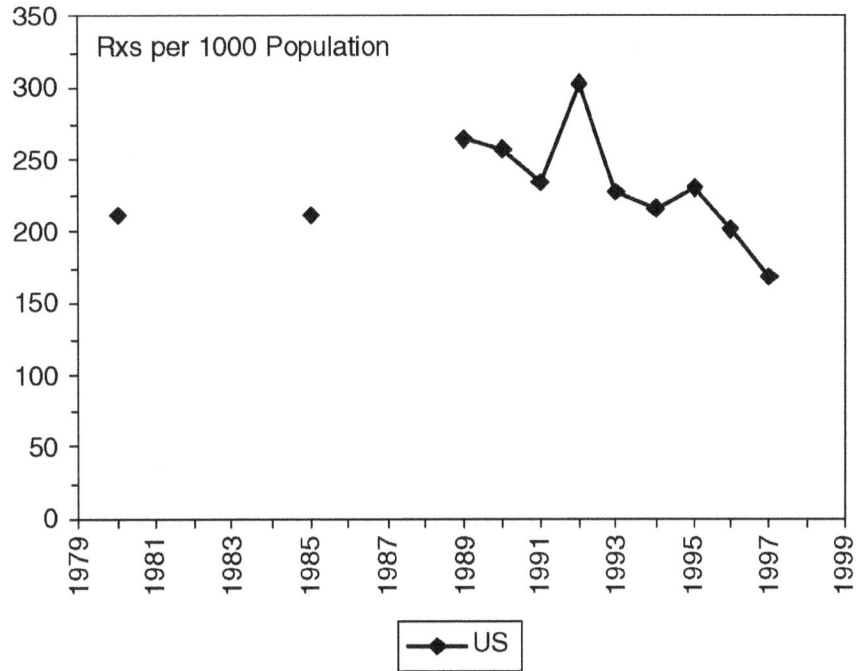

Fig. 6. Office-based beta-lactam antibiotic use in the United States.

to about 60 million prescriptions per year, or 60 percent of the roughly 100 million office-based prescriptions reported for all antibiotics combined.[15] Beta-lactam use increased significantly from 1985 to 1989, then gradually declined during the 1990s (with the exception of 1992); by 1996, it had declined to below where it was in 1980 and 1985.[16]

For Spain, published data exist on the use of all antibiotics combined, but too little on beta-lactam use *per se* to allow us to draw any firm conclusions. Consumption of all antibiotics by outpatients peaked sometime during 1966–1976, then decreased by about 40 percent from 1976 to 1993, about 40 percent of that decrease coming from 1983 to 1988. As of 1993, about 50 percent of the consumed antibiotics were penicillins and another 13 percent were cephalosporins, but such a breakdown of use for prior years is not provided (Baquero *et al.* 1996).

A bacterial population model of antibiotic resistance

Our system dynamics model attempts to explain what we have learned about nationwide pneumococcal resistance to beta-lactams in a way that is straightforward and potentially generalizable, yet is able to fit the historical

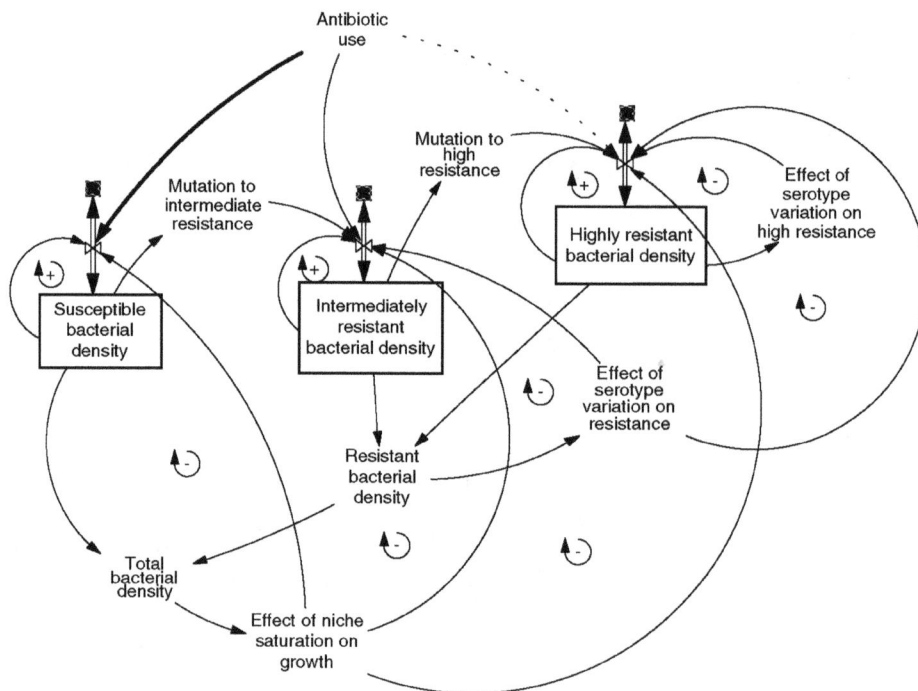

Fig. 7. Model causal structure

evidence. The model's causal structure is presented graphically in Figure 7 and in equation form in Table 1. The three levels of the model represent the three subsets of a bacterial population as defined by accepted standards of resistance measurement—namely, antibiotic-susceptible, intermediately resistant, and highly resistant. These subpopulations are modeled not in absolute terms of number of bacteria, but in terms of their relative density, where 100 represents the carrying capacity for the bacterial species or genus in question within the entire human host population under study.

Thus, the model considers the dynamics of a widely dispersed bacterial population, and the impact of antibiotic use on that population, but without direct reference to human population dynamics. This sets it apart from previous epidemiological models of resistance that have attempted to model antibiotic resistance, not by categorizing bacteria, but by categorizing their human hosts as either noncolonized, colonized entirely by susceptible bacteria, or colonized entirely by resistant bacteria (Austin *et al.* 1997; Levin *et al.* 1997). We believe that the bacterial population perspective is a useful one, especially in light of the fact that pathogenic bacteria are highly mobile, and that individual humans may be found to harbor both susceptible and resistant strains of the same bacterial species.

Table 1. Model
equations

$$\frac{\mathrm{d}S}{\mathrm{d}t} = (S \times 365 \div t_e) \times (P_S(S) - 1) \tag{1}$$

$$\frac{\mathrm{d}I}{\mathrm{d}t} = ((I \times 365 \div t_e) \times (P_I(I) - 1)) + (S \times m_{SI}) \tag{2}$$

$$\frac{\mathrm{d}H}{\mathrm{d}t} = ((H \times 365 \div t_e) \times (P_H(H) - 1)) + (I \times m_{IH}) \tag{3}$$

$$D = S + I + H \tag{4}$$

$$R_T = (I + H) \div D \tag{5}$$

$$R_H = H \div D \tag{6}$$

$$P_S(S) = P_S^{\mathrm{MAX}} \times (1 - b_S)^{\mathrm{AB}/200} \times f(D) \tag{7}$$

$$P_I(I) = P_I^{\mathrm{MAX}} \times (1 - b_I)^{\mathrm{AB}/200} \times f(D) \times g_1(R_T) \tag{8}$$

$$P_H(H) = P_H^{\mathrm{MAX}} \times (1 - b_H)^{\mathrm{AB}/200} \times f(D) \times g_1(R_T) \times g_2(R_H) \tag{9}$$

where:

S, I, H:	Bacterial densities: susceptible, intermediately resistant, highly resistant (0–1)
D:	Total bacterial density (0–1)
R_T, R_H:	Resistance fractions: total, high (0–1)
t_e:	Average elimination time (days)
m_{SI}, m_{IH}:	Mutation rates: S to I, I to H (fraction/year)
P_S, P_I, P_H:	Proliferation ratios (dimensionless: ratio of proliferation to elimination)
$P_S^{\mathrm{MAX}}, P_I^{\mathrm{MAX}}, P_H^{\mathrm{MAX}}$:	Maximum proliferation ratios (dimensionless)
AB:	Antibiotic use (prescriptions/thousand persons/year)
b_S, b_I, b_H:	Factors for antibiotic inhibition of proliferation (0–1)
$f(D)$:	Nonlinear effect of niche saturation on proliferation (0–1)
$g_1(R_T)$:	Nonlinear effect of serotype variation on proliferation of total resistance (0–1)
$g_2(R_H)$:	Nonlinear effect of serotype variation on proliferation of high resistance (0–1)

Before moving on to the specifics of the model, one must consider whether, in the case of pneumococcal resistance, the model's bacterial population represents only pneumococci or also includes the commensal viridans group streptococci with which pneumococci freely interact in the upper respiratory tract. On the one hand, pneumococci and the commensals share the same niche in the human body, both are inhibited by beta-lactam use, and the two share genes freely with each other. On the other hand, unlike the commensals, pneumococcal colonies are temporary guests rather than long-term residents in a human host, and consequently may tend to become less antibiotic resistant than the commensals.[17] Because the two populations have aspects of overlap as well as aspects of non-overlap, they should perhaps be modeled separately. However, national time series data on resistance in commensals do not exist and probably never will. Pathogens are the typical concern of laboratories and our primary concern as well. Thus, for the sake of simplicity and verifiability, our current model walks a fine line. Our explicit modeling is of a population of pneumococci, but we assume that resistance within the commensal populations moves in parallel with that of the pneumococci.

Each of the three bacterial population levels in Figure 7 is shown with a two-way flow that represents the net difference between additions to that level due to normal reproduction, gene exchange, and mutation, and reductions due to elimination; see Eqs (1), (2), and (3) in Table 1. Mutation (or, perhaps, the absorption of DNA from another species of bacteria in the same niche) may create a new intermediately resistant bacterium from the subpopulation of susceptible bacteria, or a new highly resistant bacterium from the subpopulation of intermediately resistant bacteria. In this way, mutation may introduce resistance genes that were not previously present in the population. But growth in resistance occurs through the self-perpetuating processes of reproduction and gene exchange within the bacterial population, described in the model with the single-term proliferation.

Proliferation is formulated through the use of proliferation ratios, one for each bacterial population level, which specify the rate of proliferation relative to that of bacterial elimination; see Eqs (7), (8), and (9). A maximum proliferation ratio describes the rate of proliferation relative to the rate of elimination in the absence of antibiotic use or constraints such as niche size. By setting the maximum proliferation ratio of susceptible bacteria to exceed that of intermediately or highly resistant bacteria, one may specify the fitness cost of resistance.

Three factors are specified in the model that may inhibit or constrain proliferation. First, antibiotic use, expressed as prescriptions per thousand persons per year, significantly inhibits the proliferation of susceptible bacteria, but has less effect on the proliferation of intermediately resistant bacteria, and very little effect on the proliferation of highly resistant bacteria.[18] Second, with or without antibiotic use, all three proliferation ratios are constrained equally (that is, by the same table function) when the total bacterial density climbs toward its limit of 100. This effect of niche saturation reflects the idea that, as the bacteria approach the limits of their natural niche in their human hosts, their ability to proliferate any further is constrained by problems of survival in the inhospitable surroundings.

Let us pause for a moment to consider the combined effects of antibiotic use and niche saturation. Start with a state of equilibrium in which antibiotic use is relatively low and the susceptible bacterial density is high—high enough to reduce net population growth to zero due to niche saturation. Now increase antibiotic use. The first thing that happens is that the net growth rate turns negative and the bacterial density therefore decreases. If there are no resistant bacteria present, the susceptible bacterial population will simply decline to a new lower equilibrium, one where the antibiotic effect is exactly balanced out by a less restrictive effect of niche saturation. If there are some resistant bacteria present, and antibiotics do relatively little to inhibit their proliferation, the easing of the niche saturation effect will allow them to proliferate quickly in a way they could not do before the increase in antibiotic use. As a result, resistance may now emerge rapidly until the niche limit is once again approached. In sum, antibiotic use sets the stage for the emergence of resistance, both by giving resistant bacteria the reproductive advantage over susceptibles and by clearing a space for them to proliferate within their human hosts.

The third and final type of constraint on proliferation to consider is that related to serotype variation, a subject which may be somewhat unique to pneumococci because of their remarkable diversity of serotypes. There are two of these effects in the model, one affecting both intermediately and highly resistant bacteria, and the other affecting only highly resistant bacteria. The first of these effects says that as total (intermediate plus high) resistance grows beyond some critical level (say, 40 percent) it becomes more and more difficult for resistance to spread to additional serotypes and therefore to additional bacteria. This concept reflects the rather good evidence, discussed previously, that a significant proportion of common pneumococcal serotypes have a low propensity for becoming resistant.

The second of the serotype variation effects is specific to highly resistant bacteria and says that as high resistance grows beyond some critical level (say, 10 percent) it becomes more and more difficult for high resistance *per se* to spread to additional serotypes and therefore to additional bacteria. Unlike the first effect, we currently have no direct evidence to support this second effect. In fact, this second effect is something that was added to the model only when we could find nothing else to explain the plateau of high PRP at 15 percent in Spain in the late 1980s. Not only is the effect a speculative one at present, but it may actually be contradicted by the data from Hungary and Asia, cited previously, suggesting the possibility of percentages of high PRP well in excess of 15 percent.

Model calibration and historical reference testing

The current model has proved capable of reproducing many, though not all, of the behavioral features of the PRP time series data for four countries seen in Figures 2 and 3. In calibrating the model, a conservative approach was taken: it is assumed that the countries differ only in terms of bacterial density initial conditions (as of 1979) and in terms of antibiotic use over time. Thus, all of the constants and nonlinear functions specified in Table 1 are assumed to be identical for all four countries, despite the fact that there is some variation by country in the distribution of serotypes that in theory could affect these parameter values. This common set of parameter values used for all historical reference testing is presented in Table 2.

The calibration process was primarily focused on the United States, since that was the one country for which we had at least a partial time series on beta-lactam use (Figure 6). We filled in the missing U.S. beta-lactam data by assuming no change in use from 1979 through 1987, and then interpolated linearly from 1987 to 1989 to get a value for 1988. In addition, in order to best

Table 2. Parameter values used in historical reference testing

Average elimination time (days)
$t_e = 36$

Mutation rates (fraction/year)
$m_{SI} = 0.0005$
$m_{IH} = 0.0025$

Maximum proliferation ratios
(Resistance reduces reproductive fitness: $P_S^{MAX} > P_I^{MAX} \geq P_H^{MAX}$)
$P_S^{MAX} = 1.230$
$P_I^{MAX} = 1.159$
$P_H^{MAX} = 1.151$

Antibiotic inhibition factors
(Resistance reduces antibiotic effect: $b_S > b_I > b_H$)
$b_S = 0.090$
$b_I = 0.030$
$b_H = 0.004$

Effect of niche saturation on proliferation

D:	0	0.1	0.2	0.3	0.4	0.5	0.6	0.7	0.8	0.9	1
$f(D)$:	1	0.997	0.992	0.986	0.977	0.963	0.941	0.901	0.821	0.625	0

Effect of serotype variation on proliferation of total resistance

R_T:	0	0.1	0.2	0.3	0.4	0.5	0.6	0.7	0.8	0.9	1
$g_1(R_T)$:	1	1	1	1	1	0.98	0.96	0.94	0.92	0.90	0.88

Effect of serotype variation on proliferation of high resistance

R_H:	0	0.05	0.1	0.15	0.2	0.25	0.3	0.35	0.4	0.45	0.5
$g_2(R_H)$:	1	1	1	0.96	0.92	0.88	0.85	0.82	0.80	0.78	0.76

fit historical PRP data for 1998, we assumed a value of 210 Rxs/1000/year for that year. Finally, for 1999 and future years, we assumed that beta-lactam use would settle at a value of 200, the average of what it had been during the period 1995–1997. The resulting beta-lactam use curve for the U.S.A. may be seen in Figure 8. Given this curve as input, the model is able to do a rather good job of reproducing the historical growth patterns of both total PRP and high PRP in the U.S.A. for the period 1979 to 1998, as shown in Figure 9.

The U.S. data were helpful for calibrating all of the model's parameters except for the nonlinear effects of serotype variation on proliferation. In order to calibrate the serotype effects, we looked to the PRP time series from Spain, where plateaus in both total PRP and high PRP are evident. As far as beta-lactam use for Spain is concerned, we took a conservative approach, which was taken with South Africa and Hungary as well. In particular, the beta-lactam use curve was assumed to be flat at some initial value with one possible period of ramp-wise change to a new final value. Given this simplifying assumption, the beta-lactam use curve that gave PRP results most like Spain was a decline from 260 to 200 Rxs/1000/year starting in 1984 and ending in 1990. The simple curve that gave results most like South Africa was a constant value of 200. The simple curve that gave results most like Hungary was a decline from 260 to 220 starting in 1980 and ending in 1984. These three curves appear in Figure 8, where they may be compared with the much more erratic U.S. curve based on actual data.

Fig. 8. Beta-lactam use
assumptions for four
historical reference
simulations

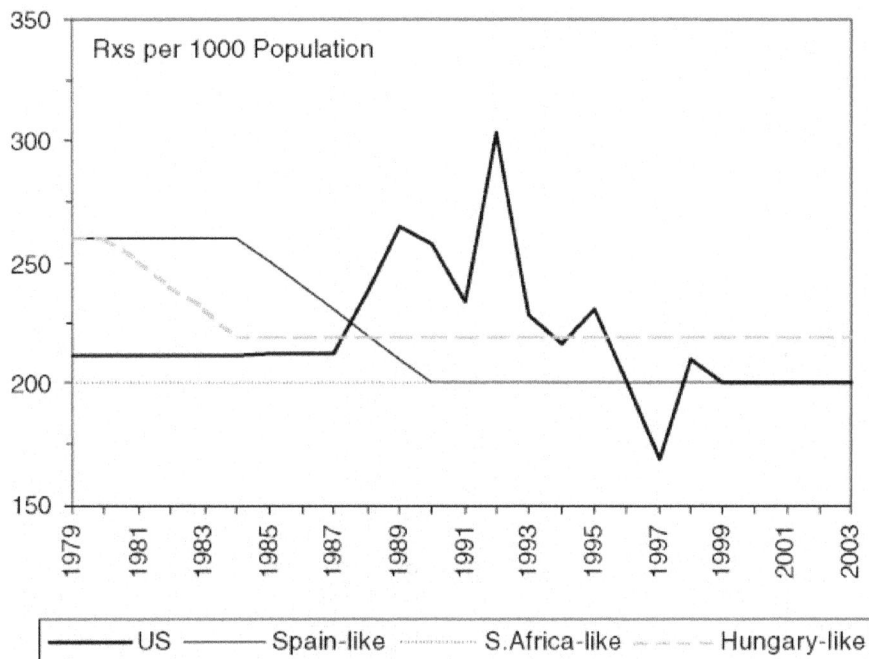

Fig. 8. Beta-lactam use assumptions for four historical reference simulations

Figure 10 presents the results, in terms of total PRP, of simulating the model with the four beta-lactam use curves in Figure 8. These four reference simulation results may be compared with their actual data counterparts in Figure 2. For all four countries, the model is able to reproduce historical growth trends and plateau levels, but misses the sometimes sizable fluctuations apparent in the historical data. This lack of fluctuation in simulated PRP to some extent reflects the fact that smooth curves rather than actual data were used for beta-lactam use for the three non-U.S. countries as well as for the U.S.A. between 1979 and 1987. For the years 1987 to 1997 in the U.S.A., one may see how erratic year-to-year swings in actual beta-lactam use translate into a bumpy rather than smooth simulated pattern of PRP growth during those years. Moreover, experiments not shown here indicate that more erratic hypothetical use curves may be constructed for each country that are able to generate some of the smaller fluctuations seen in the PRP data. However, given our current model, no plausible beta-lactam use curves can explain fluctuations in PRP as large as those seen in Hungary between 1981 and 1985, or in Spain between 1980 and 1983. One may only surmise that another driving factor is at work, such as year-to-year changes in the rate at which pneumococci are spread through the population—perhaps translating in the current model to periodic shifts in bacterial proliferation ratios.[19]

183

Fig. 9. PRP in the
United States,
simulated vs data

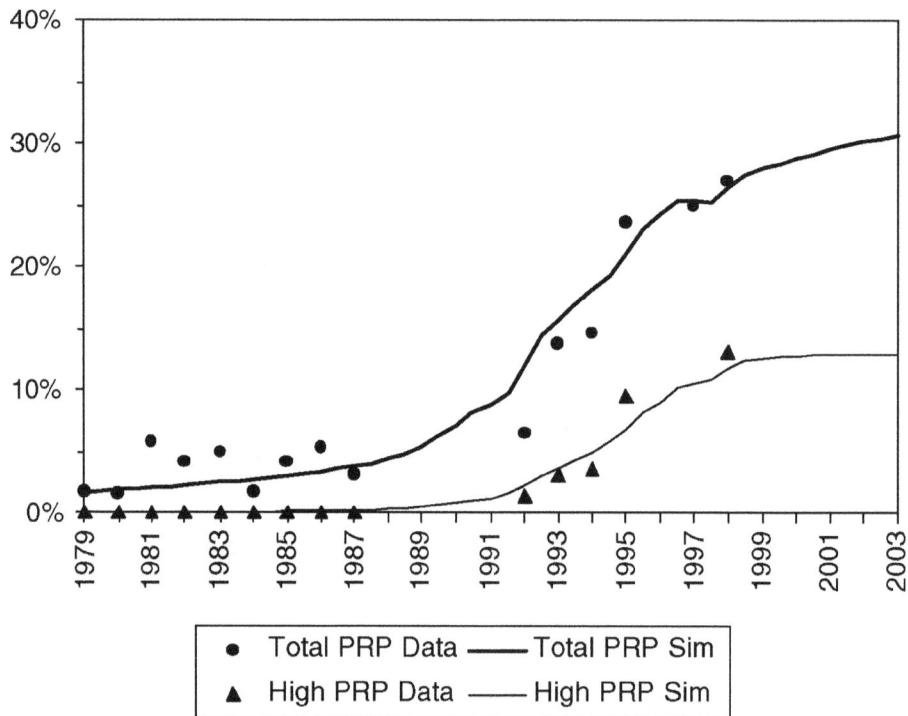

Testing the effect of reducing antibiotic use

Our modeling work was prompted by questions concerning the impact of antibiotic use reduction—in particular, the magnitude and timing of that reduction—on the emergence of antibiotic resistance. A variety of tests of antibiotic use reduction have been done, focusing on PRP emergence in the United States. The key conclusions are as follows:

- A threshold of beta-lactam use exists, beyond which first intermediate PRP and then high PRP will emerge. If beta-lactam use is reduced below this threshold before the significant emergence of high PRP, then that emergence may be prevented. Otherwise, high PRP may emerge and drive total PRP sharply upward.
- A reduction in beta-lactam use lowers the final plateau level of PRP. This can mean either a slowing or a reversal of PRP growth, depending on how far emergence has already progressed at the time of use reduction, and on the extent of use reduction. An actual reversal in PRP growth prior to full emergence typically requires a reduction in beta-lactam use below the threshold for emergence.

Fig. 10. Total PRP
results from four
historical reference
simulations

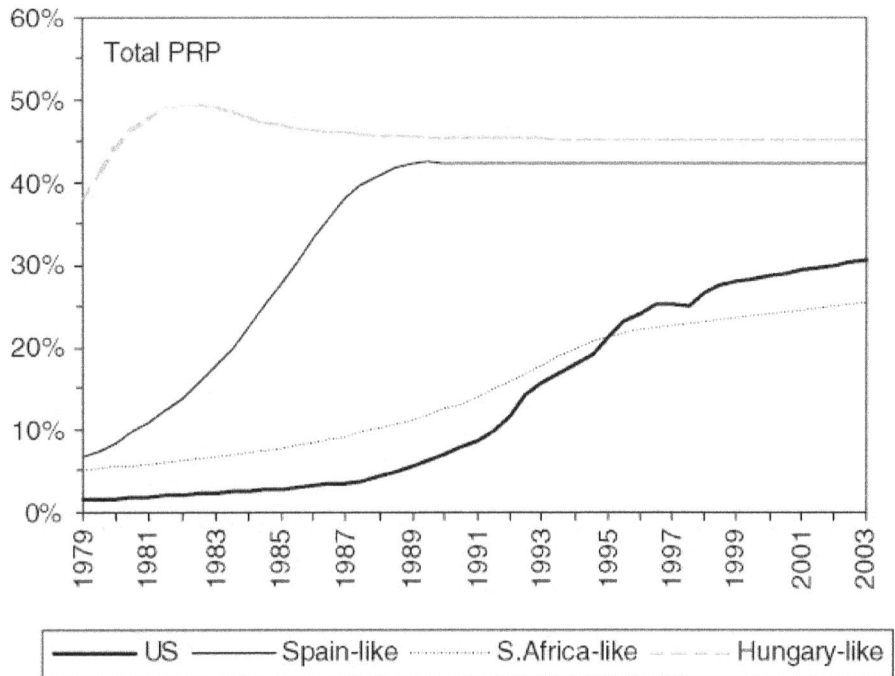

- If the reduced level of use is below the threshold for emergence, total PRP will decline substantially but not necessarily toward zero. While such a reduction in use does drive intermediate PRP toward zero, high PRP may be little affected.
- High PRP may be driven downward only by reducing beta-lactam use well below the threshold for emergence. In particular, there is a second and lower threshold, what might be called a threshold for elimination, below which even high PRP can be driven toward zero.

A typical sequence of tests is illustrated in Figures 11 and 12; Figure 11 presents results for total PRP in the U.S.A., while Figure 12 presents results for high PRP *per se*. The first of these tests is a baseline run, extended to the year 2030 based on an assumption that beta-lactam use remains at the level of 200 Rxs/1000/year for the entire period 1999–2030. In a second test, beta-lactam use is reduced to a constant value of 150 as of 1992. In a third test, beta-lactam use is reduced to 150 as of 2000. In a fourth test, beta-lactam use is reduced to an even lower value of 125 as of 2000. A usage value of 150 represents a 25 percent reduction below the 1995–1997 average of 200. This is a magnitude of reduction that seems feasible given the high percentages of inappropriate use of antibiotics for upper respiratory illnesses in the U.S.A. (see Note 1). A usage value of 125

Fig. 11. Total PRP in the United States simulated to 2030: sensitivity to antibiotic use reduction and its timing

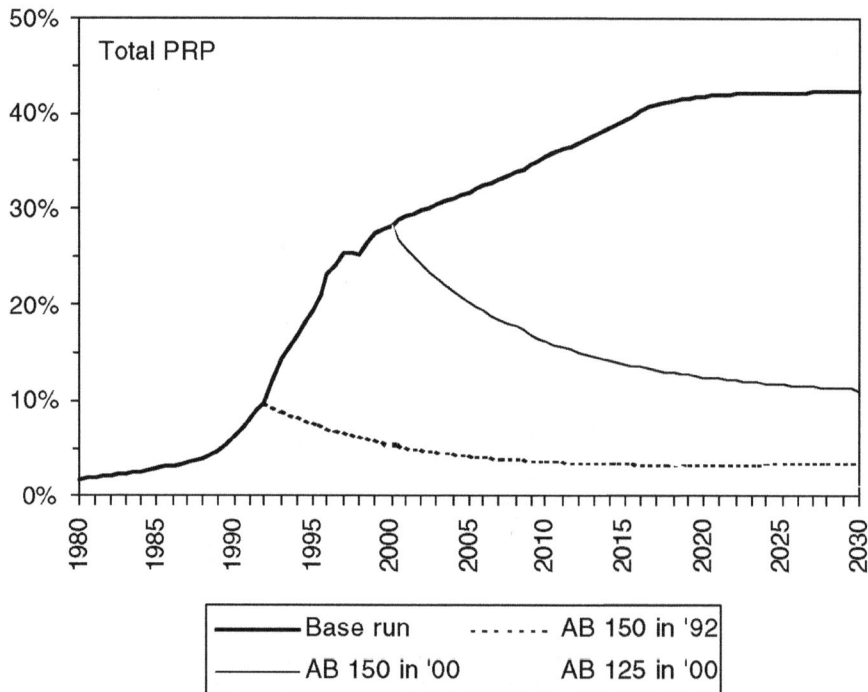

Fig. 11. Total PRP in the United States simulated to 2030: sensitivity to antibiotic use reduction and its timing

represents a 37.5 percent reduction below 200, a magnitude of reduction that might be considerably more difficult, though not impossible to achieve in real life. (Recall the example of Finland where macrolide consumption was reduced 40 percent nationwide in a short period of time.)

Given all of the numerical assumptions in the current model, the beta-lactam use threshold for PRP emergence is about 170 Rxs/1000/year, and the threshold for elimination is about 140. Below the emergence threshold, susceptible bacteria are able to proliferate more quickly than intermediately resistant ones, as a result of the fitness cost of resistance, leading to a decline in overall resistance. Figure 11 shows how total PRP can be reduced substantially, even after emergence, if beta-lactam use is cut to 150 Rxs/1000/year. If beta-lactam use is cut even further, to 125, then total PRP can be sent on a trajectory toward zero.

Figure 12 shows that if the reduction in use does not occur early enough in the simulation, high PRP emerges rapidly in the 1990s, up to a baseline plateau level of 13 percent. If use is reduced to a value of 150 in 2000, high PRP declines somewhat but still remains at a significant level of 10 percent. This behavior reflects the fact that, when beta-lactam use is at a value of 200 or even 150, highly resistant bacteria are naturally able to proliferate more

Fig. 12. High-level
PRP in the United
States simulated to
2030: sensitivity to
antibiotic use
reduction and its
timing

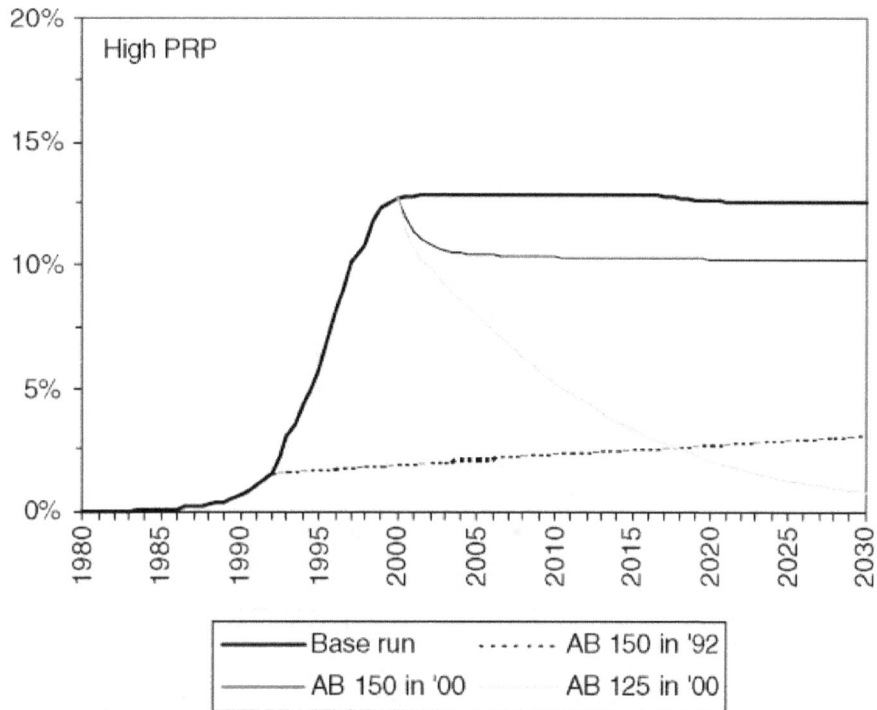

quickly than both susceptible bacteria and intermediately resistant bacteria, because they are inhibited less by antibiotics. Indeed, the only reason that highly resistant bacteria do not become entirely dominant is because of the effect of serotype variation, which is assumed to become binding when high PRP exceeds 10 percent. But if beta-lactam use is reduced to a value of 125, below the threshold for elimination, then the proliferation advantage of highly resistant bacteria is reversed, and high PRP is driven toward zero.

Sensitivity testing

The policy simulations described above suggest that the amount of reduction in antibiotic resistance that can be expected from a reduction in antibiotic use is largely determined by where the new usage rate lies relative to thresholds for emergence and elimination. To gain a fuller understanding of how steady-state levels of resistance are affected by antibiotic use and by other factors, several batteries of sensitivity tests were performed. Key results of this testing in terms of total PRP are shown in Figure 13; results in terms of high PRP are shown in Figure 14.

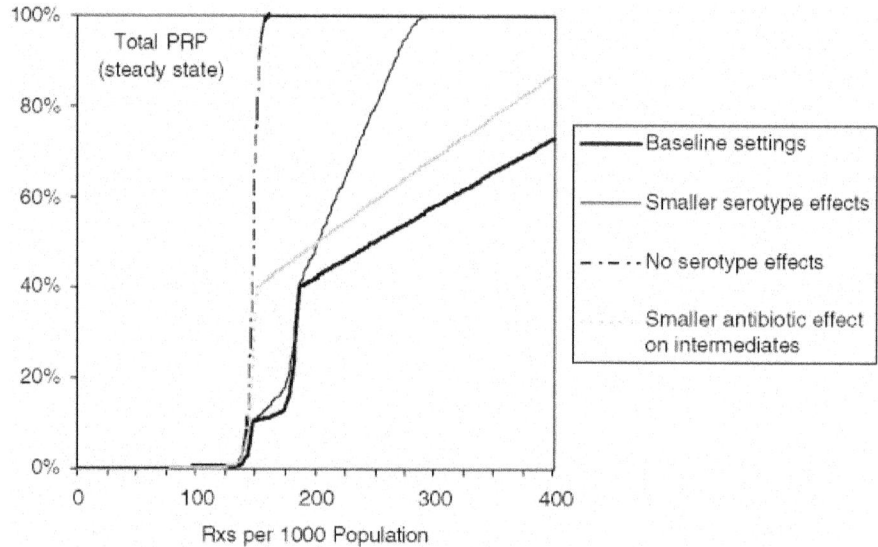

Fig. 13. Total PRP in steady state as a function of antibiotic use: sensitivity to assumptions about serotype variation and antibiotic effect

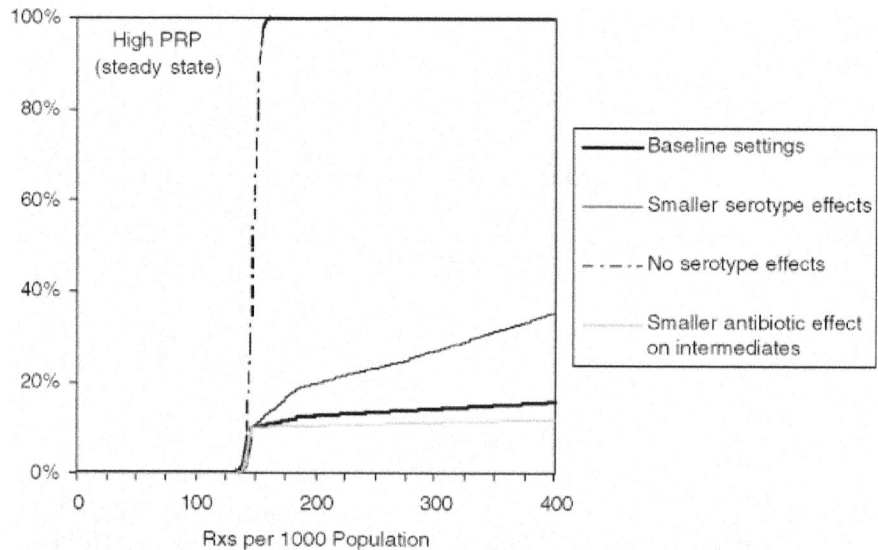

Fig. 14. High-level PRP in steady state as a function of antibiotic use: sensitivity to assumptions about serotype variation and antibiotic effect

In each sensitivity test, beta-lactam use was fixed at a constant whole-number value between 0 and 400 Rxs/1000/year; a full battery consisted of 401 tests in all. In order to ensure that true steady-state values were achieved, both intermediate and high PRP were initialized at high levels of 33 percent, and the simulation was run out one hundred years to 2080. What are shown in

Figures 13 and 14 are the values of PRP as of the year 2080, and how these steady-state values change as a function of antibiotic use.

Baseline sensitivity results

The first battery of sensitivity tests employs the baseline parameter settings used in all of the reference simulations and policy simulations discussed previously and shown in Table 2. Focusing on the "baseline settings" line in Figure 13, one sees that the steady-state value of total PRP is at or near zero for values of antibiotic use of 140 or less, rests in the vicinity of 10 percent when antibiotic use is between 145 and about 170, shoots up to a value of 40 percent when antibiotic use is at about 185, and then increases with a constant slope from there, reaching 70 percent when antibiotic use is at 400.

Figure 14 shows that the first 10 percent step in total PRP seen in Figure 13 is virtually all high-level PRP. This corresponds to a region of antibiotic use above the threshold for elimination but below the threshold for emergence, a region of use where highly resistant bacteria have a proliferation advantage but intermediately resistant bacteria do not. Because of the strongly constraining serotype variation effect of high PRP, steady-state high PRP climbs only very gradually with further increases in antibiotic use, rising from 12 percent when antibiotic use is at 185 to 16 percent when use is at 400. Thus, the vast majority of the increase in steady-state total PRP that in Figure 13 is predicted for values of antibiotic use exceeding the emergence threshold is an increase in intermediate PRP and not high PRP. This is a region where intermediately resistant bacteria enjoy the proliferation advantage over both antibiotic-susceptible and highly resistant bacteria, as a result of the assumed serotype variation effect of high PRP.

Reducing or neutralizing the effects of serotype variation

The impact on steady-state resistance of alternative assumptions about the effects of serotype variation may be seen in Figures 13 and 14 by comparing the results for two other batteries of tests with those of those of the baseline battery. In the "smaller serotype effects" scenario, the table functions for both of the serotype variation effects (seen in Table 2 as $g_1(R_T)$ and $g_2(R_H)$) are assumed to be only one quarter as steep as in the baseline. In "no serotype effects", the effects are neutralized entirely. One may think of these alternate scenarios as applying to hypothetical situations or locations in which serotype heterogeneity with respect to antibiotic resistance is much less of a factor or is nonexistent; ultimately, we may find these scenarios best apply to apply to pathogens other than pneumococcus.

The results of "smaller serotype effects" are qualitatively similar to the baseline, in the sense that the growth of high PRP is still limited beyond 10 percent. This constraint on high PRP, though much reduced, is still enough that intermediate resistance takes the proliferation advantage when the emergence threshold is exceeded, as it does in the baseline scenario. The constraint on the growth of intermediate PRP is also less than in the baseline scenario, so that with high enough rates of usage (300 and above), steady-state total PRP rises to 100 percent.

The results of "no serotype effects" are qualitatively different from those of the baseline. Now unconstrained by growth, highly resistant bacteria take a permanent proliferation advantage for all values of antibiotic use exceeding the elimination threshold. This advantage allows high PRP to climb to 100 percent for usage rates of 160 and above, driving out both antibiotic-susceptible and intermediately resistant bacteria.

Reducing the inhibitory effect of antibiotics on intermediately resistant bacteria

In all previous runs, the antibiotic inhibition factors were assigned the values seen in Table 2; namely, 9 percent for susceptibles, 3 percent for intermediately resistant bacteria, and 0.4 percent for highly resistant bacteria. (These three values specify inhibition at a usage rate of 200.) Although these values appeared to work well for the four countries whose PRP history was mimicked by our reference simulations, the possibility exists that the inhibition factors may be different in different countries or locations. In particular, we have thought about how noncompliance, a situation in which patients take inappropriately low dosages or cut short their course of treatment, might affect the inhibitory effect.

We speculate that the main effect of noncompliance would be to reduce the inhibition factor for intermediately resistant bacteria. Even if a dosage level is less than optimal, it may still be sufficient to suppress susceptible bacteria effectively, but fail to suppress many of the intermediately resistant bacteria that would have been suppressed at a full dosage level. In the fourth scenario in Figures 13 and 14, the inhibition factor for intermediates is cut in half relative to the baseline, from 3 percent to 1.5 percent.

As a result of this parameter change, intermediates take the proliferation advantage at lower usage rates than they do in the baseline, and the threshold for emergence is thus reduced. The net result in Figure 13 is essentially a shifting of the total PRP curve to the left for all usage rates exceeding the emergence threshold. For example, whereas in the baseline scenario it takes a usage rate of about 185 to generate steady-state total PRP of 40 percent, in the scenario representing lower patient compliance the same result is produced by a usage rate of only 150.

Open questions

At various points in this article we have touched on pieces of empirical evidence that may raise questions as to the adequacy of our current model and indicate the need for further research and model improvement. The most pressing of these open questions are as follows:

- Can high PRP grow and remain above the 15 percent seen in Spain?
- Can total PRP grow and remain above the 50 percent seen in Hungary?
- What explains the fluctuations in total PRP seen in all four countries we studied?

To answer these questions, we will want to start by filling gaps in the existing data. In addressing all three questions, we would obviously like to find data on beta-lactam use in countries other than the U.S.A. In addition, for the first two questions, we would like to collect more complete time series data on high PRP in Spain, and more data from various countries on the serotype distribution of PRP in general and high PRP in particular. And for the third question, we would like to collect time series on the incidence of pneumococcal disease, to see whether it fluctuates in a way that could help explain the fluctuations of PRP.

If we find that the current model is indeed inadequate to explain all of the evidence, even allowing for plausible adjustments in parameter values, then the model's structure will have to be modified in some way. A couple of opportunities for disaggregation come to mind, though it is not obvious how they would solve the problem. One possibility is to further disaggregate the bacterial population by explicitly representing invasive pneumococci (the most common objects of PRP measurement), noninvasive pneumococci, and the commensal streptococci with which the pneumococci interact. Another possibility is to disaggregate the human host population in some way, as other epidemiological models have done (Austin *et al.* 1997, Levin *et al.* 1997).

Ultimately, after the above questions have been adequately addressed, we would like to investigate the applicability of the model to cases other than PRP. These future investigations should address the following questions:

- Can the model be successfully applied to study pneumococcal resistance to antibiotic classes other than the beta-lactams?
- Can the model be successfully applied to study antibiotic resistance in pathogens other than pneumococcus?

Conclusion

This article reports on the current status of an ongoing modeling effort. Focusing on the case of pneumococcal resistance to beta-lactams, the objective has been to develop a model that reflects the full breadth of evidence and theory on the subject, including data from a variety of countries. Although we have made good progress, it is important that we continue to develop the model so that it has the widest possible applicability and provides valid insights.

Aside from continued model development, we must also think about how best to facilitate the use of model-based findings by the policymakers with whom our team continues to work. A systematic approach to such facilitation (see, for example, Ritchie-Dunham and Rabbino 2001) involves returning to the larger policy context of the problem and thinking further about the resources and actions available to policymakers, as well as the perceptions, incentives, and likely responses of the different stakeholders affected.

The development of a new dynamic theory of a phenomenon as complex as antibiotic resistance is never an easy thing, particularly when key data are missing or difficult to gather. Despite the difficulties, it is a process that can reward researchers and policymakers with new ideas as the work progresses. These ideas, in turn, serve to concentrate the process of information gathering and may evolve into firm policy findings when the model matures to the point that it can stand up to the preponderance of empirical evidence and reviewers' critiques. For this reason, it is important to take stock periodically as a modeling project moves forward, and to report interim ideas and findings as we have done here.

Acknowledgements

The authors would like to acknowledge Beverly Ray, Julie Rawlings, Mike Kelley, Jane Mahlow, and Annette Riggio of the Texas Department of Health for their support of this project and Benjamin Schwartz of the U.S. Centers for Disease Control and Prevention for his generous assistance and advice.

Notes

1. An analysis of nationwide survey data on office visits compared actual prescribing behavior with established standards. This analysis suggests that 30 percent of the 23 million prescriptions for ear infections are inappropriate, 50 percent of the 13 million prescriptions for sore throat, 50 percent of the 13 million prescriptions for sinusitis, and 80 percent of the 16 million prescriptions for bronchitis (Okie 1997).
2. Hygiene and infection control measures are particularly important in settings where human-borne bacteria are plentiful, such as day-care centers and hospitals. But there is also growing concern that antibiotic resistant bacteria may develop agriculturally and then be passed on to humans. In several recent cases, drug-resistant strains of *Salmonella* in humans have been traced to the use of antibiotics in livestock (Ferber 2000). More than 40 percent of the antibiotics used in the U.S.A. are for livestock, of

which 40 to 80 percent has been judged unnecessary for animal well-being. Antibiotics are also widely used in aerosols sprayed on fruit trees, aerosols that can easily carry to other plants as well (Levy 1998, Bright 1999).

3. The specific pathway for interspecies gene exchange is known as transposon-mediated exchange. Other pathways for gene exchange include plasmid conjugation, transformation, and transduction (Neu 1992).

4. The new 7-valent pneumococcal conjugate vaccine (PCV, active against seven of the most common serotypes) is recommended for all children under the age of two (FitzGerald 2000). In addition, a 23-valent pneumococcal polysaccharide vaccine (PPV) has been available since the early 1980s, but it is not effective in children and is recommended primarily for the elderly (CDC 1996; Long 1999).

5. One study found the mean duration of pneumococcal carriage in disease cases to be 36 days, with a range of 3 to 267 days, and the mean duration of carriage in non-disease cases to be 26 days, with a range of 4 to 192 days (Ekdahl *et al.* 1997).

6. CDC surveys are reported in Spika *et al.* (1991), Breiman *et al.* (1994), Butler *et al.* (1996), and CDC (1999a; 1999b). Surveys directed by the university-based researchers Gary Doern and Angela Brueggemann are reported in Doern *et al.* (1996a; 1999).

7. The National Committee for Clinical Laboratory Standards (NCCLS) recommends an initial screening for PRP by the oxacillin disk technique, which is highly sensitive and specific and detects almost all isolates with any resistance to penicillin or other beta-lactams. Isolates found to be nonsusceptible by this method should then be subjected to quantitative testing of minimum inhibitory concentration (MIC), using accepted concentration intervals and breakpoints for specific antibiotics. For penicillin, the breakpoints are Susceptible: ≤ 0.06 μg/mL; Intermediately resistant: ≥ 1.0 μg/mL; Highly resistant: ≥ 2.0 μg/mL (CDC 1996).

8. Data sources for Figure 2: United States: Breiman *et al.* 1994; Butler *et al.* 1996; Doern *et al.* 1996a; 1999; CDC 1999a; 1999b; Spain: Fenoll *et al.* 1991; 1998; South Africa: Koornhof *et al.* 1992; Hungary: Marton 1992. Data sources for Figure 3: United States: Breiman *et al.* 1994; Butler *et al.* 1996; Doern *et al.* 1996a; 1999; CDC 1999a; 1999b; Spain: Fenoll *et al.* 1991.

9. The U.S. time series have a gap during 1988–1991, when no national hospital surveys were done. These data were culled from multiple articles representing the work of two different research groups with some changes over time in the hospitals they surveyed, so that some data are not strictly comparable. In the case of Hungary, the sample includes both invasive and noninvasive isolates, mixed together in the data without identification, whereas in the other three countries the sample includes only invasive isolates. The Spanish authors (Fenoll *et al.* 1998) also present resistance data for noninvasive isolates for 1983–1996, not presented here. These

noninvasive isolates are consistently more resistant than the invasive ones, with total PRP in the range of 50–60 percent throughout the period studied.

10. The U.S. data for Figure 6 come from 19 hospitals submitting 5,469 isolates to the CDC over the period 1979–1987 (Spika *et al.* 1991). The Spanish data for Figure 7 come from 26 hospitals submitting 2,197 isolates to the Centro Nacional de Microbiología over the period 1979–1989 (Fenoll *et al.* 1991).

11. In the U.S. data, four of the common serotypes (each with over 100 isolates in the sample) account for 36 percent of total isolates but 72 percent of resistant isolates. These serotypes, in descending order of resistance (from 38 percent down to 5 percent), are #19A, #14, #23F, and #6B. In the Spanish data, five of the common serotypes (each with over 100 isolates) account for 36 percent of total isolates but 84 percent of resistant isolates. These serotypes, in descending order of resistance (from 89 percent down to 41 percent), are #23, #6, #9, #14, and #19. (Lettered subvariants were not identified in the article from Spain.) The resistance of serotype #9 is only 2 percent in the U.S. data, making this the only common serotype to have relatively low resistance in the U.S. data but relatively high resistance in the Spanish data.

12. For Hungary, a sample from 1988–1989 (135 isolates) exhibited total PRP of 58 percent and high PRP of 35 percent. For Asia, samples are from 1996–1997, and reflect 996 invasive isolates from 14 medical centers in 11 countries. After Korea, the highest PRP percentages are for Japan (65 percent total, 27 percent high), Vietnam (61 percent total, 33 percent high), and Thailand (58 percent total, 22 percent high). These are followed by Sri Lanka and Taiwan (about 40 percent total PRP), Indonesia and Singapore (about 20 percent), and Malaysia, India and China (10 percent or less).

13. These resistance genes alter the penicillin-binding proteins naturally present on streptococcal species, thereby reducing the affinity of these proteins for beta-lactams and effectively making it more difficult for beta-lactams to bind to the affected bacteria (Neu 1992).

14. Figure 6 reflects NAMCS data for 1980, 1985, 1989, and 1992 reported by McCaig and Hughes (1995), as well as data we have extracted directly from the annual NAMCS data sets for 1990 to 1997. In McCaig and Hughes, rates of use were reported for a variety of antibiotics, of which the beta-lactams include amoxicillin, ampicillin, the penicillins, and the cephalosporins. The beta-lactams for which we found mentions in the 1990–1997 NAMCS data sets, under both generic names (lower case) and brand names (upper case), are as follows: amoxicillin, Amoxil, Augmentin, ampicillin, penicillin, penicillin V, Keflex, cephalexin, Ceclor, and Ceftin. The only year of overlap in these two sources is 1992. For 1992, McCaig and Hughes give a total rate of beta-lactam use of 303 prescriptions per thousand population, 19 percent higher than the rate of 254 that we calculated. We do not know what the source of this discrepancy might be

(there are several one might consider) and have decided to resolve it for now simply by shifting all of our calculated rates for 1990–1997 up by 19 percent, in deference to McCaig and Hughes.

15. The total for office-based and hospital-based prescriptions combined is reported to be about 150 million per year (McCaig and Hughes 1995).

16. Detailed analysis of the NAMCS data shows that most of the decline is accounted for by a reduction on the order of 50 percent in the use of beta-lactams for otitis media from 1990 to 1996. The majority of that reduction occurred in young children ages zero to two, who had in 1990 accounted for over 20 percent of all office-based beta-lactam use, but by 1996 accounted for only 17 percent.

17. In a sample of 352 viridans group isolates from 43 medical centers in the U.S.A. during 1993–1994, total penicillin resistance was 56 percent, including 13 percent high resistance *per se* (Doern *et al.* 1996b). This rate of resistance among oral streptococci is much higher than those found in any of the national surveys of PRP in the U.S.A. to date.

18. A simple exponential function is used to describe the effect of antibiotic usage on proliferation, as seen in Eqs (7), (8), and (9). This function includes a parameter (one for each population level) specifying the inhibitory effect when antibiotic usage is at a level of 200 prescriptions (Rxs) per thousand persons per year.

19. Some epidemic diseases such as measles have for decades experienced repeated oscillations that appear to be driven by population growth and changes in vaccination rates. These cycles are dynamically complex and may exhibit periods of one or more years (Earn *et al.* 2000).

References

Austin DJ, Kakehashi M, Anderson RM. 1997. The transmission dynamics of antibiotic-resistant bacteria: the relationship between resistance in commensal organisms and antibiotic consumption. *Proceedings of the Royal Society of London Series B* **264**: 1629–1638.

Austin DJ, Kristinsson KG, Anderson RM. 1999. The relationship between the volume of antimicrobial consumption in human communities and the frequency of resistance. *Proceedings of the National Academy of Science USA* **96**: 1152–1156.

Ballow CH, Schentag JJ. 1992. Trends in antibiotic utilization and bacterial resistance: report of the National Nosocomial Resistance Surveillance Group. *Diagnostic Microbiology of Infectious Diseases* **15**: 37S–42S.

Baquero F, Baraibar R, Campos J *et al.* 1996. Antibiotic resistance in Spain: what can be done? *Clinical Infectious Diseases* **23**: 819–823.

Breiman RF, Butler JC, Tenover FC *et al.* 1994. Emergence of drug-resistant pneumococcal infections in the United States. *Journal of the American Medical Association* **271**(23): 1831–1835.

Bright C. 1999. Super-bacteria arrive. *World-Watch* (March/April 1999): 9–11.

Butler JC, Hofmann J, Cetron MS *et al*. 1996. The continued emergence of drug-resistance *Streptococcus pneumoniae* in the United States: an update from the Centers for Disease Control and Prevention's Pneumococcal Sentinel Surveillance System. *Journal of Infectious Diseases* **174**: 986–993.

Centers for Disease Control and Prevention (CDC). 1996. Defining the public health impact of drug-resistant *Streptococcus pneumoniae*: report of a working group. *Morbidity and Mortality Weekly Report* **45**(RR-1): 1–20.

Centers for Disease Control and Prevention (CDC). 1999a. Geographic variation in penicillin resistance in *Streptococcus pneumoniae*—selected sites, United States, 1997. *Morbidity and Mortality Weekly Report* **48**(30): 656–661.

Centers for Disease Control and Prevention (CDC). 1999b. Active Bacterial Core Surveillance (ABCs) Report, Emerging Infections Program Network, *Streptococcus pneumoniae*—1997 and 1998. On-line reports at http://www.cdc.gov/ncidod/dbmd/abcs.

Cetron MS, Jernigan DB, Breiman RF *et al*. 1996. Action plan for drug-resistant *Streptococcus pneumoniae*. *Emerging Infectious Diseases* **1**(2): 64–65.

Chadwick DJ, Goode J (eds.) 1997. *Antibiotic Resistance: Origins, Evolution, Selection and Spread*. Wiley: New York.

Doern GV, Brueggemann A, Holley Jr HP, Rauch AM. 1996a. Antimicrobial resistance of *Streptococcus pneumoniae* recovered from outpatients in the United States during the winter months of 1994 to 1995: results of a 30-center national surveillance study. *Antimicrobial Agents and Chemotherapy* **40**(5): 1208–1213.

Doern GV, Ferraro MJ, Brueggemann AB, Ruoff KL. 1996b. Emergence of high rates of antimicrobial resistance among viridans group streptococci in the United States. *Antimicrobial Agents and Chemotherapy* **40**(4): 891–894.

Doern GV, Brueggemann AB, Huynh H *et al*. 1999. Antimicrobial resistance with *Streptococcus pneumoniae* in the United States, 1997–98. *Emerging Infectious Diseases* **5**(6): 757–765.

Earn DJD, Rohani P, Bolker BM, Grenfell BT. 2000. A simple model for complex dynamical transitions in epidemics. *Science* **287**(28 January): 667–670.

Ekdahl K, Ahlinder I, Hansson HB *et al*. 1997. Duration of nasopharyngeal carriage of pencillin-resistant *Streptococcus pneumoniae*: experiences from the South Swedish Pneumococcal Intervention Project. *Clinical Infectious Diseases* **25**: 1113–1117.

Ekdahl K, Hansson HB, Mölstad S *et al*. 1998. Limiting the spread of pencillin-resistant *Streptococcus pneumoniae*: experiences from the South Swedish Pneumococcal Intervention Project. *Microbial Drug Resistance* **4**(2): 99–104.

Fenoll A, Martín Bourgon C, Muñóz R *et al*. 1991. Serotype distribution and antimicrobial resistance of *Streptococcus pneumoniae* isolates causing systemic infections in Spain, 1979–1989. *Review of Infectious Diseases* **13**: 56–60.

Fenoll A, Jado I, Vicioso D *et al*. 1998. Evolution of *Streptococcus pneumoniae* serotypes and antibiotic resistance in Spain: update (1990 to 1996). *Journal of Clinical Microbiology* **36**(12): 3447–3454.

Ferber D. 2000. Superbacteria on the hoof? *Science* **288**(5 May): 792–794.

FitzGerald S. 2000. A new tool to protect baby from infections. *Philadelphia Inquirer* 18 February 2000.

Frimodt-Møller N, Espersen F, Jacobsen B *et al*. 1997. Problems with antibiotic resistance in Spain and their relation to antibiotic use in humans elsewhere. *Clinical Infectious Diseases* **25**: 939–940.

Garrett L. 1994. *The Coming Plague: Newly Emerging Diseases in a World Out of Balance*. Penguin Books: New York.

Geslin P, Buu-Hoi A, Frémaux A, Acar JF. 1992. Antimicrobial resistance in *Streptococcus pneumoniae*: an epidemiological survey in France, 1970–1990. *Clinical Infectious Diseases* **15**: 95–98.

Guillernot D, Carbon C, Balkau B *et al*. 1998. Low dosage and long treatment duration of beta-lactam: risk factors for carriage of penicillin-resistant *Streptococcus pneumoniae*. *Journal of the American Medical Association* **279**(5): 365–370.

Hardin G. 1968. The tragedy of the commons. *Science* **162**(13 December): 1243–1248.

Homer JB, St Clair C. 1991. A model of HIV transmission through needle sharing. *Interfaces* **21**(3): 26–49.

Koornhof HJ, Wasas A, Klugman K. 1992. Antimicrobial resistance in *Streptococcus pneumoniae*: a South African perspective. *Clinical Infectious Diseases* **15**: 84–94.

Kristinsson KG. 1997. Effect of antimicrobial use and other risk factors on antimicrobial resistance in pneumococci. *Microbial Drug Resistance* **3**(2): 117–123.

Lee H-J, Park J-Y, Jang S-H *et al*. 1995. High incidence of resistance to multiple antimicrobials in clinical isolates of *Streptococcus pneumoniae* from a university hospital in Korea. *Clinical Infectious Diseases* **20**: 826–835.

Levin BR, Lipsitch M, Perrot V *et al*. 1997. The population genetics of antibiotic resistance. *Clinical Infectious Diseases* **24**(Suppl. 1): S9–16.

Levy SB. 1992. *The Antibiotic Paradox: How Miracle Drugs are Destroying the Miracle*. Plenum Publishers: New York.

Levy SB. 1998. The challenge of antibiotic resistance. *Scientific American*, (March 1998): 46–53.

Long SS. 1999. Current status of pneumococcal vaccines. *Clinical Updates in Pediatric Infectious Diseases* **2**(6): 1–4.

Marton A. 1992. Pneumococcal antimicrobial resistance: the problem in Hungary. *Clinical Infectious Diseases* **15**: 106–111.

McCaig LF, Hughes JM. 1995. Trends in antimicrobial drug prescribing among office-based physicians in the United States. *Journal of the American Medical Association* **273**(3): 214–219.

Morell V. 1997. Antibiotic resistance: road of no return. *Science* **278**: 575–576.

National Center for Health Statistics (NCHS). 1997. *1997 National Ambulatory Medical Care Survey*. NCHS CD-ROM Series 13 No. 21 (machine-readable data set). Also: CD-ROMs from NCHS Series 13 for 1990 (No. 2), 1991 (No. 3), 1992 (No. 4), 1993 (No. 6), 1994 (No. 8), 1995 (No. 11), and 1996 (No. 14). National Center for Health Statistics: Hyattsville, Maryland.

Neu HC. 1992. The crisis in antibiotic resistance. *Science* **257**: 1064–1073.

Okie S. 1997. Experts urge steps to stem antibiotic resistance. *Washington Post* (Health section), 26 August 1997.

Rahal JJ, Urban C, Horn D *et al*. 1998. Class restriction of cephalosporin use to control total cephalosporin resistance in nosocomial *Klebsiella*. *Journal of the American Medical Association* **280**(14): 1233–1237.

Reichmann P, König A, Liñares J *et al*. 1997. A global gene pool for high-level cephalosporin resistance in commensal *Streptococcus* species and *Streptococcus pneumoniae*. *Journal of Infectious Diseases* **176**: 1001–1012.

Ritchie-Dunham JL, Méndez Galvan JF. 1999. Evaluating epidemic intervention policies with systems thinking: a case study of dengue fever in Mexico. *System Dynamics Review* **15**(2): 119–138.

Ritchie-Dunham JL, Rabbino HT. 2001 (forthcoming). *GRASP: The Art and Practice of Strategic Clarity*. Wiley: Chichester.

Seppälä H, Klaukka T, Vuopio-Varkila J *et al.* 1997. The effect of changes in the consumption of macrolide antibiotics on erythromycin resistance in Group A streptococci in Finland. *New England Journal of Medicine* **337**(7): 441–446.

Song J-H, Lee NY, Ichiyama S *et al.* 1999. Spread of drug-resistant *Streptococcus pneumoniae* in Asian countries: Asian Network for Surveillance of Resistant Pathogens (ANSORP) Study. *Clinical Infectious Diseases* **28**: 1206–1211.

Spika JS, Facklam RR, Plikaytis BD *et al.* 1991. Antimicrobial resistance of *Streptococcus pneumoniae* in the United States, 1979–1987. *Journal of Infectious Diseases* **163**: 1273–1278.

Sterman JD. 2000. *Business Dynamics: Systems Thinking and Modeling for a Complex World*. McGraw-Hill: Boston.

9

Maps and Models in System Dynamics: A Response to Coyle (2001)

Homer J, Oliva R. *System Dynamics Review*, 17(4): 347-355

Maps and models in system dynamics: a response to Coyle

Jack Homer[a]* and Rogelio Oliva[b]

Jack Homer is a system dynamics modeling consultant and has directed Homer Consulting since 1989. He was formerly an assistant professor at the University of Southern California and received a PhD in management from MIT for a thesis modeling the diffusion of new medical technologies. He also holds BS and MS degrees in applied mathematics and statistics from Stanford University, and received the Jay W. Forrester Award in 1997.

Rogelio Oliva is an assistant professor in the Technology and Operations Management Unit at the Harvard Business School. He holds a BS in industrial and systems engineering from ITESM (Mexico), an MA in systems management from Lancaster University, and a PhD in operations management and system dynamics from MIT. His current research interests include service operations, and the transition that product manufacturers are making to become service providers.

Abstract

Geoff Coyle has recently posed the question as to whether or not there may be situations in which computer simulation adds no value beyond that gained from qualitative causal-loop mapping. We argue that simulation nearly always adds value, even in the face of significant uncertainties about data and the formulation of soft variables. This value derives from the fact that simulation models are formally testable, making it possible to draw behavioral and policy inferences reliably through simulation in a way that is rarely possible with maps alone. Even in those cases in which the uncertainties are too great to reach firm conclusions from a model, simulation can provide value by indicating which pieces of information would be required in order to make firm conclusions possible. Though qualitative mapping is useful for describing a problem situation and its possible causes and solutions, the added value of simulation modeling suggests that it should be used for dynamic analysis whenever the stakes are significant and time and budget permit. Copyright © 2001 John Wiley & Sons, Ltd.

Syst. Dyn. Rev. **17**, 347–355, (2001)

A recent research problem by Geoff Coyle (2000) poses the question, "How much value does quantified modeling add to qualitative analysis?" And more pointedly: "The research issue is whether or not there are circumstances in which the uncertainties of simulation may be so large that the results are seriously misleading to the analyst and the client." By uncertainties, Coyle is largely concerned about soft variables, which may be a challenge to formulate as equations and for which numerical data may be lacking. He suggests that an appropriate response in the face of substantial uncertainties is to use causal-loop (influence) diagrams alone, without simulation, to assist in policy analysis. He believes that this approach will remain appropriate at least until better techniques for modeling soft variables have been developed.

This line of reasoning emerges from a stream of work on qualitative analysis, much of it by Coyle and Eric Wolstenholme, that dates back to the early 1980s (see, especially, Wolstenholme and Coyle 1983). Summarizing a major theme of this work, Wolstenholme (1999) states, "[T]he idea [of using stand-alone causal loop diagrams] was aimed at providing insight into managerial issues by inferring, rather than calculating, the behaviour over time of the system represented."

Our purpose in this paper is to review the roles of qualitative maps and simulation models in system dynamics, and to argue that simulation—following established system dynamics methodology—can add value beyond mapping

[a] Voorhees, New Jersey, U.S.A.

[b] Harvard Business School, Boston, Massachusetts, U.S.A.

* Correspondence to: Jack Homer, 36 Covington Lane, Voorhees, New Jersey 08043, U.S.A.

System Dynamics Review Vol. 17, No. 4, (Winter 2001): 347–355
DOI: 10.1002/sdr.224
Copyright © 2001 John Wiley & Sons, Ltd.

Received March 2001
Accepted July 2001

alone in most cases. This argument rests upon two points that have been demonstrated repeatedly in the laboratory and in real life: first, that one cannot reliably draw inferences from complex causal maps without simulation and, second, that system dynamics is well suited for addressing issues associated with soft variables and incomplete data.

Despite these demonstrations, as George Richardson (1999) has observed, "We are perceiving a growing tendency to use qualitative mapping approaches by themselves, without quantitative simulation." It may be that some systems practitioners see mapping and modeling as similar, with simulation primarily offering greater numerical precision than mapping, and even that only in some cases. It is just such a blurring of the lines that we wish to argue against here, with the hope that the unique value of simulation may become better appreciated.

Why we map

Causal-loop diagrams have long been used in standard system dynamics practice for two purposes connected with simulation modeling. They were initially employed after simulation, to summarize and communicate model-based feedback insights (see, for example, Forrester 1968). A few years later, they also started to be used prior to simulation analysis, to depict the basic causal mechanisms hypothesized to underlie the reference mode of behavior over time—that is, for articulation of a dynamic hypothesis (Randers 1973; Randers 1980).

The dynamic hypothesis is a cornerstone of good system dynamics modeling practice. It "explains the dynamics as endogenous consequences of the feedback structure" (Sterman 2000), and explicitly states how structure and decision policies generate behavior (Richardson and Pugh 1981). Moreover, "The inclusion of basic mechanisms from the outset forces the modeler to address a meaningful whole at all stages of model development." (Randers 1973). That is, a dynamic hypothesis is the key to ensuring that the analysis is focused on diagnosing problematic behavior and not on enumerating the unlimited details of a "system".

With the advent of qualitative analysis in the 1980s, the causal-loop diagram started to be used for purposes not necessarily related to simulation modeling, namely, for detailed system description and for stand-alone policy analysis. Wolstenholme (1999) puts the case clearly:

'Causal loop' qualitative system dynamics enhances linear and 'laundry list' thinking by introducing circular causality and providing a medium by which people can externalise mental models and assumptions and enrich these by sharing them. Furthermore, it facilitates inference of modes of behaviour by assisting mental simulation of maps.

We agree that system description can be a useful activity. Systems thinking consultants everywhere can attest to the utility of word-and-arrow diagrams for helping clients to understand the interconnected nature of their problems and the many possible side effects of their decisions. These maps generally go well beyond the conciseness of a dynamic hypothesis, and may often be better described as "hurricane diagrams" (Thompson 1999). Their usual intent is to improve the process of thinking about the structure underlying a problem—including feedback loops and perhaps time delays, accumulations, and nonlinear effects—rather than to hypothesize the root causes of reference mode behaviors. For this purpose, it does not much matter whether one uses causal-loop diagrams, or stock-and-flow diagrams, or Wolstenholme and Coyle's modular approach (Wolstenholme and Coyle 1983; Wolstenholme 1990; Coyle 1996), or the "rich picture" technique of soft systems methodology (Checkland 1981; Coyle and Alexander 1997) for describing a problem situation.

Although system description has value, it is important to recognize that what is described is structure and not dynamics. When a diagram is built to capture the full complexity of a problem situation, it will invariably include some details not relevant to the generation of reference mode behavior. Indeed, the richer the system description becomes, the more it departs from the concept of a dynamic hypothesis—and the more difficult it becomes to see how structure may be linked to behavior.

Why we model

Even when a causal map is kept relatively simple, its implied behavior may not be so simple or obvious. In fact, causal-loop diagrams are notoriously unreliable tools for behavioral inference (Richardson 1996a). A simulation model corresponding properly to the diagram may fail to behave as expected, even if the causal map contains loops with the polarities and delays one believes are necessary to recreate the reference modes of behavior. This is why we call our initial ideas about causal mechanisms a dynamic *hypothesis* and not an explanatory model. Experimental studies have shown repeatedly that people do a poor job of mental simulation even when they have complete knowledge of system structure and even when that structure is quite simple (Sterman 1994a, 2000).

Simulation modeling provides a tool for formally testing the dynamic hypothesis and determining its adequacy. If the testing is done properly, the flaws in a model will generally make themselves evident when the model's behavior is compared with that of the real world. Often, one finds that a simulation model may reproduce some aspects of reference mode behavior and other pieces of evidence but not all of them. In that case, the model, and perhaps the dynamic hypothesis itself, must be revised and tested again and again until it is found to be fully consistent with available evidence and logical considerations (Homer 1996). Even then, one must be careful not to confuse the model with reality, but always be aware of its limitations, and keep an eye open for additional evidence that may call into doubt some aspect of the model.

Although simulation may not always reveal the shortcomings of a dynamic hypothesis, the risk of leaving the flaws unrecognized is much greater without simulation. As John Sterman (1994b) has said, "Without modeling, we might think we are learning to think holistically when we are actually learning to jump to conclusions." This is not to say that it is *impossible* to test the inferences drawn from a causal map without computer simulation. One might, for example, gather additional data—new or historical—that could be compared against map-based mental simulations of how the system should behave under specified conditions. Or one might, in some cases, carry out a small-scale, pilot implementation of map-based policy recommendations, to see whether the indicated improvements come to pass. But these empirical approaches to hypothesis testing often have serious drawbacks, including lack of reproducibility, inadequate breadth, and high costs, risks, or time requirements. These are the very drawbacks of traditional social science and policy analysis that simulation modeling is generally able to overcome.

Dynamic analysis without modeling?

Are there instances in which it is appropriate to use a map alone without a simulation model for analyzing dynamic behavior? In theory, the problem may be so straightforward or familiar that dynamic conclusions may be drawn reliably from a causal-loop diagram alone. As George Richardson (1996a) has stated:

> The history [of feedback thought in the social sciences] suggests that [properly] inferring dynamic behavior from word-and-arrow maps involving circular causality comes only in two situations, when the map is a recognized structure previously verified by simulation and when the map is extremely simple.

When might these two situations—a previously-simulated map, such as a generic structure (Paich 1985; Lane and Smart 1996), or a transparently simple one—actually arise in the real world? Infrequently. In both cases, the assumption is that the client has introduced no significant messy details that complicate matters. But problems that have proved so nettlesome as to cause an organization to call in a consultant rarely fit neatly into any straightforward structure. Once one starts to consider the real-world details and the multiple perspectives on a problem, the system description nearly always grows and becomes more elaborate. For this reason, Paich (1985) concludes that "generic structures are not appropriate for solving specific problems. Rather, generic structures are educational tools for learning about the fundamentals of complex systems.''

Suppose that, despite the messy details and multiple perspectives, somebody persuasive—maybe a member of the client team, maybe the consultant—proposes that a familiar or simple structure, lying at the core of the messiness, really is sufficient to explain the problem and its possible solution. And suppose that the team, after some discussion, unanimously supports this proposal. Is it appropriate to stop at that point and declare the problem solved? Some might say yes, that the team is now "on the same page" and can move forward, secure in its new sense of unity and purpose.

Unfortunately, the new insights may be off the mark, no matter how confident the team may feel. Overconfidence is one of the most prevalent and insidious biases in judgment and decision making, while groupthink is a common barrier to learning in the face of complexity (Sterman 1994a, 2000). If the proposed core structure is to be reliably used as the basis for policy analysis, its primacy or dominance over all other loops and pieces of structure that have been described by the team must be demonstrated somehow. Such demonstration is one of the strengths of computer simulation. Just as simulation makes possible the reliable inference of behavior from a given system structure, so too it facilitates the isolation of the core structure underlying a problem behavior.

Dealing with uncertainty

Are there instances, as Coyle suggests, in which the problem description is so rife with uncertainty, and soft variables in particular, that a reliable simulation model cannot be built? Soft variables—hard to measure and often subject to multiple influences—have been a central feature of system dynamics models since Forrester (1961) first described models depicting management decision-making and consumer response. Any reasonably complete text or course on system dynamics should cover principles and guidelines for modeling decision making and human behavior, and for formulating nonlinear relationships, including soft variables. For example, Sterman (2000) devotes entire sections to these subjects (chapters 13, 14, and 15). The ability of system dynamics simulation to handle soft variables has been demonstrated repeatedly, and across a remarkable breadth of applications in the management and social sciences (see, for example, Forrester 1969, 1971; Roberts 1978; Sterman 1985; Levine and Fitzgerald 1992; Richardson 1996b; Oliva and Sterman 2001).

It is true that there may be cases in which insufficient reference mode data or other types of information are available to determine conclusively whether the dynamic hypothesis is adequate to explain the problem behavior, or to assess adequately the various possible implications of policy change. But such cases are less common than one might think, for two reasons. First, even when numerical data are relatively lacking, mental databases are extremely rich and helpful in modeling (Forrester 1980). Second, even when parameter values are uncertain, sensitivity testing often reveals that behavior modes and policy conclusions are not affected by these uncertainties (Forrester 1969; Richardson and Pugh 1981).

Even when there is too little information to reach firm conclusions from a simulation model, it is still not *more* misleading to simulate than to map without simulation. If there is too little information to be able to make reliable inferences about behavior from a simulation model, then certainly the same must be said about a qualitative causal map. One may choose to map simply for the sake of system description and not for behavioral inference; but, in that case, as we have said, one is studying only the structure of a system and not its dynamics.

Moreover, how can one determine that too little information exists to reach firm conclusions *except* through simulation? One generally does not know how powerful a dynamic hypothesis is, or how sensitive to uncertainty, until a simulation model is built and tested. Model testing, including sensitivity analysis, can help to identify which pieces of information would be required in order to determine the adequacy of the dynamic hypothesis or to make firm policy conclusions possible. Consider three examples:

1. In a study of cocaine prevalence in the U.S.A., the behavior of an initial model was found to be sensitive to the strength of assumed causal links from wholesale price to retail price and supply. Extensive analysis of data on undercover cocaine purchases and seizures found little or no strength in these causal links, and led to a thorough reorientation of the dynamic hypothesis and the model (Homer 1996).
2. In a study of automobile leasing strategy by General Motors, sensitivity testing suggested those areas in which further model disaggregation and incorporation of market research data might make a difference to policy conclusions and other areas where it would not (Sterman 2000).
3. In an ongoing study of antibiotic resistance, discrepancies between model behavior and some pieces of empirical evidence have suggested specific areas for further data collection and possible model modification (Homer *et al.* 2000).

Conclusion

The question has been posed as to whether or not there may be situations in which simulation modeling adds no value beyond that of qualitative mapping, or may even be more misleading than mapping by itself. We have argued that simulation nearly always adds value to policy analysis, even in the face of significant uncertainties and soft variables.

Still, we appear to have a situation today in which some systems practitioners may use mapping alone to draw behavioral and policy inferences, without going on to simulation or empirical testing. Some of these analysts may be competent simulation modelers who believe, or whose clients believe, that testing may not add enough value to justify its costs and time requirements. Others may know enough to draw causal-loop diagrams but do not know how to develop proper simulation models.

Such uses of stand-alone mapping are, perhaps in most cases, better than nothing—that is, better than policy discussions that lack feedback thinking. There is clear value in helping a client recognize the existence of delayed responses and self-reinforcing side effects and possible sources of policy resistance that, without a process of mapping, would remain mere unarticulated concerns, vague observations, or simply ignored altogether. Moreover, there probably are certain situations—for example, when a decision must be made quickly or when the stakes are low—in which computer modeling is just too time-consuming or costly to justify its use.

Yet, it remains one thing to explore feedback mechanisms with a client, and quite another to draw conclusions about their dynamic implications. The mechanisms one emphasizes in an untested causal-loop diagram may or may not be the ones the client really ought to be most concerned about. In other words, the map may be misleading, and without simulation or other formal tests of the dynamic hypothesis there is no way to know whether that is so or not.

Consequently, we believe that all systems practitioners should understand and clearly describe to their clients the dangers of inferring too much from causal-loop diagrams, as stimulating as these diagrams may be. Of course, clients should also understand the limitations of simulation models, particularly in the realm of prediction, and the many requirements of proper model testing. But the fundamental distinction remains: Only through formal testing can one solidly bridge the gap from structure to behavior.

References

Checkland P. 1981. *Systems Thinking, Systems Practice*. Wiley: Chichester.

Coyle RG. 1996. *System Dynamics Modelling: A Practical Approach*. Chapman and Hall: London.

Coyle RG. 2000. Qualitative and quantitative modelling in system dynamics: some research questions. *System Dynamics Review* **16**(3): 225–244.

Coyle RG, Alexander MDW. 1997. Two approaches to qualitative modelling of a nation's drugs trade. *System Dynamics Review* **13**(3): 205–222.

Forrester JW. 1961. *Industrial Dynamics*. MIT Press: Cambridge, MA; reprinted by Pegasus Communications, Williston, VT.

Forrester JW. 1968. Market growth as influenced by capital investment. *Industrial Management Review* **9**(2): 83–105; Also in *Collected Papers of Jay W. Forrester*.

1975. Wright-Allen Press: Cambridge, MA; reprinted by Pegasus Communications, Williston, VT.

Forrester JW. 1969. *Urban Dynamics*. MIT Press: Cambridge, MA; reprinted by Pegasus Communications, Williston, VT.

Forrester JW. 1971. *World Dynamics*. Wright-Allen Press: Cambridge, MA; reprinted by Pegasus Communications, Williston, VT.

Forrester JW. 1980. Information sources for modeling the national economy. *Journal of the American Statistical Association* **75**: 555–574.

Homer JB. 1996. Why we iterate: scientific modeling in theory and practice. *System Dynamics Review* **12**(1): 1–19.

Homer JB, Ritchie-Dunham J, Rabbino H, Puente LM, Jorgensen J, Hendricks K. 2000. Toward a dynamic theory of antibiotic resistance. *System Dynamics Review* **16**(4): 287–319.

Lane DC, Smart C. 1996. Reinterpreting 'generic structure': evolution, application and limitations of a concept. *System Dynamics Review* **12**(2): 87–120.

Levine R, Fitzgerald H. 1992. *Analysis of Dynamic Psychological Systems (2 vols)*. Plenum Press: New York.

Oliva R, Sterman JD. 2001. Cutting corners and working overtime: Quality erosion in the service industry. *Management Science* **47**(7): 894–914.

Paich M. 1985. Generic structures. *System Dynamics Review* **1**(1): 126–132.

Randers J. 1973. Conceptualizing dynamic models of social systems: lessons from a study of social change. PhD dissertation, MIT Sloan School of Management, Cambridge, MA.

Randers J. 1980. Guidelines for model conceptualization. In *Elements of the System Dynamics Method*, Randers J (ed.). MIT Press: Cambridge, MA; reprinted by Pegasus Communications, Williston, VT.

Richardson GP. 1996a. Problems for the future of system dynamics. *System Dynamics Review* **12**(2): 141–157.

Richardson GP. (ed.) 1996b. *Modelling for Management: Simulation in Support of Systems Thinking (2 vols)*. Dartmouth: Aldershot.

Richardson GP. 1999. Reflections on the future of system dynamics. *Journal of the Operational Research Society* **50**: 440–449.

Richardson GP, Pugh AL III. 1981. *Introduction to System Dynamics Modeling*. MIT Press: Cambridge, MA; reprinted by Pegasus Communications, Williston, VT.

Roberts EB (ed.). 1978. *Managerial Applications of System Dynamics*. MIT Press: Cambridge, MA; reprinted by Pegasus Communications, Williston, VT.

Sterman JD. 1985. The growth of knowledge: testing a theory of scientific revolutions with a formal model. *Technological Forecasting and Social Change* **28**(2): 93–122.

Sterman JD. 1994a. Learning in and about complex systems. *System Dynamics Review* **10**: 291–330.

Sterman JD. 1994b. Beyond training wheels. In *The Fifth Discipline Fieldbook*. Senge PM, Kleiner A, Roberts C, Ross RB, Smith BJ (primary authors). Doubleday: New York.

Sterman JD. 2000. *Business Dynamics: Systems Thinking and Modeling for a Complex World*. Irwin/McGraw-Hill: Boston.

Thompson JP. 1999. Consulting approaches with system dynamics: three case studies. *System Dynamics Review* **15**(1): 71–95.

Wolstenholme EF. 1990. *System Enquiry: A System Dynamics Approach*. Wiley: Chichester.

Wolstenholme EF. 1999. Qualitative vs. quantitative modelling: the evolving balance. *Journal of the Operational Research Society* **50**: 422–428.

Wolstenholme EF, Coyle RG. 1983. The development of system dynamics as a methodology for system description and qualitative analysis. *Journal of the Operational Research Society* **7**: 569–581.

10

Models for Collaboration: How System Dynamics Helped a Community Organize Cost-Effective Care for Chronic Illness (2004)

Homer J, Hirsch G, Minniti M, Pierson M. *System Dynamics Review*, 20(3): 199-222

Related writing

Hirsch G, Homer J. Modeling the Dynamics of Health Care Services for Improved Chronic Illness Management. *Proceedings of the 22nd International System Dynamics Conference*, Oxford, England, 2004.

Models for collaboration: how system dynamics helped a community organize cost-effective care for chronic illness

Jack Homer[a]*, Gary Hirsch[b], Mary Minniti[c] and Marc Pierson[d]

Jack Homer has been an independent system dynamics consultant to private and public organizations since 1989. In the public arena, his work focuses on public health and health care policy. He was formerly an assistant professor at the University of Southern California and received a PhD in management from MIT. He also holds BS and MS degrees in applied mathematics and statistics from Stanford University, and received the Jay W. Forrester Award in 1997.

Gary Hirsch is an independent consultant with 35 years of experience in applying system dynamics to complex problems in health care, education, social services and industry. He holds SB and SM degrees from MIT's Sloan School of Management in system dynamics. His work in health care has focused on improved management of chronic illness, strategic planning for health care delivery

Abstract

Chronic illness is a large and growing problem throughout the world. Experts agree that the U.S. health care system is poorly organized to care for chronic illnesses and, as a result, is wasteful and unresponsive to the needs of patients. This article describes a program to improve chronic care in a county of Washington State, and how system dynamics models focusing on diabetes and heart failure supported the planning of that program. The models project the program's costs and benefits over 20 years and have given its leadership the ability to do resource planning, set realistic expectations, determine critical success factors, and evaluate the differential impacts on affected parties. Relying upon model projections, the leadership is seeking ways to address concerns about financial "winners" and "losers" so that all parties are willing to participate in and support the program. Copyright © 2004 John Wiley & Sons, Ltd.

Syst. Dyn. Rev. **20**, 199–222, (2004)

Magnitude of chronic illness as a problem

Chronic illness—that which lasts more than three months and is not self-limiting in nature—is a huge and growing problem in the U.S.A. and throughout the world. Chronic illnesses are the leading cause of illness, disability, and death and are responsible for at least 70% of all health care expenditures in the U.S.A. (Hoffman *et al.* 1996; Institute for Health and Aging 1996). More than half of the adult population in the U.S.A. has at least one chronic illness and the number of the chronically ill is expected to grow from 125 million in 2000 to 157 million by 2020. About half of those with a chronic illness have more than one such affliction, and are responsible for the great majority of the total cost of chronic illness. The aging of the population will drive growth in the prevalence of chronic illness, as one third of the chronically ill are over 65. People over 65 currently make up 13 per cent of the population, but that segment will reach 20 per cent by 2030 after the entire "baby boom" population has reached age 65. Chronic illness is much more prevalent among the older population, as is the likelihood of having two or more chronic conditions, and will grow in importance and cost as the population ages (IoM 2001; Partnership for Solutions 2004).

A recent report by the Institute of Medicine of the National Academy of Sciences (IoM 2001) suggests that the health care system in the U.S.A. is not up

[a] Homer Consulting, Voorhees, NJ 08043, U.S.A. E-mail: jhomer@comcast.net
[b] Creator of Learning Environments, Wayland, MA 01778 USA. E-mail: gbhirsch@comcast.net
[c] PeaceHealth/Healthcare Improvement Division, Bellingham, WA 98225, USA. E-mail: mminniti@peacehealth.org
[d] PeaceHealth/St. Joseph Hospital, Bellingham, WA 98225 USA. E-mail: mpierson@peacehealth.org
* Correspondence to: Dr. Jack Homer.

System Dynamics Review Vol. 20, No. 3, (Fall 2004): 199–222
Published online in Wiley InterScience
(www.interscience.wiley.com). DOI: 10.1002/sdr.295
Copyright © 2004 John Wiley & Sons, Ltd.

Received August 2003
Accepted April 2004

systems, and development of interactive learning environments for helping managers and providers better understand those systems.

Mary Minniti has worked in the health care quality improvement profession for the last eight years, and is Project Manager for the Pursuing Perfection Project of Whatcom County in Washington State. She is a Certified Professional in Healthcare Quality, with a BS in sociology from the University of Oregon. She has worked for over 20 years in the social service field, promoting and initiating innovative community programs for improving education and employment opportunities for at-risk populations, and coordinating the involvement of a wide variety of organizations to ensure that the programs benefit all stakeholders.

Marc Pierson, MD, was a practicing emergency medicine physician for 18 years and has worked for 16 years implementing community-based

to the challenge posed by chronic illness, describing it as "poorly organized". It states that, "The prevailing model of health care delivery is complicated, comprising layers of processes and handoffs that patients and families find bewildering and clinicians view as wasteful." Involvement of patients and their families in care is especially important in chronic illness where they can provide much of the care and make a difference between good outcomes and deteriorating health. The report indicates that fundamental change is needed to effectively deal with chronic illness.

The Pursuing Perfection Program

This article describes a program to improve the care of chronic illness in Whatcom County, Washington, and the role played by a pair of system dynamics models in support of that program. The county is semi-rural and its largest town is Bellingham, about two hours north of Seattle. It has a population of 171 thousand with 14 per cent living below the poverty line. The program is a collaborative effort of healthcare providers in Whatcom County and includes two of the leading insurers active there and a primary Medicaid (government-funded coverage for people with low income) insurer that has recently joined the effort. The program has received $1.9 million in funding from the Robert Wood Johnson Foundation (RWJF) as one of seven sites in a larger program called Pursuing Perfection ("P2") that is designed to improve the care of chronic illness (Pursuing Perfection Learning Network 2004). Whatcom County's P2 program built on a foundation of cooperation that had already been established in the county.

Pursuing Perfection in Whatcom County is focused on the following problems:

- *Poor cooperation among organizations.* More competition between organizations than cooperation on behalf of patients.
- *Poor patient care.* Care is often unsafe, unscientific, filled with delays and inefficiencies, not seamless, not transparent, broken up into silos of care, and delivered inequitably.
- *Lack of focus on chronic care.* Although chronic care is responsible for the majority of healthcare utilization and costs, the current system is designed more around acute care than chronic care.
- Consequently, *chronically ill patients carry the burden* of an inadequate health care system.

In line with the Institute of Medicine's recommendations, the mission of the Whatcom County P2 program is to create a community-based system of chronic care that it is patient-centered, evidence-based, effective, safe, timely, and equitable (Patient Powered 2004).

systems to improve health care quality in Whatcom County. He is a long-standing member of the executive team at Whatcom County's St. Joseph Hospital, and Project Executive for the Pursuing Perfection Project. He has also led the implementation of a community-wide clinical information network and medical record availability securely across the community.

The program is initially concentrating on two chronic illnesses as prototypes for improved care: Type 2 diabetes and heart failure. Both of these illnesses affect a great many people in the U.S.A. and other countries. About 17 million people have Type 2 diabetes (NIDDK 2004a) and nearly 5 million have heart failure in the U.S.A. alone (AHA 2000). Total costs of diabetes in the U.S.A. in 2002 were estimated to be $132 billion, with $92 billion of that in direct medical expenditures and the other $40 billion in indirect costs due to disability and premature mortality (NIDDK 2004a). Heart failure was estimated to cost $38 billion in 1991 in health care costs alone (O'Connell and Bristow 1993). The prevalence of both diseases is growing rapidly as the population ages and the number of people above age 65 increases. Obesity, the prevalence of which has more than doubled in the past 20 years in the U.S.A. (Flegal *et al.* 2002), has also contributed to the growth of diabetes prevalence and heart disease, as well as being a risk factor for several other chronic diseases (NIDDK 2004b). The following sections describe how the disease processes of diabetes and heart failure were modeled, with special attention to diabetes, and how the results were used to further the goals of the P2 program.

Role of system dynamics in Pursuing Perfection

The Whatcom County P2 program had two critical needs for making decisions about potential interventions for improving the care of chronic illnesses such as diabetes and heart failure:

- Getting a sense of the overall impact of these interventions on incidence and prevalence of diabetes and heart failure, health care utilization and cost, and mortality and disability rates in the community.
- Understanding the impact of the various interventions on individual health care providers in the community and on those who pay for care—insurers, employers, and individuals. There was a concern that the costs and benefits of the program be shared equitably and that providers who helped produce savings not be overly penalized by a loss of revenue that might result.

These needs could not be met with common quantitative tools such as spreadsheet models that project impacts in a simple, linear fashion. Interventions in chronic illness do not have simple direct impacts. The aging of the population, incidence of new cases and progression of disease, deaths, and the interventions themselves all create a constantly changing situation. For example, interventions ideally reduce mortality rates, leaving more people with the disease alive and requiring care at a later point in time. Similarly, people prevented from advancing to a more serious stage of the illness will have fewer health care requirements at a later point in time.

People kept from developing the disease altogether have even fewer needs and better prognoses. What mix of preventive programs and more active treatment of those who already have the disease yields the best results for the community? How might screening programs that identify these illnesses at an earlier stage improve outcomes? To fully evaluate these interventions, it is necessary to be able to track the effects of interventions over time.

The stock-and-flow structure of system dynamics models is ideal for this purpose. The models that were developed for diabetes and heart failure track flows of patients across several stages of severity of illness, calculating health care requirements and mortality and disability rates for patients at each stage. Interventions slow the rates of progression across these stages as well as preventing the disease in the first place.

The models offered additional capabilities that would help to advance the goals of Pursuing Perfection:

- They would support sensitivity analyses to help deal with uncertainty in the available data. With the models, we could create a range of projections to illustrate possible impacts, from worst-case to best-case scenarios. Conservative (worst-case) scenarios would be helpful for for those reluctant to take risks who worried that certain benefits of the programs might not materialize.
- The models would also provide a framework in which to assess controversial issues and get a better understanding of them in the context of the larger system. For example, the literature on both diabetes and heart failure contains an active debate about the value of screening at-risk patients to find those who are at an early, asymptomatic stage of the disease. Both models would provide a framework for testing different screening strategies and understanding their costs and benefits.
- It would also be possible to compare different implementation paths and understand their consequences for resource requirements and impacts. Providers and insurers initially involved in P2 represented only a fraction of the community's health care system. The manner in which others were assumed to get involved or whether they got involved at all would have a major impact on the magnitude and timing of the project's benefit to the community.

Modeling of chronic illness in this project draws on an extensive body of system dynamics work on specific chronic illnesses such as cardiovascular disease (Luginbuhl *et al.* 1981; Hirsch and Wils 1984), on dental care and oral health (Hirsch and Killingsworth 1975), and on a microworld dealing with community health status in which chronic illness is a central focus (Hirsch and Immediato 1998; 1999).

The community will eventually want to do model-based analyses for all of the other major chronic illnesses. Diabetes and heart failure are a starting point and serve as prototypes. Expanding the range of chronic illnesses will eventually let us model the synergies of treating risk factors that lead to multiple chronic illnesses and the downstream benefits of treating one such as diabetes that can be a risk factor for others such as heart failure. The models will eventually be able to show how creating a treatment and prevention infrastructure to do this will have beneficial effects on multiple illnesses.

A modeling framework

Pictured in Figure 1 is the conceptual framework we used for modeling the costs and benefits of a program to address any particular chronic illness. Program adoption by providers of care occurs against a backdrop of the community's demographics, prevalence of the disease, and the prevailing approach to caring for the disease. The program may have significant infrastructure costs, including costs of program personnel (administrators, consultants, clinical care specialists) and the costs of information systems that allow providers and patients to record and share data electronically. Adoption of the program leads to a shift in care patterns, typically toward greater intensity of planned, non-urgent, care, which, in turn, directly affects health care costs. This shift in care is intended to reduce the incidence and progression of disease and consequent complications and deaths. Reductions in the health care costs associated with diseases, as well as productivity losses due to disability, ideally would offset the added costs of infrastructure and greater intensity of planned care, resulting in a net savings for the community as well as improving outcomes for patients.

Fig. 1. A framework for modeling chronic illness program impacts

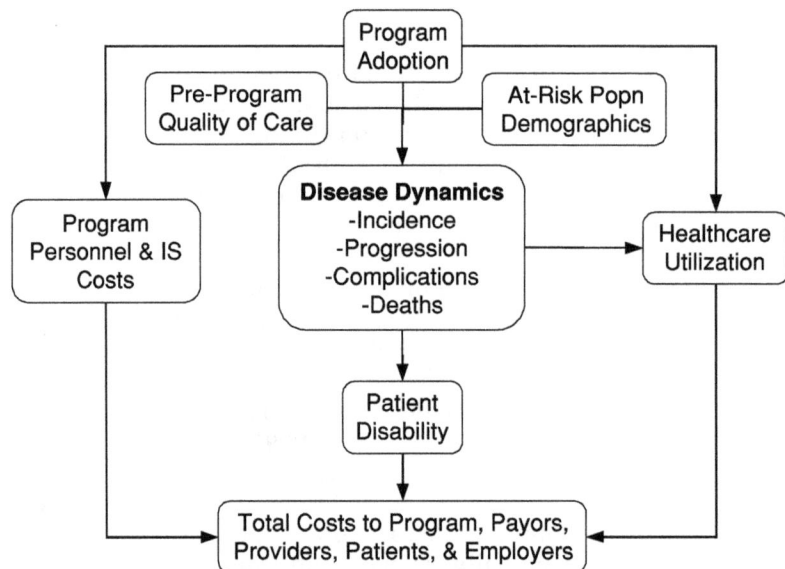

214

Applying the modeling framework to diabetes

Application of the modeling framework to a specific chronic illness and intervention program requires the specification of four things:

- the patient stock-and-flow structure for the illness and its calibration to reflect data for a typical patient population;
- the types, amounts, and unit costs of healthcare utilization associated with the patient stock-and-flow structure;
- how the program would affect patient flows;
- how the program would directly affect infrastructure and health care costs.

In the remainder of this article examples of model structure and behavior will be presented for Type 2 diabetes. Space constraints here do not allow for a similar presentation of heart failure, which may, however, be found in a previous version of this article (Homer *et al.* 2003).

Figure 2 presents a somewhat simplified view of the stock-and-flow structure used in modeling Type 2 diabetes. The actual model has two separate

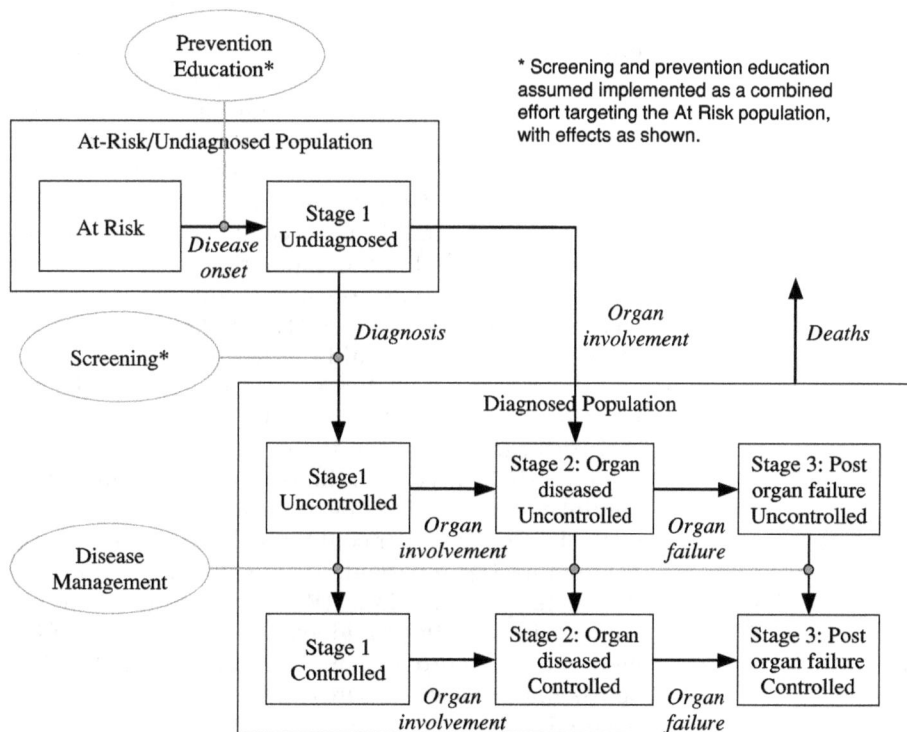

Fig. 2. Diabetes stages and intervention points

structures like those shown in Figure 2, one for the 18-to-64 age group and one for the 65-and-older age group, which are linked by flows of patients turning 65. The model also calculates an inflow of population turning 18, death out-flows from each stock based on patient age and stage of illness, and flows of migration into and out of the county. The three stages of diabetes portrayed in this figure were identified through discussions with clinicians in Whatcom County.

About 30 per cent of the general population is at risk for developing diabetes primarily by virtue of being overweight and physically inactive, or having a family history of diabetes. Diabetes develops through several stages. Increased sugar in the blood of a pre-diabetic leads to changes that weaken the body's ability to maintain blood sugar control, and take a patient from pre-diabetic to Stage 1 diabetes, unless treated with oral medications and/or changes in diet and activity. Most Stage 1 diabetics, up to two-thirds of all diabetics, have no outward symptoms, and more than half are undiagnosed.

If Stage 1 diabetics go untreated, most will eventually progress to Stage 2, marked by organ disease. In Stage 2 diabetes (about 18 per cent of diabetics in Whatcom County), blood flow disturbances impair the functioning of organ systems and potentially lead to heart attack, stroke, kidney disease, peripheral vascular disease, loss of sensation in the extremities, or eye disease (retinopathy, glaucoma, cataracts). At this stage it is still possible to reduce complications through glycemic and blood pressure control. A patient who has suffered irreversible organ damage, or organ failure, is said to be in Stage 3 (about 14 per cent of diabetics in Whatcom County); this would include patients post-heart attack, post-stroke, post-amputation, with end-stage renal disease, or with blindness. These patients are at the greatest risk of further complications leading to death. Despite the advanced state of their disease, even Stage 3 diabetics may benefit from glycemic and blood pressure control.

Several studies have demonstrated that the incidence, progression, and complications of diabetes can be reduced significantly through concerted in-tervention (Wagner et al. 2001; ADA/NIDDK 2002). As indicated in Figure 2, primary prevention would consist of efforts to screen the at-risk population and educate pre-diabetics about the lifelong diet and activity changes they need to prevent progression to diabetes. Intensive preventive programs for pre-diabetics can reduce the incidence of diabetes by 50–60 per cent. For con-firmed diabetics, a comprehensive disease management approach, such as that employed by the P2 program, can increase the fraction of patients under control from the 40 per cent typically seen without a program up to nearly 100 per cent for those patients who make the required lifestyle changes and take the required medications. (We have estimated more conservatively, based on conversations with providers, that a maximum of 80 per cent of known dia-betics could be brought under control under P2.) The benefits of control are substantial: disease progression is reduced by perhaps two-thirds, and the hospitalization rate at each stage of the disease cut by about half.

The diabetes model was implemented in Vensim,[1] using its array capability to replicate the structure shown in Figure 2 for the 18–65 and over-65 age groups, several different primary care providers participating in P2, and for patients covered by a number of different insurers. As indicated earlier, this disaggregation was important for being able to see how the costs and benefits of P2 would fall on different health providers and insurers. The model also includes an extensive set of calculations of health care utilization and costs for each of the subgroups defined by this array structure.

Figure 3 presents a 20-year *status quo* projection of diabetes prevalence by stage in Whatcom County, which assumes no intervention program. The total number of diabetics grows from about 8,000 in 2001 to nearly 13,000 in 2021, an average growth rate of 2.2 per cent per year. During this same period the total county 18-and-over population grows by only 1.5 per cent per year. As a result, diabetics increase from 6.5 per cent of the population to 7.5 per cent over the 20 years. The reason for this growth of diabetes, more rapid than that of the overall population, is that the prevalence of diabetes is much greater among the faster-growing elderly population (with about 17 per cent prevalence of diabetes) than among the slower-growing non-elderly (less than 5 per cent prevalence of diabetes). Note in Figure 3 that the distribution of diabetics by stages remains about the same throughout the simulation. This reflects an assumption that there are no significant advances in diabetes diagnosis and care, such as those contemplated by P2, and no further increases in the fraction

Fig. 3. *Status quo* projection of diabetes in Whatcom County, 2001–2021

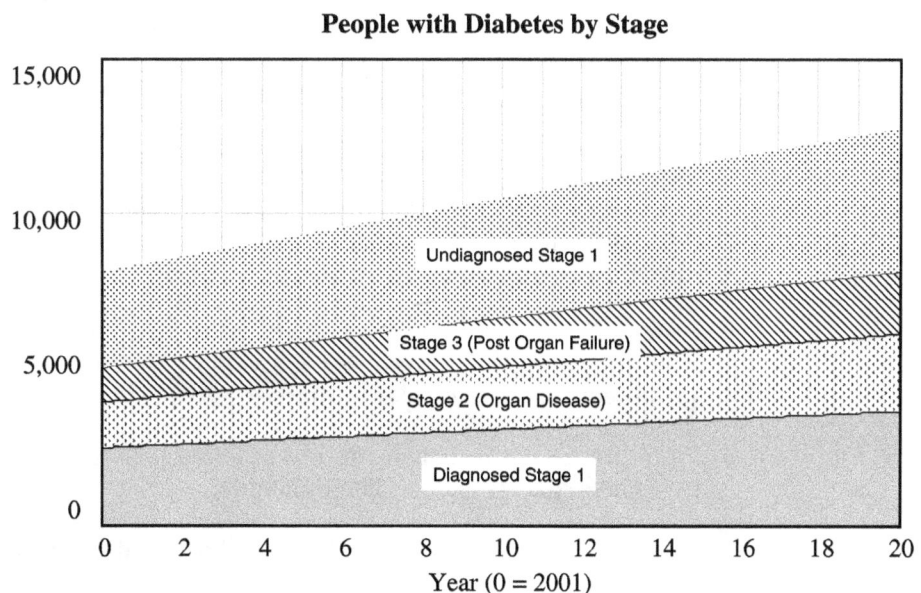

People with Diabetes by Stage

Undiagnosed Stage 1

Stage 3 (Post Organ Failure)

Stage 2 (Organ Disease)

Diagnosed Stage 1

Year (0 = 2001)

System Costs for Diabetes

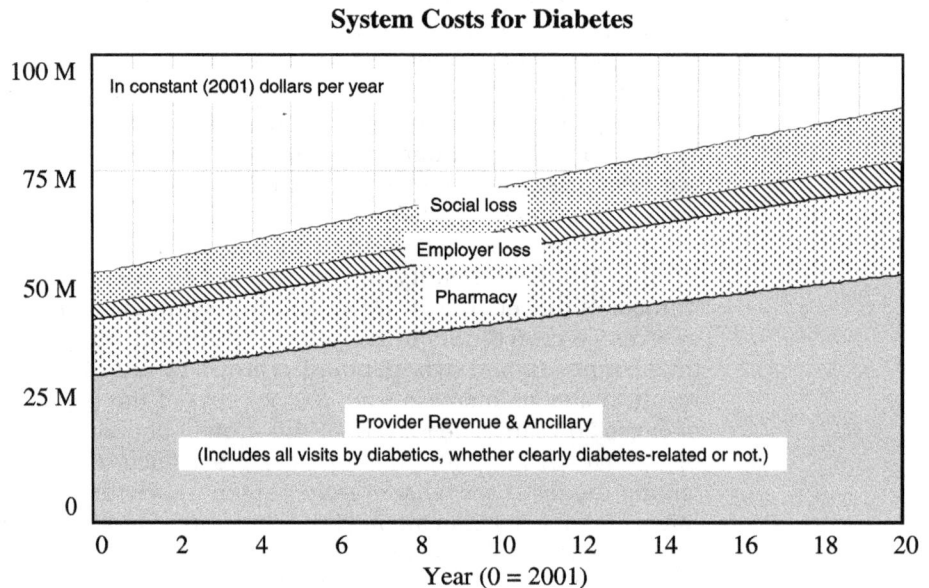

In constant (2001) dollars per year

Social loss

Employer loss

Pharmacy

Provider Revenue & Ancillary
(Includes all visits by diabetics, whether clearly diabetes-related or not.)

Year (0 = 2001)

of the population at risk for diabetes. (This may be a conservative assumption in view of recent data on the growing prevalence of obesity in the US.)

Figure 4 presents a *status quo* projection of diabetes-related costs, broken into four major categories. These costs are presented in constant 2001-dollar terms, excluding inflation in the general economy and in healthcare *per se*. The observed growth in costs is a direct reflection of the growth in the diabetic patient population, and especially growth in the number of the Stage 2 and 3 patients, who generate most of the costs. The largest cost category is Provider Revenue and Ancillary. Within this category, hospital costs account for 74 per cent of the total, ancillary costs (e.g., for laboratory tests) for 14 per cent, specialist MD visits for less than 10 per cent, and primary care physician (PCP) visits for less than 3 per cent. Pharmacy is the cost of all drugs used by diabetics. Somewhat less tangible but no less important for the community are the losses of productivity due to disability, shown in Figure 4 as Employer loss (value of the employee's productivity to the employer) and Social loss (loss of income to the employee and the value of having that person participate in the economy). (Testa and Simonson 1998 describe how these costs are distinguished.)

Calibration of the diabetes and heart failure (HF) models was made possible by diverse sources of data. These include:

- county population projections from the Washington State Office of Fiscal Management;

- illness prevalence by age from the National Center for Health Statistics (diabetes) and the literature (HF), which were in close agreement with member data from Group Health Cooperative (GHC), a major payor in the area offering both commercial and government plans;
- distribution of illness by stage from GHC based on diagnostic codes (diabetes) and the literature (HF);
- in-control/out-of-control fractions (diabetes) from laboratory data provided by the Family Care Network (FCN) primary care group;
- hospital utilization and financial data from St. Joseph Hospital;
- PCP utilization and financial data from PCPs currently participating in P2 (FCN, SeaMar, and CSH);
- specialist utilization and financial data from GHC (diabetes) and the North Cascade Cardiology group (HF);
- pharmacy costs from GHC;
- effect of control (diabetes) and ideal care (HF) on utilization and costs from the medical literature and expert judgment of clinicians in the community.

Small focus groups, made up of a cross-section of physicians, nurses, other health providers, and a patient representative were invaluable in helping to develop parameters for which numerical data were not available. They gave generously of their time and drew heavily on many years of experience with diabetes and heart failure patients to help estimate important model parameters.

Representing program impacts with the diabetes and heart failure models

The basic clinical intervention components of the P2 program in Whatcom County include screening and prevention education for diabetes, risk management for heart failure, and disease management for both. The models reflect detailed information on the personnel, information systems, and healthcare costs that the P2 interventions are expected to entail. The models also describe how the clinical interventions would affect patients flows (as illustrated for diabetes in Figure 2), and specify two possible factors that could mitigate the ability of the program to bring patients successfully under control (diabetes) or ideal care (heart failure) via disease management. The first of these issues is drug affordability, particularly for elderly patients who lack sufficient drug coverage under Medicare. (Medicare is the government-sponsored insurance plan for people over age 65 in the US.) The second issue is the possible insufficiency of clinical care specialists (CCSs) to keep up with the demand for their services.

Figure 5 illustrates the causal structure used in the diabetes model to model the movement of patients from uncontrolled to controlled status, for any given stage in the three-stage chain of diagnosed diabetics shown in Figure 2. (An

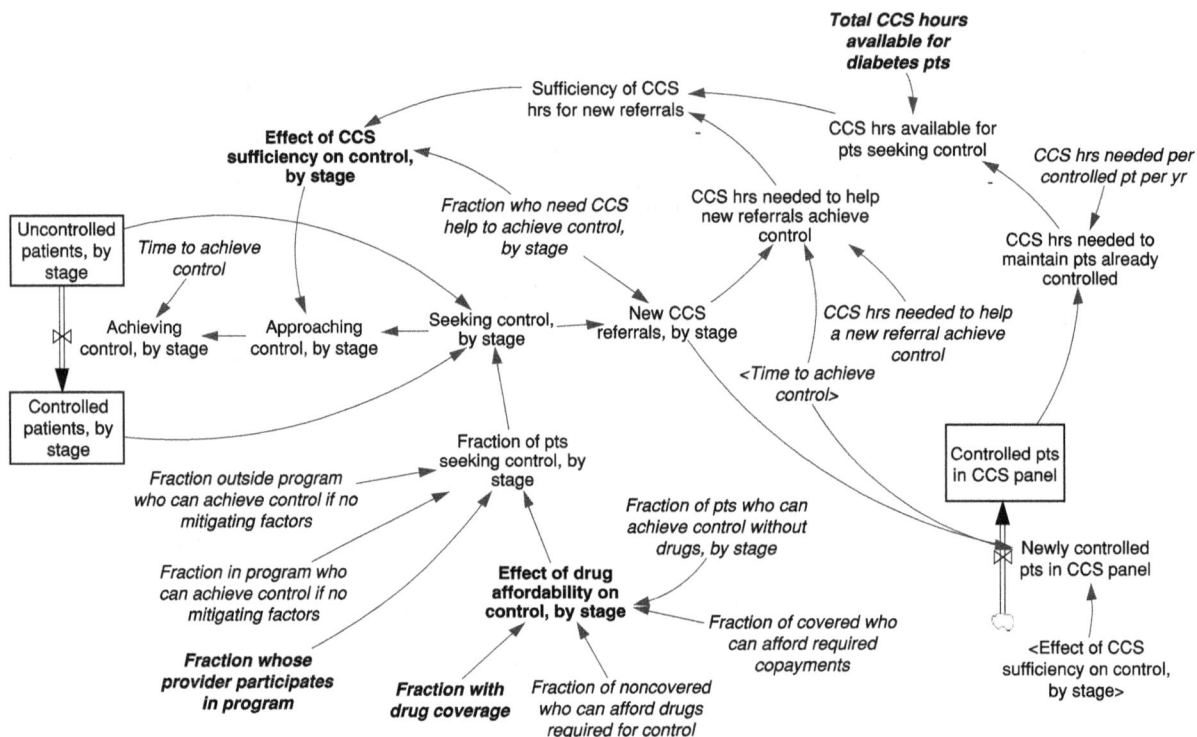

Fig. 5. Causal structure describing control of diabetes, with drug affordability and clinical care specialist (CCS) sufficiency as potential mitigating factors

identical structure is used in the heart failure model to model the movement of patients from usual care into ideal care.) The P2 program raises the fraction of patients who will achieve control, absent any problems with drug affordability or CCS sufficiency.

As shown in Figure 5, four key factors combine directly to determine the effect of drug affordability, namely, the fractions of:

- patients with drug coverage;
- those without coverage who can afford to self-pay for the additional drugs required for disease management;
- patients with coverage for drugs who can afford the co-payments for these drugs required by most U.S. plans (for example, $10 per prescription);
- patients in the given disease stage who can achieve control without the need for additional drugs but rather through diet and exercise alone. (This last fraction becomes smaller as one progresses to more advanced stages of disease.)

Unlike the structurally straightforward effect of drug affordability seen in Figure 5, the structure for the effect of CCS sufficiency on control involves some interesting dynamic complexity. In particular, CCS hours available to assist diabetes patients are increasingly siphoned off to meet the maintenance demand of a growing panel of clients already under control, leaving fewer hours available to new referrals. These maintenance demands are relatively minor on a per-patient basis—only about 1½ hours per patient per year on the average, compared with about 6 hours required to help bring new referrals under control over the course of a few months. However, these maintenance demands can accumulate, within just a few years' time of program initiation, to double and then triple in aggregate the number of hours needed for bringing new referrals under control, according to our simulations. In addition to the time requirements just mentioned, there are two other key factors affecting CCS sufficiency:

- the fraction of patients in a particular stage who need CCS help to achieve control (which becomes larger as one progresses to more advanced stages of disease);
- total CCS hours available for diabetes patients—which, in turn, are determined by the number of clinical care specialists. The number of CCSs should be expected to grow over time to accommodate demand and is treated in the model as an exogenous decision variable.

Table 1 presents a group of scenarios for evaluating program impacts for diabetes and heart failure. These scenarios, and several others not seen here, were presented to program participants and other community stakeholders

Table 1. Selected Scenarios Tested with Diabetes and Heart Failure Models

Status quo
- No program implementation, costs or benefits

Full program adoption
- Administrative costs incurred starting Year 0 (2001)
- PCP & specialist adoption grows to 100% during Years 1 to 4 (2002–2005):
  - (Yr 1) 2 FCN sites, SeaMar, and CSH; (Yr 2) Other 6 FCN sites;
  - (Yr 3) Half of other PCPs; (Yr 4) All remaining PCPs
  - Specialists: Ramp-up in parallel with PCP adoption
- Clinical care specialists hired to meet demand as projected by model
  - Start with 2 CCSs, grow to 7 by Year 4, up to 10 by Year 18
- Program components included:
  - Diabetes: Community-based screening and prevention education for At-Risks (with referral to PCP if test positive), and disease management for known diseased
  - Heart Failure: Risk management for At-Risks, and disease management for known diseased, but no additional screening of At-Risks beyond what MDs do already (mostly post-infarction)

Disease management only
- Diabetes: No screening or prevention education of At-Risk beyond *status quo* amount
- Heart failure: No risk management beyond *status quo* amount

Full program plus comprehensive Medicare drug coverage (for people over age 65)

as part of the P2 planning process. The key scenario for comparison with the *status quo* is "Full program adoption." This scenario, which represents the fully realized vision of P2, assumes that all of the county's office-based physicians will participate in the program by 2005. It also assumes comprehensive disease management for both diabetes and heart failure, similarly rigorous risk management for heart failure, and a community-based mass screening and preventive education program for diabetes. Finally, it assumes a ramping up of the number of CCSs sufficient to meet the demand for their services projected by the model.

Figure 6 shows the growth of direct program costs (in constant 2001 dollars) under all scenarios in which there is full adoption of the program by physicians, and CCS growth to meet the corresponding demand. (These conditions apply to all scenarios in Table 1 other than *status quo*.) The largest category is personnel costs, which include administrative, process and organizational development consultants for redesigning office practices, and the CCSs. The consultants drop out during Year 6 (2007), after they have completed their final office practice redesign and implementation. There are seven CCSs by Year 4, growing to ten by Year 18, at an annual cost of $74,000 each. The information systems cost about $1,500 per physician per year, leading by Year 4 to an annual cost of over $400,000 county-wide.

The direct cost of the diabetes mass screening and prevention effort is tiny in comparison, involving about 5,000 subjects per year at a cost of only about eight dollars per person. We assume that half of the county's at-risk population of 30,000 will be screened in this way, once every three years as recommended

Fig. 6. Program costs under full adoption scenarios

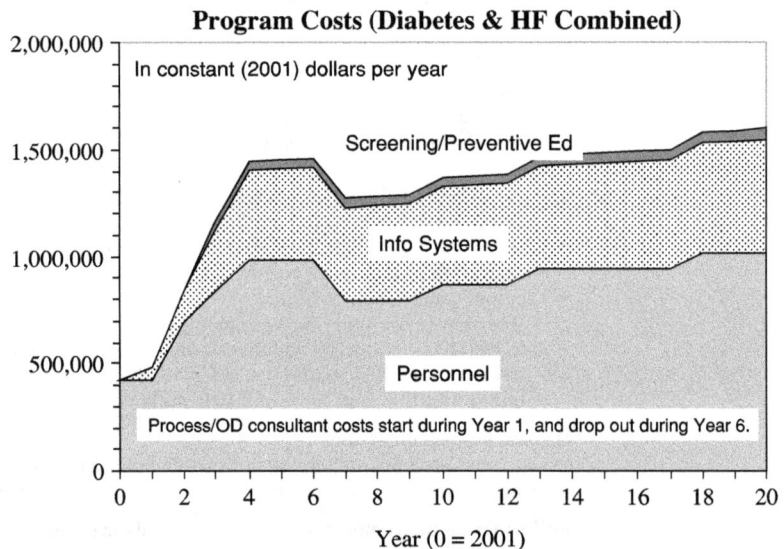

Program Costs (Diabetes & HF Combined)

by guidelines. A quarter of those tested will get a positive reading for diabetes or prediabetes (impaired glucose tolerance), and will be referred to their physician for additional testing and counseling.

Simulated Results of Alternative Program Scenarios

Figures 7 to 10 present graphical output from the diabetes model allowing comparison of the four scenarios described in Table 2 over the full 20-year time horizon. (Similar graphs from the heart failure model are presented in Homer *et al.* 2003.) The following is a summary of these results.

Fig. 7. Program impact on fraction of diabetics under control

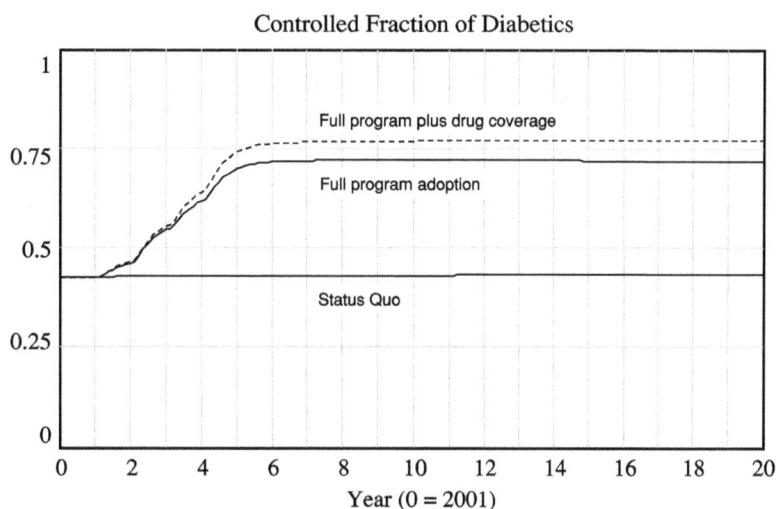

Controlled Fraction of Diabetics

Full program plus drug coverage

Full program adoption

Status Quo

Year (0 = 2001)

FRACTION OF DIABETES PATIENTS UNDER CONTROL (FIGURE 7). The fraction of known diabetics under control starts at a *status quo* value of 43 per cent. It is assumed that 80 per cent of known diabetics could be brought under control if they could afford the drugs and there were sufficient CCS support. With full program adoption, the fraction under control is increased to 72 per cent, and with adoption plus full Medicare drug coverage it is raised to 77 per cent. This final value is short of 80 per cent not because of any mitigating factors, but because there is a steady influx of newly diagnosed diabetics who require some months to achieve control.

DEATHS FROM DIABETES COMPLICATIONS (FIGURE 8). Under the *status quo*, the number of diabetes-related deaths grows continuously along with the size of the diabetic population. Full program adoption reduces these deaths by 40 per cent, and adoption plus drug coverage by 54 per cent, in line with the greater

Fig. 8. Program
impact on deaths from
diabetic complications

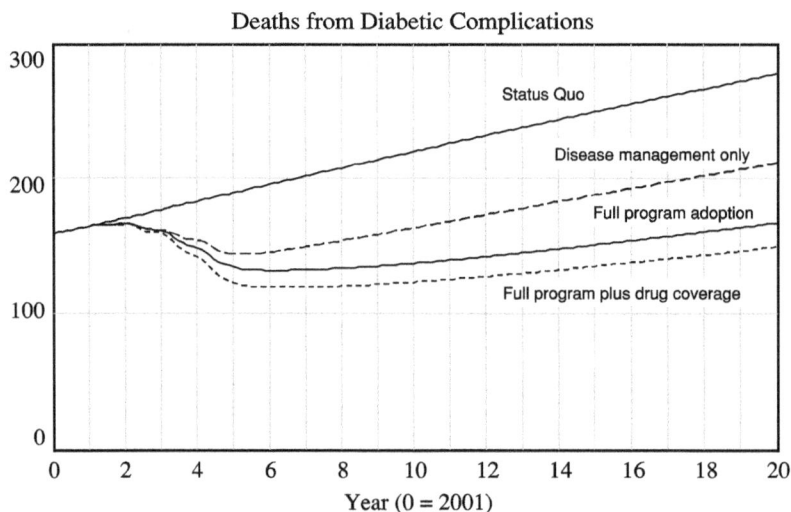

Fig. 8. Program impact on deaths from diabetic complications

Deaths from Diabetic Complications

fractions of diabetics being brought under control. A program with disease management only (no screening and prevention component) is effective at reducing deaths early on, but becomes less and less effective as time progresses.

PATIENTS WITH ADVANCED DIABETES (FIGURE 9).   Under the *status quo*, the number of patients with later stage (Stage 2 or 3) diabetes grows continuously at an average rate of 2.7 per cent per year. The full program—with or without drug coverage— ends up reducing the number of later stage diabetics by about 20 per cent relative

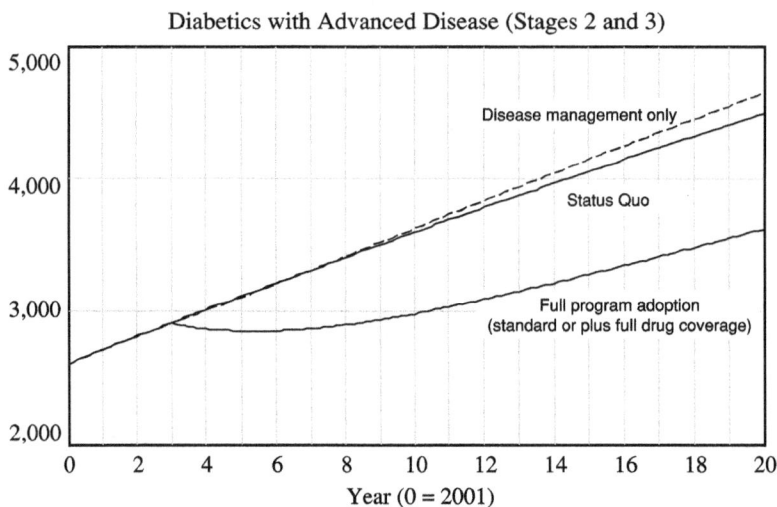

Fig. 9. Program impact on prevalence of advanced diabetes (Stages 2 and 3)

Diabetics with Advanced Disease (Stages 2 and 3)

to the *status quo*. These two scenarios share the same screening and prevention component, a component that markedly reduces the incidence and progression of diabetes to later stages. A program with disease management only actually leads to an increase in later stage diabetics relative to the *status quo*, because (lacking a screening component) it does more to keep later stage diabetics alive longer than it does to reduce the progression from Stage 1 to Stage 2.

TOTAL SYSTEM COSTS (FIGURE 10). A breakdown of total system costs for the *status quo*, including disability losses to employers and society-at-large, was previously shown in Figure 4. In the scenarios with full program adoption and full adoption plus drug coverage, net savings are achieved by Year 3, only two years after the program is launched. By Year 5, drug coverage generates further reductions in disability beyond those provided by the program alone. The scenario with disease management only, in contrast, achieves total net savings initially, but gives back most or all of these savings by the end of 20 years. By the end of 20 years, the full adoption approach results in a net savings of $6 million per year, or 7 per cent of the *status quo* costs, including a $4 million per year reduction in disability losses.

Fig. 10. Program impact on total system costs of diabetes (including disability costs)

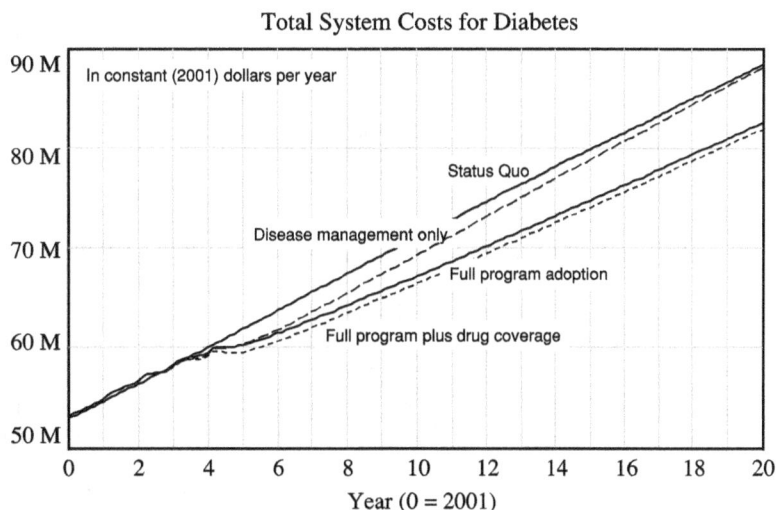

There was naturally concern that the numbers used in the models were subject to some uncertainty and that this might affect the conclusions one might draw. In addition to reviewing many of the model's parameters with providers in the community, we performed sensitivity analyses to examine the effect of varying key parameters on the models' results. These used best-case and worst-case assumptions for the impact of P2 on disease progression and rates of complications from disease. These simulations suggest that, while there is some uncertainty about the exact magnitude of impact, P2 is likely to result in significant health benefits at acceptable cost, even with worst-case assumptions. However, the crossover point for net savings does not occur as soon as it does in the mid-range or best-case simulations.

The P2 program planners and stakeholders in Whatcom County appreciated the long-term view afforded by the preceding graphs, but also required a more detailed sense of the program's impacts over the shorter term. Table 2 presents numbers covering the period 2003–2008 (model Years 2–7) that describe the impact of the full program adoption scenario relative to the *status quo* for diabetes and heart failure combined. A table like this one but with somewhat more detail has served in Whatcom County as a tool for identifying likely "winners" and "losers" over the next several years. It has also helped with developing ideas for program funding and mechanisms for the redistribution of savings so that all stakeholders might have a financial interest in program participation.

The table has also helped to convince stakeholders that the cost of the program is worthwhile, even if one ignores disability savings and longer-term benefits. The final section at the bottom of Table 2 suggests that over the 2003–2008 time period program-related outlays (program and payor costs) will generate health benefits that rival or beat those of other accepted health interventions on the basis of cost–benefit ratio. The model suggests that the P2 program will result in an average outlay per life-year saved of less than $30,000 over the 2003–2008 period, an amount that compares very favorably with accepted medical technologies. This cost-benefit ratio turns negative by 2010, signifying a net saving from that year onward for diabetes and heart failure combined.

Using model results to reach a common understanding

The work described here began in July of 2002. It started with a series of community meetings designed to help P2 participants better understand the process and objectives of modeling and begin to create an approach to model diabetes, the first of the two illnesses to be modeled. Results of the pilot effort at diabetes modeling done a year earlier were shared to help participants visualize the projections and insights that would be available at the end of the current effort in the Spring of 2003. The meetings were also valuable for providing input to the design of the two models and critique as they developed.

As indicated earlier, focus groups of clinicians helped us define the flow of patients through the stages of the two illnesses being modeled and changes in care that could result in improved outcomes. Community meetings provided participants with the first set of critical insights from the modeling work. These were about the overall impact of P2 on the community and were essential for helping to build commitment to continue with the program. Key insights included:

Table 2. Six-year program impacts on diabetes and heart failure combined, comparing full program adoption scenario to *status quo*

As of start of year:	2003	2004	2005	2006	2007	2008	Total
1. Health impacts							
1.1 Disability days avoided	2,781	10,201	24,134	41,400	48,688	52,976	180,180
1.2 Inpatient days avoided	214	781	1,637	2,806	3,251	3,517	12,205
1.3 Deaths avoided (life-years saved)	6	22	54	89	104	114	388

Financials below are in constant 2001 dollars. Values are the result of subtracting status quo projections from full-program projections. Parentheses indicates decrease relative to status quo; no parentheses indicates increase.

	2003	2004	2005	2006	2007	2008	Total
2. Program costs ($000)							
2.1 Personnel and operations	689	878	1,025	1,026	1,026	835	5,479
2.2 Info systems paid for by MDs	147	279	416	423	431	438	2,134
2.3 Total	836	1,157	1,442	1,449	1,457	1,273	7,613
3. Impact on provider net income ($000)							
3.1 Primary care MDs	(65)	(111)	(118)	(54)	(44)	(37)	(428)
3.2 Specialist MDs	(71)	(174)	(291)	(343)	(375)	(394)	(1,647)
3.3 Hospital	(123)	(495)	(1,039)	(1,758)	(2,052)	(2,231)	(7,697)
4. Impact on supplier revenue ($000)							
4.1 Pharmaceuticals	513	1,716	3,794	6,128	6,519	6,591	25,261
4.2 Implanted devices	(19)	(103)	(346)	(701)	(891)	(1,020)	(3,079)
5. Impact on payor costs ($000)							
5.1 Commercial plan reimbursements	77	222	428	575	391	190	1,883
5.2 Medicaid reimbursements	59	121	474	862	883	843	3,241
5.3 Medicare reimbursements	(154)	(607)	(1,531)	(2,995)	(3,753)	(4,204)	(13,245)
5.4 Patient out-of-pocket payments	206	674	1,574	2,609	2,787	2,838	10,688
5.5 Total	189	410	945	1,050	307	(334)	2,567
6. Impact on disability losses ($000)							
6.1 Employer loss	(116)	(478)	(1,016)	(1,641)	(1,904)	(2,062)	(7,217)
6.2 Social loss	(246)	(922)	(2,142)	(3,638)	(4,269)	(4,642)	(15,859)
6.3 Total	(362)	(1,400)	(3,158)	(5,278)	(6,174)	(6,704)	(23,076)
7. Impact on combined costs ($000)							
7.1 Outlay (program+payor)	1,025	1,566	2,387	2,500	1,764	938	10,180
7.2 Total (program+payor+disability)	663	166	(771)	(2,779)	(4,409)	(5,765)	(12,896)
8. Cost-benefit ratios ($)							**Average**
8.1 Outlay per disability day avoided	369	154	99	60	36	18	56
8.2 Outlay per inpatient day avoided	4,800	2,006	1,458	891	543	267	834
8.3 Outlay per life-year saved	173,479	70,491	44,370	28,155	16,954	8,263	26,216

- Complete implementation of P2 involving all providers in the community would produce more extensive benefits than partial implementation involving only those providers already participating.
- Knowing the magnitude and growth of P2 costs over time enabled participants to budget for their shares of those costs.

- Total system costs for the P2 program are less than *status quo* costs without P2 even though reduced mortality rates keep more people alive. Some health care payors have expressed concern that improved care for older people will lead to higher costs because they will be kept alive longer and continue to require care. However, the P2 program produced a net reduction in cost in the simulations by keeping people in the less severe stages of the diseases for a longer period of time and reducing the complications of these diseases that require expensive hospitalizations. Given the sensitivity of payors who were already bearing high costs, this was an important insight to help motivate their continued participation.

In addition to these overall results, the impacts on particular providers and those paying for care yielded additional insights that were important to these conversations:

- Benefits of P2 in terms of savings are likely to fall unevenly among those paying for care for at least the first several years of the program. Medicare is likely to be the biggest "winner" from the start; see Table 2. Commercial insurers, on the other hand, would actually pay out more under P2 relative to the *status quo* through 2008, after which time they too start to realize net savings due to the accumulated achievements of primary prevention under the program. Medicare patients are older and are, on the average, at more severe stages of the two diseases. As a result, they have higher rates of acute complications and hospitalizations that can be prevented by the more rigorous care available under P2. For these patients, it is possible to achieve immediate and substantial savings despite higher costs for prescription drugs and primary care. Commercial insurers cover younger patients whose disease is not typically as advanced and who are therefore less likely to have acute complications and require hospitalization. Savings from reduced hospitalization are not enough to offset higher costs of care under P2 and prescription drug costs for the first six years of the program. Medicare, in fact, benefits from the investment made by these commercial insurers that helps keep patients healthier when they are younger and require less care once Medicare becomes responsible for them. This insight highlighted the importance of "bringing Medicare to the table" to help pay part of the increased costs of P2 since it would be the biggest recipient of the savings generated for payors.
- Employers in the community and the community at large are also winners in terms of the reduction of disability losses resulting from diabetes and heart failure. Employers may be willing to fund some of the additional costs created by P2, so that they might reap the bulk of this loss reduction. For example, employers may be willing to pay higher premiums to commercial insurers to help cover P2, so that these insurers will have a greater financial interest in supporting the program.

- The other big winner is the pharmaceutical industry; see Table 2. This windfall for the pharmaceutical industry suggested that the drug companies be brought to the table and asked to help fund P2.
- The hospital and physicians (particularly specialists) are likely to see reductions in their net income as a result of P2, as seen in Table 2. This reduction is, of course, a concern to those providers. The hospital depends on "bread and butter" chronic illnesses to generate income that helps to subsidize other services such as mental health that are poorly covered by insurers. The good news is that hospital income contributions from diabetes and heart failure under P2 are not projected to fall below their 2001 values at any time, even though these contributions are lower than they would have been without P2. Reductions in hospital utilization from diabetes and heart failure are also not bad news in the sense that the community has undergone consolidation of its hospitals and can use the excess capacity to provide services that might otherwise be lacking as the population ages and requires more care. This is also true of a perceived physician shortage the community is experiencing. Having less severe chronic illness with fewer complications means that the limited number of physicians can spend more time keeping patients healthier.
- The model also provided a framework in which to examine mechanisms for redressing any perceived inequities in the distribution of program costs and benefits. One of these might be a payment scheme in which the hospital's payments from insurers are kept relatively constant despite reductions in the number of admissions. This might be justified by the need to support the broader role that the hospital plays in the community's health care system and the fact that some services subsidize others. One approach would be to have Medicare, the largest insurer and also the largest beneficiary of program-generated savings, pay a fixed annual amount per patient with a chronic illness (regardless of hospital use) rather than on a per-admission basis. Tests of the diabetes model demonstrated the feasibility of using such a severity-adjusted per-capita payment mechanism for leveling hospital revenues. The model demonstrates that the mechanism would effectively shift some of the windfall Medicare stands to receive under the program in order to "make the hospital whole", but without causing a net increase in Medicare payments relative to the *status quo* projection. The ability to use the model to do such testing permitted differences of opinion about equity to be pursued constructively rather than becoming stumbling blocks for the program.

From common understanding to collaboration

The Whatcom County P2 Leadership Board met on March 17–18, 2003, to learn about model-based findings and to discuss next steps for the program. The members of this board are leaders of the P2 participant organizations, representing the hospital, primary and specialist care providers, and major local insurers, plus a patient representative (also involved in all of our community meetings) who has both diabetes and heart failure. Much of the day's discussion focused on financial support for P2, and model findings proved helpful in this regard.

Financial support during transition period

A key concern addressed at the meeting was how the P2 program in Whatcom County would be supported during the nine-month period after the RWJF funding was due to run out and before other anticipated sources of funds could take over. Much of the discussion about this was purely practical: how much money was required to continue making progress during the transition period, and how much each of the participants could contribute. In this regard, the model contributed in two ways:

- The model helped the insurer, GHC, see that it could have a direct return from P2 during and soon after the nine-month transition period as a result of savings from the care of its older, sicker patients. While the program might cause GHC to pay more than it would have otherwise for its younger, non-Medicare patients, savings on the Medicare patients it manages would outweigh these higher costs and result in a net savings. After showing GHC what these net savings were projected to be through 2004, they agreed to contribute more for the interim funding than they had offered earlier in the discussion.
- The model also helped participants understand the value of preventive care and risk management in controlling the long-term cost and health impacts of the two diseases. The long-term nature of the impacts of these activities and the short-term financial needs might have made it tempting to postpone any spending on prevention until after long-term funding was assured. However, based on insights from the model, a community-based screening program for diabetes was retained in the program budget, and the importance of getting ideal care to hypertensives and hyperlipidemics at risk for heart failure was underscored.

Identification of other funding sources

While some additional funding might still come from RWJF starting mid-2004, it was clear that P2 had to develop other sources of outside funding. As indicated earlier, the model showed that employers in the community would enjoy a substantial reduction in cost due to disability from these two diseases among people who were still working. Pharmaceutical companies would benefit from substantial increases in the volumes of drugs prescribed. Other insurers in the community who managed programs for Medicare patients as GHC does would also benefit from significant savings. These insights helped shape the strategy for pursuing additional funding sources. The discussion also identified additional potential sources such as disability insurance carriers.

Going public, going forward

Following soon after the Leadership Board meeting in March 2003, meetings outside of Whatcom County have taken place that illustrate what it will take, and what questions must be answered, in order to spread the Whatcom County P2 approach to other communities and gain needed support from government and other major institutions.

- On April 14, a "Policy Summit" organized by the P2 staff and Leadership Board was held in Seattle. This was a well-publicized, all-day event, attended by some 200 representatives from government, foundations, and industry and community organizations. Morning presentations described Whatcom County's success in forging agreement around P2, early successes of the program, and an overview of the system dynamics approach and findings. Facilitated small group discussions in the afternoon generated further ideas on cost-effective improvements in health care, as well as support for taking the P2 approach beyond Whatcom County.
- On April 28–29, we discussed the use of system dynamics in Whatcom County at a meeting of the Pursuing Perfection Partners, hosted by the Institute for Healthcare Improvement, and attended by representatives of all seven of the RWJF P2 grantee institutions in the U.S.A., as well as groups from England, the Netherlands, and Sweden. One question raised in discussion was about whether and how easily the models might be adapted to other communities and other countries. Another question was about what greater benefits and cost savings one might expect to see as a one- or two-illness P2 approach is expanded ultimately to include all of the major chronic illnesses, and about how one might model these multi-illness synergies. Interest was also expressed in modeling the impacts of a program like P2 on patient access to caregivers and the dynamics of physician supply in a community.
- On May 9, we presented the P2 modeling work to the team from the American Hospital Association responsible for developing policy positions used in lobbying for or against proposed legislation on behalf of hospitals nationwide. Much of the discussion revolved around what the models might say about different approaches to Medicare reform being discussed in Congress. On one side of the debate was a proposal to provide expanded drug benefits to seniors under privately run disease management programs. On the other side was a more ambitious (and initially more costly) "case management" approach bearing a likeness to P2, involving not only drug benefits but also multi-disciplinary provider teams.

Every audience we have presented to has agreed that there is a role for modeling in support of program planning and policy evaluation in the complex area of chronic illness care. The leaders of P2 in Whatcom County are convinced that the models have given them the ability to do resource planning, set realistic expectations, determine critical success factors, and evaluate the differential impacts on affected parties. They feel that the models have led them to conclusions and decisions they likely would not have reached otherwise. They are now seeking ways to address concerns about financial winners and losers so that all parties are willing to participate and support the P2 program.

Note

1. The models contain proprietary financial information and are not available to the public.

References

American Diabetes Association and National Institute of Diabetes and Digestive and Kidney Diseases (ADA/NIDDK). 2002. The Prevention or Delay of Type 2 Diabetes. *Diabetes Care* **25**: 742–749.

American Heart Association (AHA). 2000. *2001 Heart and Stroke Statistical Update*. AHA: Dallas, Tx.

Flegal KM, Carroll MD, Ogden CL, Johnson CL. 2002. Prevalence and trends in obesity among US adults, 1999–2000. *Journal of the American Medical Association* **288**: 1723–1727.

Hirsch GB, Immediato CS. 1998. Design of simulators to enhance learning: examples from a health care microworld. *Proceedings of the International Conference of the System Dynamics Society*, Quebec City, Quebec.

Hirsch GB, Immediato CS. 1999. Microworlds and generic structures as resources for integrating care and improving health. *System Dynamics Review* **15**(3): 315–330.

Hirsch GB, Killingsworth WR. 1975. A New Framework for Projecting Dental Manpower Requirements. *Inquiry* **12**(2): 126–142.

Hirsch GB, Wils W. 1984. Cardiovascular disease in the Dutch population: a model-based approach to scenarios. *Ministry of Health: Conference on Health Care Scenarios*. The Hague, Netherlands, August.

Hoffman C. Rice DP, Sung H-Y. 1996. Persons with chronic conditions: their prevalence and costs. *Journal of the American Medical Association* **276**(18): 1473–1479.

Homer J, Hirsch G, Minniti M, Pierson P. 2003. Models for collaboration: how system dynamics helped a community organize cost-effective care for chronic illness. *Proceedings of the 21st International System Dynamics Conference*, New York, NY.

Institute for Health and Aging, University of California, San Francisco. 1996. *Chronic Care in America: A 21st Century Challenge*. Robert Wood Johnson Foundation: Princeton, NJ. Available at: http://www.rwjf.org/publications/publicationsPdfs/Chronic_Care_in_America.pdf. [Last accessed 21 July 2004]

Institute of Medicine Committee on Quality of Health Care in America (IoM). 2001. *Crossing the Quality Chasm: A New Health System for the 21st Century.* National Academy Press: Washington, DC.

Luginbuhl W, Forsyth B, Hirsch G, Goodman M. 1981. Prevention and rehabilitation as a means of cost-containment: the example of myocardial infarction. *Journal of Public Health Policy* **2**(2): 103–115.

National Institute of Diabetes and Digestive and Kidney Diseases (NIDDK). 2004a. *National Diabetes Statistics.* Available at: http://diabetes.niddk.nih.gov/dm/pubs/statistics/index.htm. [Last accessed 21 July 2004]

——. 2004b. *Statistics Related to Overweight and Obesity.* Available at: http://www.niddk.nih.gov/health/nutrit/pubs/statobes.htm#preval. [Last accessed 21 July 2004]

O'Connell JB, Bristow MR. 1993. Economic impact of heart failure in the United States: time for a different approach. *Journal of Heart and Lung Transplantation* **13**(suppl): S107–S112.

Partnership for Solutions, Better Lives for People with Chronic Conditions. 2004. Johns Hopkins University and the Robert Wood Johnson Foundation. Available at: http://www.partnershipforsolutions.org. [Last accessed 21 July 2004]

Patient Powered: Patient Centered Healthcare in Whatcom County. 2004. Whatcom County Pursuing Perfection Project. Available at: http://www.patientpowered.org.

Pursuing Perfection Learning Network. 2004. Institute for Healthcare Improvement. Available at: http://www.ihi.org/IHI/Programs/PursuingPerfection. [Last accessed 21 July 2004]

Testa MA, Simonson DC. 1998. Health economic benefits and quality of life during improved glycemic control in patients with Type 2 diabetes mellitus: a randomized, controlled double-blind trial. *Journal of the American Medical Association* **17**: 1490–1496.

Wagner EH, Sandhu N, Newton KM, McCulloch DK, Ramsey SD, Grothaus LC. 2001. Effect of improved glycemic control on health care costs and utilization. *Journal of the American Medical Association* **285**(2): 182–189.

11

Understanding Diabetes Population Dynamics Through Simulation Modeling and Experimentation (2006)

Jones A, Homer J, Murphy D, Essien J, Milstein B, Seville D. *American Journal of Public Health*, 96(3): 488-494

Related writings

Milstein B, Jones A, Homer J, Murphy D, Essien J, Seville D. Charting Plausible Futures for Diabetes Prevalence in the United States: A Role for System Dynamics Simulation Modeling. *Preventing Chronic Disease*, 4(3), July 2007. Available at: http://www.cdc.gov/pcd/issues/2007/jul/06_0070.htm.

Homer J, Jones A, Seville D, Essien J, Milstein B, Murphy D. The CDC's Diabetes Systems Modeling Project: Developing a New Tool for Chronic Disease Prevention and Control. *Proceedings of the 22nd International System Dynamics Conference*, Oxford, England, 2004.

Understanding Diabetes Population Dynamics Through Simulation Modeling and Experimentation

| Andrew P. Jones, MS, Jack B. Homer, PhD, Dara L. Murphy, MPH, Joyce D. K. Essien, MD, MBA, Bobby Milstein, MPH, and Donald A. Seville, MS

Health planners in the Division of Diabetes Translation and others from the National Center for Chronic Disease Prevention and Health Promotion of the Centers for Disease Control and Prevention used system dynamics simulation modeling to gain a better understanding of diabetes population dynamics and to explore implications for public health strategy.

A model was developed to explain the growth of diabetes since 1980 and portray possible futures through 2050. The model simulations suggest characteristic dynamics of the diabetes population, including unintended increases in diabetes prevalence due to diabetes control, the inability of diabetes control efforts alone to reduce diabetes-related deaths in the long term, and significant delays between primary prevention efforts and downstream improvements in diabetes outcomes. (*Am J Public Health.* 2006;96:488–494. doi:10.2105/AJPH.2005.063529)

DIABETES MELLITUS IS A

growing health problem worldwide. In the United States, the number of people with diabetes has grown since 1990 at a rate much greater than that of the general population; it was estimated at 20.8 million in 2005. Total costs of diabetes in the United States in 2002 were estimated at $132 billion, with $92 billion of that amount in direct medical expenditures and the other $40 billion in indirect costs because of disability and premature mortality.[1]

There are no quick or easy fixes for addressing the health and cost burdens of diabetes. Like other dynamically complex problems, diabetes is characterized by long delays between causes and effects, and the public health effort to address it is characterized by multiple concurrent goals that may conflict with one another. For example, although planners have called for reductions both in the prevalence of diabetes and in deaths because of its complications,[2] the fact is that fewer deaths, other things being equal, would lead to increased, not decreased, prevalence. Given such interconnections, a satisfactory solution will be found not in focusing on just 1 aspect of the overall health system—such as disease management, or detection, or risk factor reduction—but rather in addressing all major components together *as a system.*

We report results of simulation experiments with a system

dynamics model developed to explore the past and future burden of diabetes—its morbidity, mortality, and costs—in the United States. Model development was sponsored by the Division of Diabetes Translation and the Division of Adult and Community Health at the Centers for Disease Control and Prevention (CDC). For background on system dynamics methodology and applications, see Sterman's comprehensive textbook.[3]

MODEL STRUCTURE AND CALIBRATION

Figure 1 displays the basic causal structure of the system dynamics model. The full structure also includes an inflow of adult population growth and outflows of non–diabetes-related deaths. This structure reflects the knowledge and policy concerns of project team participants and is grounded in the scientific literature on diabetes, obesity, and related topics. Like all models, this one is a simplification: it omits many details in order to enhance understanding and includes assumptions that are uncertain to some degree. The model has evolved through a collaborative and iterative process that still continues.

At the core of the model is a chain of population stocks (appearing as boxes) and flows (appearing as double-thick arrows with valve symbols) portraying the movement of people into and out of the following stages: (1) normal blood glucose

(normoglycemia); (2) prediabetes, defined as having impaired glucose tolerance, impaired fasting glucose, or both[4,5]; (3) uncomplicated diabetes—that is, meeting the testing criteria for diabetes but not yet symptomatic nor showing detectable signs of disease in the eyes, feet, kidneys, or other organs; and (4) complicated diabetes.

The prediabetes and diabetes (hyperglycemic) stages are further divided among stocks of people whose conditions are diagnosed or undiagnosed. Diagnosis has dynamic significance because it is a prerequisite for proper management and control of hyperglycemia and the often accompanying risk factors of hypertension and hyperlipidemia; and such management or control can, in turn, greatly reduce the rates of diabetes onset, progression, and death.[6–10] In addition, diagnosis affects the extent to which the prevalence of diabetes in the population is recognized and measured, as well as the amount of effort and money that are put into the clinical management of prediabetes and diabetes.

Outside the population stock–flow structure, Figure 1 shows the potentially modifiable influences in the model that affect the rates of population flow, including influences that may be directly amenable to policy intervention (indicated in italics). These flow-rate drivers include prediabetes and diabetes detection, prediabetes management, diabetes control, and (because of

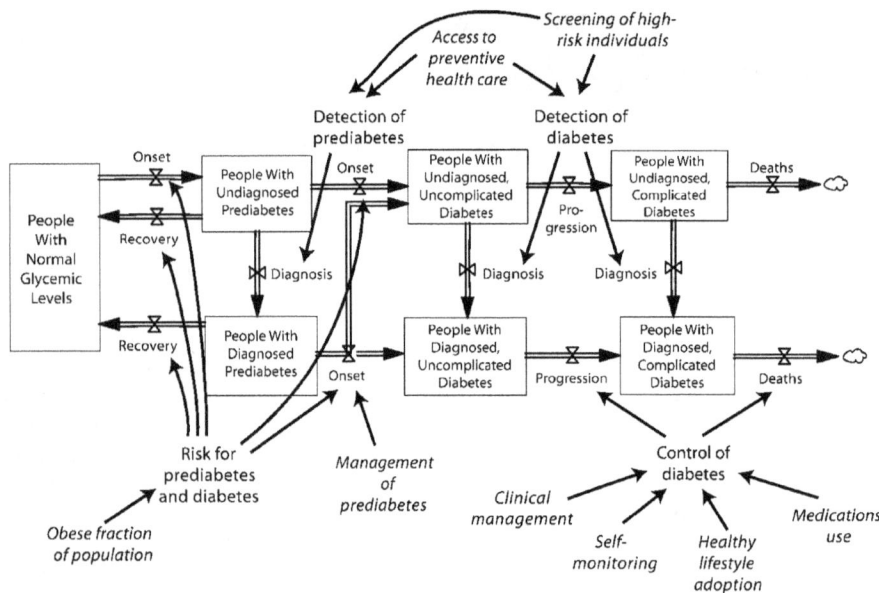

FIGURE 1—Overview of model structure, showing primary population stocks (boxes) and flows (arrows with valve symbols and cloud symbols for deaths), modifiable factors affecting flows (roman), and inputs amenable to policy intervention (italics).

and topics are summarized in Table 1.

BASELINE MODEL BEHAVIOR

Figure 2 presents selected output generated by simulating the model through the historical period starting in 1980 and then into the future through 2050 under a set of hypothetical baseline assumptions. The model requires assumptions about the future for each of the policy-related model inputs indicated in italics in Figure 1. In the baseline scenario, we assume that no further changes occur in obesity prevalence after reaching a value of 37% in 2006 (Figure 2a), and that inputs affecting the detection and control of prediabetes and diabetes remain fixed at their 2004 values through 2050. This fixed-inputs assumption is not meant to represent the project team's forecast of what is most likely to occur to policy inputs in the future, but it does make a useful and transparent starting point for policy analysis.

In addition to baseline simulation output, Figure 2 also presents historical data (Reported)

its influence on the risks of prediabetes and diabetes onset) the population prevalence of obesity. Prediabetes and diabetes detection may be improved by 2 types of interventions: those increasing the glucose screening of high-risk individuals by their providers, and those increasing access to preventive health care. Diabetes control may be improved by 4 types of interventions: those enhancing clinical management and those encouraging patients to self-monitor glucose levels, adopt healthy lifestyles, or use prescribed medications. (Another factor affecting flow rates in the model, but not indicated in Figure 1, is aging of the population, which affects death rates as well as prediabetes and diabetes onset rates. The system dynamics model, for the sake of simplicity, does not explicitly depict the

additional effects of changing racial and ethnic composition, as a Markov model by other researchers does.[11] However, by including the effects of changes in obesity prevalence, the system dynamics model does capture what may be the most salient

consequence of changes in racial and ethnic composition.)

The model's parameters were calibrated on the basis of historical data available for the US adult population, as well as estimates from the scientific literature. The primary data sources

TABLE 1—Primary Data for Model Calibration

Information Sources	Data Topics
US Census Bureau[12,13]	Population growth and death rates
	Health insurance coverage
National Health Interview Survey[14]	Diabetes prevalence
	Diabetes detection
National Health and Nutrition Examination Survey[15]	Prediabetes prevalence
	Obesity prevalence
Behavioral Risk Factor Surveillance System[16]	Glucose self-monitoring
	Eye and foot examinations
	Use of medications
	Attending diabetes self-management classes
Research literature	Effects of disease control and aging on onset, progression, death, and costs
	Direct and indirect costs of diabetes

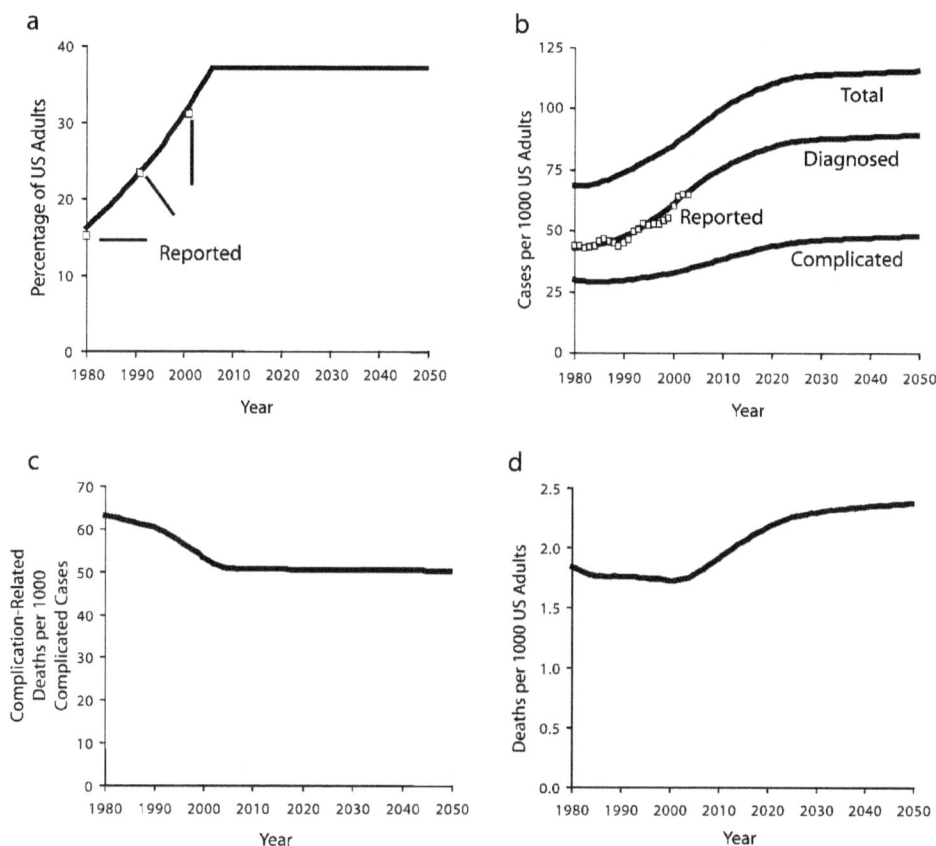

Note. Reported obesity prevalence based on National Health and Nutrition Examination Survey,[15] and reported diabetes prevalence based on National Health Interview Survey.[14] Baseline projection assumes that obesity prevalence rises to 37% in 2006 and remains fixed thereafter, and that disease detection and control efforts all remain fixed after 2004.

FIGURE 2—Selected baseline model output, 1980–2050, and comparison to historical data for obesity prevalence (a), diabetes prevalence (b), complication-related deaths per complicated cases (c), and complication-related deaths (d).

on obesity prevalence in the overall population[15] and diagnosed diabetes prevalence[20] and shows how closely the simulated output for diagnosed diabetes prevalence lies to the latter of these time series. The model is also able to reproduce available historical data on prediabetes prevalence, the diagnosed and controlled fractions of people with diabetes, population average BMI, obese fractions of people with prediabetes and diabetes, losses in health-related quality of life because of diabetes, and the

costs of both urgent/extended and preventive care of diagnosed diabetes.[17]

The 4 graphs in Figure 2 together tell the following story of diabetes prevalence and mortality for the historical period from 1980 to 2004 as indicated by model simulation. Two forces have worked in opposition to affect the number of diabetes-related deaths. The first force is a rise in the prevalence of obesity (Figure 2a). This increase in obesity led to a greater incidence of prediabetes and diabetes through

the chain of causation seen in Figure 1. Increased onset led to increased prevalence, first of uncomplicated diabetes and then of complicated diabetes (Figure 2b).

The second and opposing force is a noteworthy improvement in the control of diabetes, achieved through greater efforts to detect and manage the disease. It appears that glucose screening and clinical management of diabetes by providers, as well as self-monitoring and adoption of healthier lifestyles by people with diagnosed diabetes, all

increased significantly between 1980 and 2004. For example, we estimate that the fraction of primary care physicians who periodically test blood glucose levels in their patients at high risk for hyperglycemia rose steadily from 69% in 1980 to 95% in 2004, and that such screening has been the primary driver in increasing the fraction of patients with diabetes who have been diagnosed from 62% to 74% during the same period.[17] Model simulation suggests that progress on detection and management has reduced the rate at which people with diabetes move from uncomplicated to complicated diabetes, as well as the rate at which people with complicated diabetes die from the complications (Figure 2c).

From 1980 to 2004, the beneficial influence of increased diabetes control managed to hold mostly in check the harmful influence of increased disease prevalence: the model indicates that per capita deaths from complications of diabetes decreased by about 5% (in fact achieving a 7% decline by 2001 before giving back some of that gain from 2001 to 2004 because of some slowing in the rate of improvement in clinical management apparent in the data[16]). This result occurred because although the simulated prevalence of complicated diabetes *increased* by 17% (Figure 2b) from 1980 to 2004, the complications-related death rate for people with complicated diabetes *decreased* by 19% (Figure 2c) during the same time period.

The baseline simulation indicates a future for diabetes prevalence and diabetes-related deaths for the period 2004–2050 quite different from the past. With obesity prevalence fixed, by assumption, at its assumed high point of

37% from 2006 onward, the diabetes onset rate would be at its high point as well, and diabetes prevalence would consequently continue to grow through 2050 (Figure 2b). The rate of growth of diabetes prevalence would gradually diminish, and prevalence would become more level (from about 2025 onward) only when the outflow of deaths (because of diabetes as well as all other causes) started to catch up with the inflow of onset. The situation is comparable to the gradual filling of a bathtub that has a slow drain—in this case, the drain being deaths of people with diabetes. In fact, with the outflow of deaths being equal to only about 4% per year of the diabetes population (for example, in 2004 about 800 000 deaths out of 20 million, about half of these deaths because of complications of diabetes), more than 20 years would be required—after the assumed peaking-out of obesity in 2006—for the growth in diabetes prevalence (as a fraction of a growing adult population) finally to slow to a trickle.

With the prevalence of complicated diabetes growing by 38% from 2004 to 2050 (Figure 2b) and the death rate among the complicated cases declining by only 2% (Figure 2c; this 2% decline reflecting some continued reduction in the undiagnosed fraction of complicated cases), deaths from complications of diabetes would increase on a per capita basis by 36% (Figure 2d). Absent further improvements in disease detection, management, and control, and with obesity prevalence and diabetes onset remaining at their all-time highs, the past progress in mortality reduction would soon be undone; starting from its lowest point in 2001, the per capita complication death rate

would rebound to surpass its 1980 level by 2008.

INTERVENTION TESTS

What can be done now and in the future to reduce the number of deaths associated with diabetes complications? Simulation experiments with the system dynamics model may help shed light on this question. Here we consider just 3 of many possible policy intervention scenarios that may be tested and compared with the baseline scenario. (A scenario consists of a particular set of assumptions for the future values of all time series inputs in the model.) In each of these scenarios, a single policy-related input is changed starting in 2006 and ramping up through 2012 or 2017, remaining constant thereafter. The 3 scenarios are as follows:

• *Enhanced clinical management of diabetes.* The fraction of people with diagnosed diabetes whose providers are adequately managing their disease (doing all appropriate monitoring and adjustment of medications) is increased; specifically, this fraction is ramped up from the baseline 48% in 2006 to 67% by 2012. Real-life implementation of this strategy might involve broader adoption of clinical standards of care, better patient tracking systems, more computerized reminder systems, and greater reimbursements or other incentives for the provision of preventive clinical services.
• *Increased management of prediabetes.* The fraction of people with diagnosed prediabetes whose providers are adequately managing their disease is increased; specifically, this fraction is ramped up from the

assumed 20% in 2006 to 50% by 2012. Appropriate management of prediabetes includes monitored regimens of increased physical activity and improved diet, plus medications for control of blood glucose, blood pressure, or lipids as needed.[8–10]
• *Reduced obesity prevalence.* The obese fraction of the adult population is reduced. Specifically, this fraction is ramped down from the assumed 37% in 2006 to 26% in 2017. This reduction returns obesity prevalence to where it was in about 1995. Real-life implementation of this strategy might involve consumer education, insurance reimbursements for calorie-control and physical activity programs, and working with industry and government to bring healthier foods and improved opportunities for physical activity to a broader spectrum of communities.

Resulting output graphs for 2 variables—total diabetes prevalence and per capita deaths from complications—are shown in Figure 3. The variables used in Figure 3 are the same as those seen previously in Figures 2b and 2d but use narrower y-axis ranges so that intervention impacts can be seen clearly. For each of the 3 intervention tests, Figure 3 shows how the intervention alters the behavior of the diabetes system from 2006 to 2050 relative to the baseline scenario.

Enhanced Clinical Management of Diabetes

As a result of this intervention, the controlled fraction of the diagnosed diabetes population increases from 41% in 2006 to 47.5% by 2012. Increased control, in turn, immediately reduces

the flow rates of diabetes progression and complications deaths. These flow-rate reductions, in turn, slow the growth in the number of diabetes-related deaths (Figure 3b). Because no further improvement in clinical management is assumed to occur after 2012, and because nothing has been done to slow the growth of diabetes prevalence (Figure 3a), the rapid growth in complications deaths resumes immediately after 2012. The resumed growth follows a trajectory that parallels that of the baseline scenario, actually slightly exceeding it in terms of percentage growth from 2012 to 2050. For this and any other scenario (namely, scenarios involving improved self-monitoring, medication use, or healthier lifestyles for people with diabetes) in which the proposed intervention has the sole effect in the model of increasing the fraction of diabetes patients who are controlled, the model suggests that as long as the controlled fraction is increasing, deaths from complications will grow more slowly; but after the increase in the controlled fraction ceases, deaths will resume a faster rate of growth in line with the growth in diabetes prevalence itself.

Figure 3a indicates that the intervention improving the clinical management of diabetes ultimately leads to a small but noticeable increase in the prevalence of diabetes. This is a direct reflection of the fact that deaths from complications have been reduced relative to the baseline scenario. Returning to the bathtub analogy, the outflow drain has been made smaller whereas nothing has been done to reduce the inflow. Just as the water in a bathtub with a "backed up" drain rises further than it would otherwise, one may say that the diabetes population

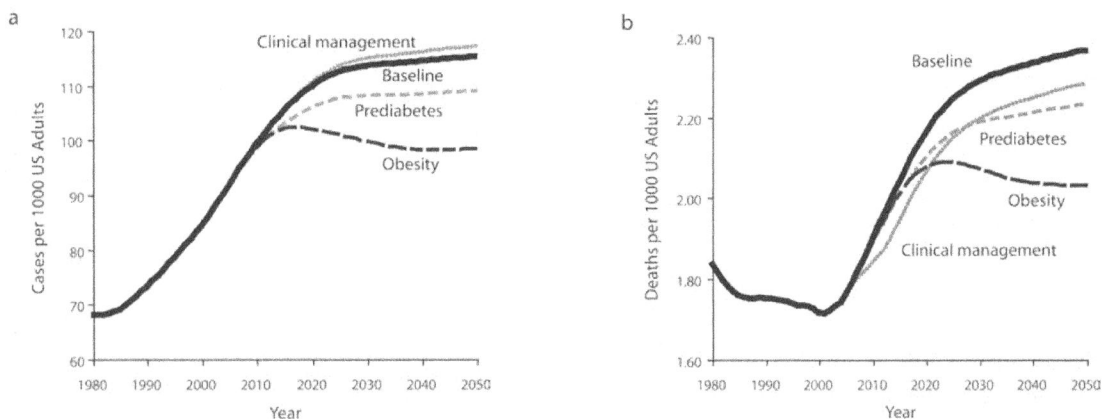

FIGURE 3—Model output for 3 intervention scenarios compared with the baseline scenario for diabetes prevalence (a) and complication-related deaths (b).

becomes backed up when the death rate is reduced.

Increased Management of Prediabetes

As a result of this intervention, many more people with diagnosed prediabetes are effectively managed. Consequently, the per capita rate of diabetes onset decreases (by about 5%), and reduced onset then leads to reduced prevalence. From 2006 to 2050, diabetes prevalence rises by 17% under the intervention, compared with a rise of 23.5% under the baseline scenario. Although the reductions in diabetes onset and prevalence are significant, they may be less than one would expect from such a large assumed increase in effective prediabetes management. This is because the intervention does not do anything to reduce the onset of prediabetes in the first place and thus allows a "backing up" of people in the prediabetes category. For many individuals, prediabetes management does not altogether prevent diabetes onset but, rather, just postpones it.

Although the reduction in diabetes prevalence under the prediabetes management intervention is less than one might have hoped, it is still sufficient to reduce deaths from complications, and is ultimately more effective at doing so than the diabetes management intervention described in the previous section. But it is not until after the year 2028 that per capita deaths under the prediabetes intervention begin to dip below those under the diabetes management scenario. Also, it should be noted that after 2028, although the growth in per capita deaths is less under this intervention than under the baseline or diabetes clinical management intervention scenarios, this growth in deaths does continue right through 2050. Although the growth in diabetes prevalence has been slowed under the prediabetes intervention, it has not been halted (Figure 3a).

Reduced Obesity Prevalence

As a result of this intervention, onset rates for prediabetes and diabetes are reduced. Also, reduced obesity allows more recovery from prediabetes back to normal glycemic levels, and the prevalence of prediabetes thus declines. Because there are fewer people with prediabetes, and fewer of them are obese, diabetes onset declines—by 15% to 19% relative to the baseline scenario. This is enough of a decline in onset to cause diabetes prevalence to peak in 2018 and then decline continuously thereafter. Overall, diabetes prevalence rises only 5.5% from 2006 to 2050, compared with the 23.5% increase in the baseline scenario.

The peak and decline of diabetes prevalence is ultimately translated into a similar peak and decline in per capita deaths from complications. Per capita deaths under the obesity reduction scenario first dip below those under the prediabetes management scenario in 2017 and first dip below those under the diabetes management scenario in 2021. The success of the reduced obesity intervention in halting and reversing the growth of diabetes prevalence and complications deaths stands in contrast to the inability of the prediabetes and diabetes management scenarios to do so. Obesity reduction leads to a lower flow rate of diabetes onset, as in the prediabetes scenario, but also reduces prediabetes prevalence and avoids the backing-up phenomenon seen in the prediabetes scenario. The model indicates that this dual action is the key to the success of the obesity reduction intervention in stemming the growth of diabetes prevalence and deaths.

CONCLUSIONS

The analyses presented in this article indicate the sorts of insights and conclusions that one may draw from simulation experiments using the system dynamics model. In particular, such experiments can improve understanding of 4 characteristic dynamics of the simulated diabetes population: (1) obesity's role in driving the growth of prediabetes and diabetes prevalence; (2) the "backing up" phenomenon—in which reduced outflow from a population stock causes a buildup in that stock—that may undercut the benefits of management and control efforts; (3) the inability of management and control efforts alone to reduce diabetes prevalence in the long term; and (4) the significant delays between primary prevention efforts and downstream improvements in diabetes outcomes. Simulation experiments suggest that these 4 characteristic dynamics in combination may often cause intervention impacts to look different in the short term than they do in the long term. For example, in addition to the experiments we have presented, we have also simulated strategies that represent a mix of increased diabetes management and reduced obesity prevalence. Comparing a mixed strategy to one that focuses entirely on diabetes management, the experiments suggest that the focused diabetes management scenario may quickly reduce diabetes-related complications and deaths but is less effective in the long term than the mixed strategy.

Such model-based insights may help the CDC and other organizations and individuals to identify more effective public health strategies and also to interact more effectively with one another in diabetes planning efforts. The fact that the model is an integrated tool interrelating all key dimensions of the burden of diabetes should be helpful in such endeavors. Although this article has focused on the dynamics of prevalence and deaths, the model also generates measures of morbidity and financial costs and allows one to simulate how they too may be affected in the future by alternative interventions.

The system dynamics model may also help in the setting of goals for diabetes management. Simulation experiments evaluating the national Healthy People 2010 objectives for diabetes (also see: Milstein, Jones, Homer et al., unpublished data, 2006) suggest that the specified goal for diagnosed prevalence reduction may be virtually impossible to achieve and moreover is inconsistent with other stated goals.

System dynamics modeling could also conceivably be used to integrate the effect of other chronic disease programs with diabetes prevention and control. One promising direction being pursued by the CDC is to develop a dynamic model of overweight and obesity capable of projecting plausible alternative futures, allowing an examination of a closer look at the roles of nutrition and physical activity programs. Another useful way to extend the work could be the development of separate models of hypertension and hyperlipidemia as well as explicit representation of them as risk factors (separate from obesity though certainly affected by it) in the diabetes model.

Aside from such extensions, more work remains in the refinement and testing of the existing diabetes model and in identifying alternative future scenarios and intervention strategies suitable for simulation. The model's assumptions, embodied in its equations and parameter estimates, are continually being refined as new information and ideas come to light. We are also working to better specify the uncertainty surrounding parameter values and performing sensitivity analyses to determine the impact of this uncertainty. Even in those cases in which the impact of the uncertainty may be great enough to affect policy conclusions, modeling may contribute by helping to prioritize empirical research agendas.

In summary, system dynamics simulation modeling and experimentation help diabetes policy planners and other stakeholders to better anticipate the multiple effects of interventions in both the short and the long term. ∎

About the Authors

Dara L. Murphy is with the Division of Diabetes Translation, Joyce D.K. Essien is with the Division of Public Health Partnerships, and Bobby Milstein is with the Division of Adult and Community Health, Centers for Disease Control and Prevention, Atlanta, Ga. Joyce D.K. Essien is also with the Center for Public Health Practice at the Rollins School of Public Health, Emory University, Atlanta. Jack B. Homer is with Homer Consulting, Voorhees, NJ. Andrew P. Jones and Donald A. Seville are with the Sustainability Institute, Asheville, NC, and Hartland, VT, respectively.

Requests for reprints should be sent to Dara L. Murphy or Kimbelian Barnes, Division of Diabetes Translation, Centers for Disease Control and Prevention, Mail Stop K-10, 4770 Buford Hwy, Atlanta, GA 30341 (e-mail: dlm1@cdc.gov; kbarnes1@cdc.gov).

This article was accepted April 7, 2005.

Contributors

A.P. Jones managed the various aspects of the modeling project, assisted with modeling, and led the writing of the article and design of the figures. J.B. Homer built the simulation model and collaborated on the writing of the article. D.A. Seville synthesized the analysis that led to model design. B. Milstein improved the scope and structure of the model. J.D.K. Essien framed and originated the study, built institutional support for it, and created the evaluation framework for it. D.L. Murphy identified the use and relevant insights of the model. All authors conceptualized ideas, framed the structure of the model, and edited drafts of the article.

Acknowledgments

This work was funded through the Association of Schools of Public Health Cooperative Agreement and the Center for Disease Control and Prevention's Division of Diabetes Translation and Division of Adult and Community Health, in collaboration with the Center for Public Health Practice at the Rollins School of Public Health at Emory University, Atlanta, Ga (grant S3181).

Note. The views expressed in this article are those of the authors and do not necessarily reflect those of the funding agencies.

References

1. National Diabetes Information Clearinghouse Web site. National Institute of Diabetes and Digestive and Kidney Diseases. National Diabetes Statistics. Available at: http://diabetes.niddk.nih.gov/dm/pubs/statistics/index.htm. Accessed November 30, 2005.

2. US Public Health Service. *Healthy People 2010.* Washington, DC: US Dept. of Health and Human Services; 2000. Available at: http://www.healthypeople.gov/document. Accessed January 26, 2005.

3. Sterman JD. *Business Dynamics: Systems Thinking and Modeling for a Complex World.* Boston, Mass: Irwin/McGraw-Hill; 2000.

4. Harris MI. Classification, diagnostic criteria, and screening for diabetes. In: National Diabetes Data Group, ed. *Diabetes in America.* 2nd ed. Bethesda, Md: National Institutes of Health; 1995: 15–35.

5. Kahn R, Genuth S, Expert Committee on the Diagnosis and Classification of Diabetes Mellitus. Follow-up report on the diagnosis of diabetes mellitus. *Diabetes Care.* 2003;26:3160–3167.

6. Diabetes Control and Complications Trial Research Group. The effect of intensive treatment of diabetes on the development and progression of long term complications in insulin dependent diabetes mellitus. *N Engl J Med.* 1993;329:977–986.

7. UK Prospective Diabetes Study Group. Tight blood pressure control and risk of macrovascular and microvascular complications in type 2 diabetes. *Lancet.* 1998;352:703–713.

8. Diabetes Prevention Program Research Group. Reduction in the incidence of type 2 diabetes with lifestyle intervention or metformin. *N Engl J Med.* 2002;346:393–403.

9. American Diabetes Association and National Institute of Diabetes and Digestive and Kidney Diseases. The prevention or delay of type 2 diabetes. *Diabetes Care.* 2002;25:742–749.

10. Bowman BA, Gregg EW, Williams DE, Engelgau MM, Jack L Jr. Translating the science of primary, secondary, and tertiary prevention to inform the public health response to diabetes. *J Public Health Manage Practice.* 2003; S8-S14.

11. Honeycutt AA, Boyle JP, Broglio KR, et al. A dynamic Markov model for forecasting diabetes prevalence in the United States through 2050. *Health*

Care Management Science. 2003;6: 155–164.

12. *Statistical Abstract of the United States: 2002.* Washington, DC: US Census Bureau; 2003. Available at: http://www.census.gov/prod/www/ statistical-abstract-02.html. Accessed January 26, 2005.

13. US Census Bureau Web site. Historical Health Insurance Tables.

Available at: http://www.census.gov/ hhes/hlthins/historic/hihistt7.html. Accessed January 26, 2005.

14. Centers for Disease Control and Prevention, National Center for Health Statistics Web site. National Health Interview Survey (NHIS). Available at: http://www.cdc.gov/nchs/nhis.htm. Accessed January 26, 2005.

15. Centers for Disease Control and

Prevention, National Center for Health Statistics Web site. National Health and Nutrition Examination Survey (NHANES). Available at: http://www. cdc.gov/nchs/nhanes.htm. Accessed January 26, 2005.

16. Centers for Disease Control and Prevention, Behavioral Risk Factor Surveillance System Web site: Prevalence Data and Trends Data. Available at:

http://www.cdc.gov/brfss. Accessed January 26, 2005.

17. Homer J, Jones A, Seville D. *Diabetes System Model Reference Guide.* Hartland, Vt: Sustainability Institute; 2004. Available at: http://sustainer.org/ pubs/diabetessystemreference.pdf. Accessed May 28, 2005.

12

System Dynamics Modeling for Public Health: Background and Opportunities (2006)

Homer J, Hirsch G. *American Journal of Public Health*, 96(3): 452-458

Related writing

Hirsch G, Homer J. System Dynamics Applications to Health Care in the United States. *Encyclopedia of Complexity and System Science*, ed. R Meyers. Springer-Verlag, Berlin, Germany, 2009. ISBN: 978-0-387-75888-6.

System Dynamics Modeling for Public Health: Background and Opportunities

| Jack B. Homer, PhD, and Gary B. Hirsch, SM

The systems modeling methodology of system dynamics is well suited to address the dynamic complexity that characterizes many public health issues. The system dynamics approach involves the development of computer simulation models that portray processes of accumulation and feedback and that may be tested systematically to find effective policies for overcoming policy resistance.

System dynamics modeling of chronic disease prevention should seek to incorporate all the basic elements of a modern ecological approach, including disease outcomes, health and risk behaviors, environmental factors, and health-related resources and delivery systems. System dynamics shows promise as a means of modeling multiple interacting diseases and risks, the interaction of delivery systems and diseased populations, and matters of national and state policy. (*Am J Public Health.* 2006;96:452–458. doi:10.2105/AJPH.2005.062059)

By applying a remedy to one sore, you will provoke another; and that which removes the one ill symptom produces others, whereas the strengthening one part of the body weakens the rest.

—Sir Thomas More, *Utopia, Part I* (1516)

DESPITE REMARKABLE

successes in some areas, the health enterprise in America still faces difficult challenges in meeting its primary objective of reducing the burden of disease and injury. Examples include the growth of the underinsured population, epidemics of obesity and asthma, the rise of drug-resistant infectious diseases, ineffective management of chronic illness,[1] long-standing racial and ethnic health disparities,[2] and an overall decline in the health-related quality of life.[3] Many of these complex problems have persisted for decades, often proving resistant to attempts to solve them.[4]

It has been argued that many public health interventions fall short of their goals because they are made in piecemeal fashion, rather than comprehensively and from a whole-system perspective.[5] This compartmentalized approach is engrained in the financial structures, intervention designs, and evaluation methods of most health agencies. Conventional analytic methods are generally unable to satisfactorily address situations in which population needs change over time (often in response to the interventions themselves), and in which risk factors, diseases,

and health resources are in a continuous state of interaction and flux.[6]

The term *dynamic complexity* has been used to describe such evolving situations.[7] Dynamically complex problems are often characterized by long delays between causes and effects, and by multiple goals and interests that may in some ways conflict with one another. In such situations, it is difficult to know how, where, and when to intervene, because most interventions will have unintended consequences and will tend to be resisted or undermined by opposing interests or as a result of limited resources or capacities.

THE SYSTEM DYNAMICS APPROACH

We believe that in many cases the challenges of dynamic complexity in public health may be effectively addressed with the systems modeling methodology of system dynamics. The methodology involves development of causal diagrams and policy-oriented computer simulation models that are unique to each problem setting. The approach was developed by computer pioneer Jay W. Forrester in the mid-1950s and first described at length in his book *Industrial Dynamics,*[8] with some additional principles presented in later works.[9–12] The International System Dynamics Society was established in 1983, and within the society a special

interest group on health issues was organized in 2003.

A central tenet of system dynamics is that the complex behaviors of organizational and social systems are the result of ongoing accumulations—of people, material or financial assets, information, or even biological or psychological states—and both balancing and reinforcing feedback mechanisms. The concepts of accumulation and feedback have been discussed in various forms for centuries.[13] System dynamics uniquely offers the practical application of these concepts in the form of computerized models in which alternative policies and scenarios can be tested in a systematic way that answers both "what if" and "why."[14–16]

A system dynamics model consists of an interlocking set of differential and algebraic equations developed from a broad spectrum of relevant measured and experiential data. A completed model may contain scores or hundreds of such equations along with the appropriate numerical inputs. Modeling is an iterative process of scope selection, hypothesis generation, causal diagramming, quantification, reliability testing, and policy analysis.[7] The refinement process continues until the model is able to satisfy requirements concerning its realism, robustness, flexibility, clarity, ability to reproduce historical patterns, and ability to generate useful insights. These numerous requirements help to ensure that a model is reliable and useful not only for studying the past, but also for exploring possible futures.[12,17]

The calibration of a system dynamics model's numerical inputs—its initial values, constants, and functional relations—merits special mention. In system dynamics modeling, variables are not automatically excluded from consideration if recorded measurements on them are lacking. Most things in the world are not measured, including many that experience tells us are important. When subject matter experts agree that a factor may be important, it is included in the model, and then the best effort is made to quantify it, whether through (in approximately this order of preference) the use of recorded measurements, inference from related data, logic, educated guesswork, or adjustments needed to provide a better simulated fit to history.[11,17,18] Uncertainties abound in model calibration, which is one of the reasons that sensitivity testing is critical. Sensitivity testing of a well-built system dynamics model typically reveals that its policy implications are unaffected by changes to most calibration uncertainties.[9,10] But even when some uncertainties are found to affect policy findings, modeling contributes by identifying the few key areas—out of the overwhelming number of possibilities—in which policymakers should focus their limited resources for metrics creation and measurement.

System dynamics modeling has been applied to issues of population health since the 1970s. Topic areas have included the following:

1. Disease epidemiology including work in heart disease,[19–21]

diabetes,[21,22] HIV/AIDS,[23–25] cervical cancer,[26] chlamydia infection,[26] dengue fever,[27] and drug-resistant pneumococcal infections[28];

2. Substance abuse epidemiology covering heroin addiction,[29] cocaine prevalence,[30] and tobacco reduction policy[31,32];

3. Patient flows in emergency and extended care[26,33–35];

4. Health care capacity and delivery in such areas as population-based health maintenance organization planning,[36] dental care,[37,38] and mental health,[38] and as affected by natural disasters or terrorist acts[39];

5. Interactions between health care or public health capacity and disease epidemiology.[40–43]

Most of these modeling efforts have been done with the close involvement of clinicians and policymakers who have a direct stake in the problem being modeled. A good example is a chronic illness study conducted in Whatcom County, Washington, that focused on diabetes and heart failure.[21] Health care providers, payers, and community representatives (supplemented by the health care literature) identified influential variables, articulated policy-related concerns, provided data, and provided experience-based estimates when measured data were not available. The models projected the potential impacts of programs on morbidity, mortality, disability, costs, and the various stakeholders and identified the programmatic investments required. Established system dynamics techniques for group model building[44] can help to harness the insights and involvement of those who deal with public health problems on a day-to-day basis.

It is useful to consider how system dynamics methodology and models compare with those of other simulation methods that have been applied to public health issues, particularly in epidemiological modeling. Other types of models include lumped population contagion models[45,46]; Markov models that distinguish among demographic categories of age, sex, race, and so forth[47–49]; and microsimulations or agent-based models at the level of individuals.[50–52] There is significant overlap among the methods, and one cannot always look at a model's equations and instantly know by what method it was developed. In general, though, one may say that system dynamics models tend to have broader boundaries than other types of models and accordingly tend to admit more variables on the basis of logic or expert opinion and for which solid statistical estimates may not available. System dynamics modelers find that a broad boundary including a variety of realistic causal factors, policy levers, and feedback loops is often what is needed for finding effective solutions to persistent, dynamically complex problems.[7,53]

CHRONIC DISEASE PREVENTION

The value of system dynamics modeling is best explained by way of illustration. We start with a challenging question: Why is it that, despite repeated calls for a greater emphasis on primary prevention of chronic disease (including a prominent recent example[54]), the vast majority of health activities and expenditures in the United States are made not for such prevention but rather for disease management and care?[55] This dominance of

"downstream" over "upstream" health activities appears to have grown ever greater during the era of modern medicine and is now seen as a pressing problem by public health agencies such as the Centers for Disease Control and Prevention (CDC).[56]

To illustrate how system dynamics simulation might shed light on this question, we have built a relatively simple model exploring how a hypothetical chronic disease population may be affected by 2 types of prevention: upstream prevention of disease onset, and downstream prevention of disease complications. The model demonstrates how upstream prevention may become inadvertently "squeezed out" by downstream prevention and suggests that a focusing of resources on life-extending clinical tools may ultimately hurt more than it helps. The model has only a single aggregated population stock, 27 differential and algebraic equations and 12 numerical inputs, and is based on general knowledge rather than on any specific case study or other hard data. If the model were intended for actual policymaking and not only for illustration or exploration, one would certainly expect to see a more detailed depiction of the population and causal factors and policies, and a more data-reliant approach to parameter estimation.

Figure 1 presents the model's essential causal structure and policy inputs. The single stock of people with disease represents the gradually changing net accumulation of 2 flows: an inflow of disease onset and an outflow of deaths. Skilled resources for prevention, consisting perhaps of all primary care providers in the region where the disease population is located, are assumed to be

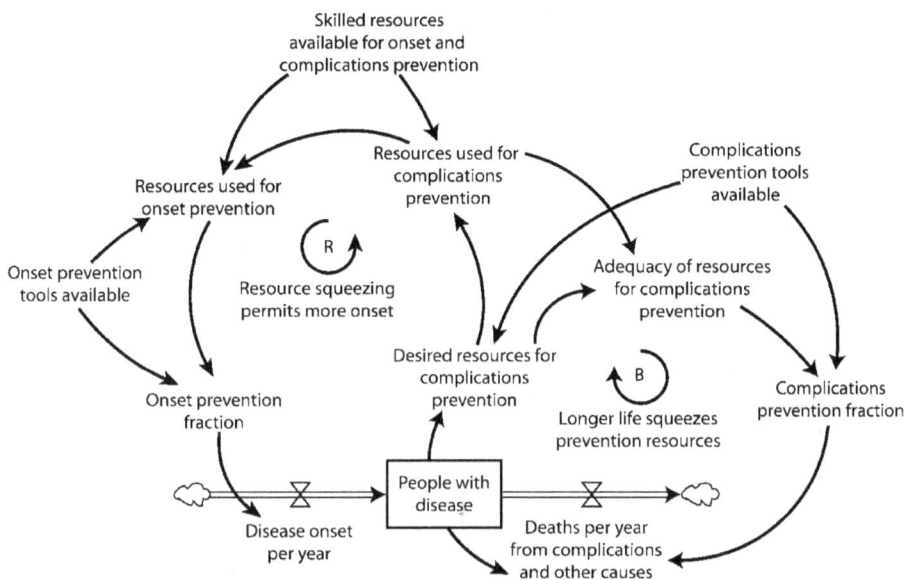

Note. The rectangle represents a stock of people; thick arrows with valves and cloud symbols represent flows of people; thinner arrows indicate causal influence; arrows with minus signs indicate inverse causal influence.

FIGURE 1—A simple model of chronic disease prevention.

on patients. Our starting assumptions are that 25% of complications are preventable, 25% of onset is preventable, and resources are sufficient to achieve both these prevention fractions, with some capacity to spare. Under these assumptions, upstream prevention activities are significant and nearly on a par with downstream prevention activities. This balanced situation may be similar to the state of affairs that prevailed in general medicine several decades ago when the tools of disease diagnosis and management were limited—and very unlike the situation today.

Figure 2 presents simulation output, over a period of 50 years, for 4 key variables (onset prevention fraction, complications prevention fraction, people with disease, and deaths from

fixed in number. Certain clinical tools (diagnostic and therapeutic) are available to these providers for complications prevention, and other tools are available for onset prevention. The greater the number of people with disease, and the greater the number of tools available for complications prevention, the more the time of providers will be devoted to complications prevention. The remainder of provider time is then available for onset prevention efforts among nondiseased patients (to the extent that available onset prevention tools allow) or is absorbed by other, nonprevention activities.

For both types of prevention, assumptions are made about the preventable fractions of cases given existing clinical tools, and also about the resource requirements per case prevented, and the time delay between the availability of new tools and their adoption by providers and impact

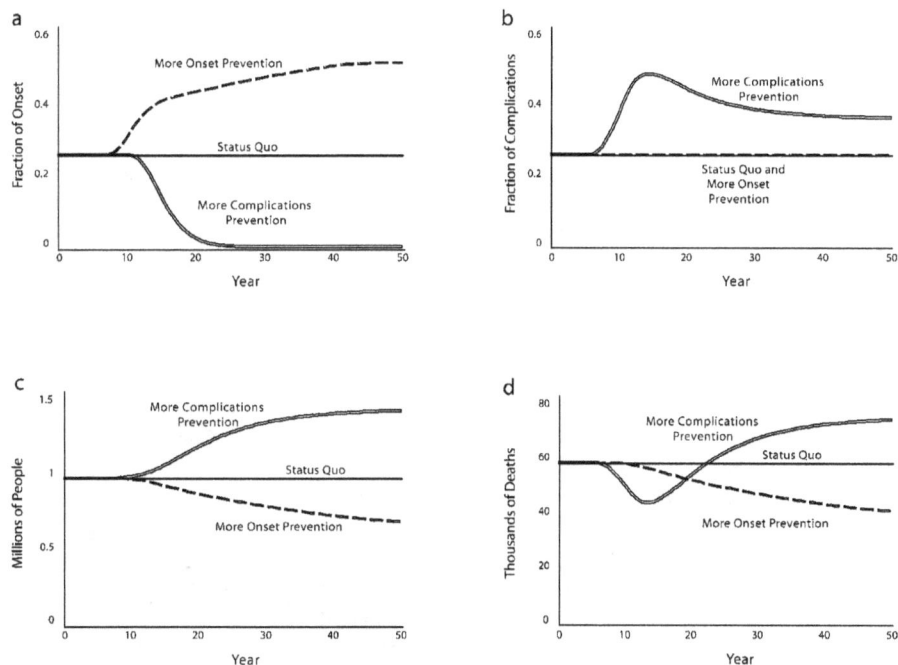

FIGURE 2—Chronic disease prevention model output for 3 scenarios over 50 simulated years, showing onset prevention fraction (a), complications prevention fraction (b), people with disease (c), and deaths from complications (d).

complications) under 3 different policy scenarios we have tested (Status Quo, More Complications Prevention, and More Onset Prevention). In all 3 scenarios, the model has been initialized in a dynamic equilibrium or steady state in which there are about 1 million people with disease, with 75 000 new cases per year and an equal number of deaths, and with 56 000 of the annual deaths from complications. In the Status Quo scenario, no new prevention tools are introduced during the simulation; consequently, the graph lines remain flat, making this scenario a convenient baseline for comparison.

In the More Complications Prevention scenario, new tools for complications prevention become available during years 5 to 10, increasing the preventable fraction of complications from 25% to 50%. The results look good early on, as a rising complications prevention fraction (Figure 2b) leads to a significant reduction in complications deaths (Figure 2d). But the reduction in deaths means a longer average stay in the diseased population stock and thus, an increase in the number of people with disease (Figure 2c). Greater disease prevalence, in turn, increases the need for resources for complications prevention.

This increased demand for limited resources has 2 negative effects. The first is that resources become inadequate to prevent complications in some patients who could have been helped otherwise. Consequently, the complications prevention fraction starts to fall from its peak, and the number of deaths starts to rebound. This effect, reflecting the balancing (B) loop seen in the right-hand portion of Figure 1, is unfortunate but by itself would

cause only a limited rebound in deaths. More problematic is the second effect of resource squeezing, which is a decline in the onset prevention fraction (Figure 2a). The drop in onset prevention allows a further increase in disease prevalence, which causes more resources to be absorbed in complications prevention, leaving even less for onset prevention. This reinforcing (R) loop, seen in the left-hand portion of Figure 1, ultimately drives out onset prevention entirely, leading to large permanent increases in both disease prevalence and complications deaths relative to their starting points.

To summarize this second scenario, although the complications prevention fraction is in fact permanently increased, the prolongation of life and the squeezing out of onset prevention ultimately cause the prevalence of disease to increase proportionately even more; the net result is an increase in deaths from complications. The squeezing out of onset prevention is a vicious cycle and a trap that the health care system may be prone to fall into, given its commitment to the best possible management of existing disease. In a system with limited prevention resources, this well-intentioned commitment may end up doing more harm than good. (The over-dependence on downstream work and squeezing out of upstream work has been observed in many domains outside of health care. This archetypical "fire fighting" dynamic has been the subject of SD models in the area of product development[57] and business process improvement.[58])

A much brighter outcome is seen in the third scenario in Figure 2, More Onset Prevention. In this scenario, new tools for onset

prevention become available during years 5 to 10, increasing the preventable fraction of onset from 25% to 50%. Using the spare resources initially available, some additional onset is prevented (Figure 2a), and the number of people with disease (Figure 2c) declines. As disease prevalence declines, even more prevention resources are freed up to do onset prevention. With disease prevalence decreasing in this scenario, the reinforcing loop in Figure 1 becomes a "virtuous cycle" rather than a vicious cycle, making possible a long-term decline in both disease prevalence and deaths from complications (Figure 2d). A similar beneficial result might be obtained by other means; for example, by changes in funding mechanisms that shift more resources toward onset prevention.

TOWARD MORE COMPLETE MODELS OF POPULATION HEALTH

The preceding example, exploring the interplay of a diseased population and the utilization of health resources, gives some indication of how a broader view of health dynamics can yield insights that would be out of the grasp of a less integrated approach. But system dynamics modeling can and should go further still to incorporate all the basic elements of a modern ecological approach that can help public health agencies achieve their goals of disease prevention, health promotion, and assurance of healthy conditions. Such a broad approach would encompass disease, health and risk behaviors, environmental factors, and resources that provide health and social services or are involved in health-related social transformation.[59,60] Figure 3

presents a system dynamics–type diagram for thinking about the dynamics of population health in these broader terms. This framework has been used at the CDC in discussions about how the agency should move forward in an era of expanded public health goals and greater health challenges.[56]

Only a few system dynamics studies to date have gone beyond diagramming to explore by means of simulation a more complete view of health like that seen in Figure 3. One such study is described by Homer and Milstein.[43] Their community health model examines the typical feedback interactions among broadly defined states of affliction prevalence, adverse living conditions, and the community's capacity to act. Like the chronic disease prevention model presented earlier, the community health model is relatively compact and was not developed on the basis of any specific case. Nonetheless, sensitivity testing of the model across many possible community and affliction characteristics has led to some conclusions about how different types of outside assistance are likely to affect a community in the short and long term. For example, the model suggests that outside assistance focused on building a community's capacity to act may be the most effective place to start in a community struggling against disease and poverty, ensuring longer-term success in a way that more direct interventions fail to do.

Hirsch and Immediato[40,61] describe another more complete view of health. Their Health Care Microworld, depicted in highly simplified form in Figure 4, simulates the health status of and health care delivered to a population. The Microworld was created

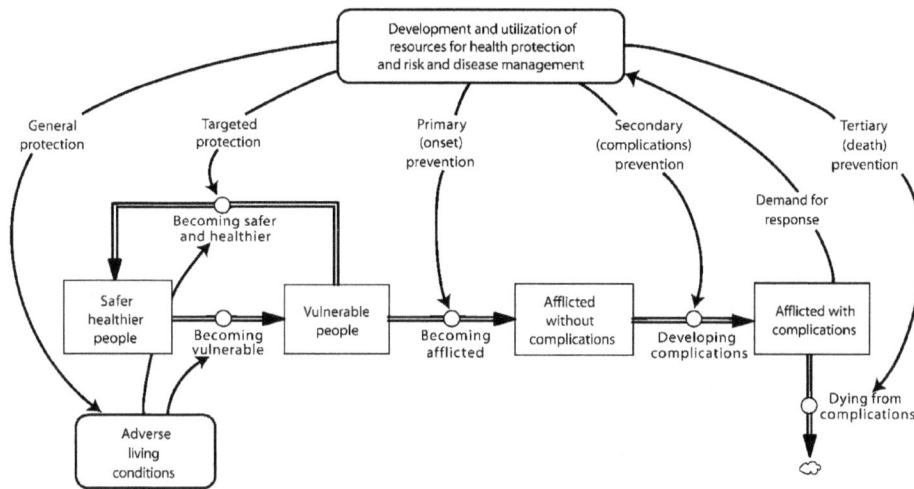

Note. Rectangles represent the stocks of people; thick arrows with circles represent flows of people; thinner arrows indicate causal influence; rounded rectangles indicate multidimensional concepts.

FIGURE 3—A broad view of population health and the spectrum of possible responses.

for a consortium of health care providers who were facing a wide range of changes in the mid-1990s and needed a means for their staff to understand the implications of those changes for how they managed. The underlying system dynamics model is quite large and was designed to reflect with realistic detail a typical American community and its providers, with data taken from public sources as well as proprietary surveys. Users of the Microworld have a wide array of options for expanding the capacity and performance of the community's health care delivery system such as adding personnel and facilities, investing in clinical information systems, and process redesign. They have a similar range of alternatives for improving health status and changing the demand for care, including screening for and enhanced maintenance care of people with chronic illnesses, programs to reduce behavioral risks such as smoking and alcohol abuse, environmental protection,

and longer-term risk reduction strategies such as providing social services, remedial education, and job training.

The Microworld's comprehensive view of health status and health care delivery can provide insights not available from approaches that focus on 1 component of the system at a time. For example, users can play roles of different providers in the community and see how attempts at creating integrated delivery systems tend to fail when participating providers care more about their own bottom lines and prerogatives than about creating a viable system. When examining strategies for improving health status, users can get a better sense of how a focus on enhanced care of people with chronic illnesses provides short-term benefits in terms of reduced deaths, hospital admissions, and costs, but how better long-term results can be obtained by also investing in programs that reduce social and behavioral health risks. When

health care delivery and health improvement are combined, users can appreciate the pitfalls of launching ambitious health improvement programs before first expanding the capacity of the delivery system to provide the medical aspects of those programs.

OPPORTUNITIES AND NEW DIRECTIONS

As long as there are dynamically complex health issues in search of answers, the system dynamics approach will have a place in the analytic armamentarium. It has already made significant contributions to addressing epidemiological issues, as well as issues of health care capacity and delivery and patient flow management. There is still much to be learned about the population dynamics of individual chronic conditions like hypertension and risk factors like obesity. system dynamics models could also address multiple interacting diseases and risks, giving a more

realistic picture of their overall epidemiology and policy implications, particularly where the diseases and risks are mutually reinforcing. For example, it has been found that substance abuse, violence, and AIDS often cluster in the same urban subpopulations, and that such "syndemics" are resistant to narrow policy interventions.[62–64] This idea could also be extended to the case of mental depression, which is often exacerbated by other chronic illnesses and may, in turn, interfere with the proper management of those illnesses. An exploratory simulation model has indicated that system dynamics can usefully address the concept of syndemics.[65]

There is also more to be learned about health-related delivery systems and capacities, with the inclusion of characteristics specific to selected real-world cases. Models combining delivery systems and risk and disease epidemiology could help policymakers and health care providers understand the nature of coordination required to put ambitious public health and risk reduction programs in place without overwhelming delivery capacities. Such models could reach beyond the health care delivery system per se to examine the potential roles of other delivery systems, such as schools and social service agencies, in health risk reduction.

The more complete view of population health dynamics advocated here may also be extended to address persistent challenges that will likely require policy changes at a national and state level, and not only at the level of local communities. Examples include the large underinsured population, high health care costs, and the shortage of nurses. System dynamics modeling can help to identify the feedback loops

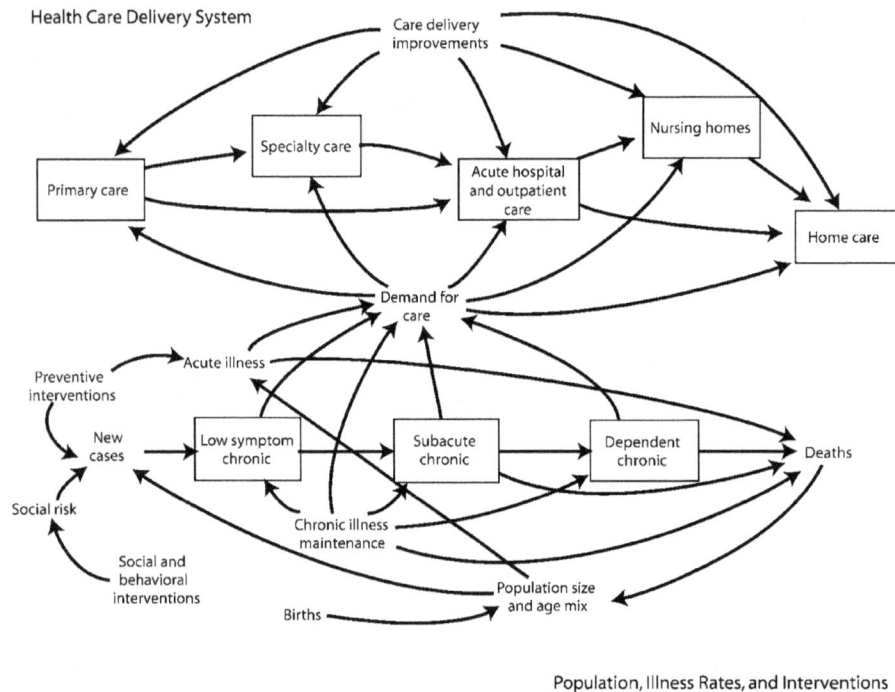

FIGURE 4—Overview of the health care microworld.

Source. Adapted from Hirsch and Immediato.[61]
Note. Episodes of illness in the bottom half of the diagram determine demand for care in the top half of the diagram. Rectangles in the bottom half represent stocks of ill patients. Rectangles in the top half represent stocks of provider capacity and workload. Arrows represent causal influences.

responsible for these problems and point the way to policies that can make a lasting difference. ∎

About the Authors

Jack Homer and Gary Hirsch are independent consultants specializing in the application of system dynamics methodology in both the public and the private spheres.

Requests for reprints should be sent to Jack Homer, 3618 Avalon Court, Voorhees, NJ 08043 (e-mail: jhomer@comcast.net).

This article was accepted March 14, 2005.

Contributors

The authors worked together equally to conceptualize, write, and edit the article.

References

1. Institute of Medicine (Committee on Quality of Health Care in America). *Crossing the Quality Chasm: A New Health System for the 21st Century.* Washington, DC: National Academies Press; 2001.

2. Institute of Medicine (Board on Health Sciences Policy). *Unequal Treatment: Confronting Racial and Ethnic Disparities in Health Care.* Washington, DC: National Academies Press; 2003.

3. Zack MM, Moriarty DG, Stroup DF, Ford ES, Mokdad AH. Worsening trends in adult health-related quality of life and self-rated health—United States, 1993–2001. *Public Health Rep.* 2004; 119:493–505.

4. Lee P, Paxman D. Reinventing public health. *Annu Rev Public Health.* 1997;18:1–35.

5. Heirich M. *Rethinking Health Care: Innovation and Change in America.* Boulder, Colo: Westview Press; 1999.

6. Schorr LB. *Common Purpose: Strengthening Families and Neighborhoods to Rebuild America.* New York, NY: Doubleday/Anchor Books; 1997.

7. Sterman JD. *Business Dynamics: Systems Thinking and Modeling for a Complex World.* Boston, Mass: Irwin/McGraw-Hill; 2000.

8. Forrester JW. *Industrial Dynamics.* Cambridge, Mass: MIT Press; 1961.

9. Forrester JW. *Urban Dynamics.* Cambridge, Mass: MIT Press; 1969.

10. Forrester JW. Counterintuitive behavior of social systems. *Technol Rev.* 1971;73:53–68.

11. Forrester JW. Information sources for modeling the national economy. *J Am Stat Assoc.* 1980;75:555–574.

12. Forrester JW, Senge PM. Tests for building confidence in system dynamics models. In: *System Dynamics, TIMS Studies in the Management Sciences.* New York, NY: North-Holland; 1980:209–228.

13. Richardson GP. *Feedback Thought in Social Science and Systems Theory.* Philadelphia, Pa: University of Pennsylvania Press; 1991.

14. Tank-Nielsen C. Sensitivity analysis in system dynamics. In: Randers J, ed. *Elements of the System Dynamics Method.* Cambridge, Mass: MIT Press; 1980: 185–202.

15. Morecroft JDW. Rationality in the analysis of behavioral simulation models. *Manage Sci.* 1985;31:900–916.

16. Sterman J. System dynamics modeling: tools for learning in a complex world. *Calif Manage Rev.* 2001; 43:8–25.

17. Homer JB. Why we iterate: scientific modeling in theory and practice. *System Dynamics Rev.* 1996;12:1–19.

18. Graham AK. Parameter estimation in system dynamics modeling. In: Randers J, ed. *Elements of the System Dynamics Method.* Cambridge, Mass: MIT Press; 1980:143–161.

19. Luginbuhl W, Forsyth B, Hirsch G, Goodman M. Prevention and rehabilitation as a means of cost-containment: the example of myocardial infarction. *J Public Health Policy.* 1981;2:1103–1115.

20. Hirsch GB, Wils W. Cardiovascular disease in the Dutch population: a model-based approach to scenarios. In: *Ministry of Health: Conference on Health Care Scenarios.* The Hague, Netherlands; 1984.

21. Homer J, Hirsch G, Minniti M, Pierson M. Models for collaboration: how system dynamics helped a community organize cost-effective care for chronic illness. *System Dynamics Rev.* 2004;20:199–222.

22. Homer J, Jones A, Milstein B, Seville D, Essien J, Murphy D. The CDC's diabetes systems modeling project: developing a new tool for chronic disease prevention and control. In: Kennedy M, Winch GW, Langer RS, Rowe JI, Yanni JM, eds. *22nd International System Dynamics Conference.* Oxford, England: System Dynamics Society; 2004.

23. Roberts C, Dangerfield B. Modelling the epidemiological consequences of HIV infection and AIDS: a contribution from operational research. *J Operational Res Soc.* 1990; 41:273–289.

24. Homer JB, St. Clair CL. A model of HIV transmission through needle sharing. *Interfaces.* 1991;21:26–49.

25. Dangerfield B, Fang Y, Roberts C. Model based scenarios for the epidemiology of HIV/AIDS: the consequences of highly active antiretroviral therapy. *System Dynamics Rev.* 2001;17:119–150.

26. Royston G, Dost A, Townshend J, Turner H. Using system dynamics to help develop and implement policies and programmes in health care in England. *System Dynamics Rev.* 1999;15: 293–313.

27. Ritchie-Dunham JL, Mendéz Galvan JF. Evaluating epidemic intervention policies with systems thinking: a case study of dengue fever in Mexico. *System Dynamics Rev.* 1999;15:119–138.

28. Homer J, Ritchie-Dunham J, Rabbino H, Puente LM, Jorgensen J, Hendricks K. Toward a dynamic theory of antibiotic resistance. *System Dynamics Rev.* 2000;16:287–319.

29. Levin G, Roberts EB, Hirsch GB. *The Persistent Poppy.* Cambridge, Mass: Ballinger; 1975.

30. Homer JB. A system dynamics model of national cocaine prevalence. *System Dynamics Rev.* 1993;9:49–78.

31. Roberts EB, Homer J, Kasabian A, Varrell M. A systems view of the smoking problem: perspective and limitations of the role of science in decision-making. *Int J Biomed Comput.* 1982;13:69–86.

32. Tengs TO, Osgood ND, Chen LL. The cost-effectiveness of intensive national school-based antitobacco education: results from the tobacco policy model. *Prev Med.* 2001;33:558–570.

33. Lane DC, Monefeldt C, Rosenhead JV. Looking in the wrong place for healthcare improvements: a system dynamics study of an accident and emergency department. *J Operational Res Soc.* 2000;51:518–531.

34. Wolstenholme EF. A management flight simulator for community care. In: Cropper S, ed. *Enhancing Decision Making in the NHS.* Milton Keynes, UK: Open University Press; 1996.

35. Wolstenholme EF. A patient flow perspective of UK Health Services: exploring the case for new "intermediate care" initiatives. *System Dynamics Rev.* 1999;15:253–271.

36. Hirsch G, Miller S. Evaluating HMO policies with a computer simulation model. *Med Care.* 1974;12:668–681.

37. Hirsch GB, Killingsworth WR. A new framework for projecting dental manpower requirements. *Inquiry.* 1975;12:126–142.

38. Levin G, Roberts EB, Hirsch GB, Kligler DS, Wilder JF, Roberts N. *The Dynamics of Human Service Delivery.* Cambridge, Mass: Ballinger; 1976.

39. Hirsch GB. Modeling the consequences of major incidents for health care systems. In: Kennedy M, Winch GW, Langer RS, Rowe JI, Yanni JM, eds. *22nd International System Dynamics Conference.* Oxford, England: System Dynamics Society; 2004.

40. Hirsch GB, Immediato CS. Microworlds and generic structures as resources for integrating care and improving health. *System Dynamics Rev.* 1999; 15:315–330.

41. Hirsch G, Homer J. Modeling the dynamics of health care services for improved chronic illness management. In: Kennedy M, Winch GW, Langer RS,

Rowe JI, Yanni JM, eds. *22nd International System Dynamics Conference.* Oxford, England: System Dynamics Society; 2004.

42. Hirsch G, Homer J. Integrating chronic illness management, improved access to care, and idealized clinical practice design in health care organizations: a systems thinking approach. In: 3rd *International Conference on Systems Thinking in Management.* Philadelphia, Pa: AFEI/University of Pennsylvania; 2004. Abstract available at: http://www.acasa.upenn.edu/icstm04/abstracts/019.rtf. Accessed November 11, 2005.

43. Homer J, Milstein B. Optimal decision making in a dynamic model of community health. In: *Proceedings of the 37th Annual Hawaii International Conference on System Sciences.* Waikoloa, Hawaii: IEEE; 2004. Available at: http://csdl.computer.org/comp/proceedings/hicss/2004/2056/03/2056toc.htm. Accessed November 4, 2005.

44. Vennix JAM. *Group Model-building: Facilitating Team Learning Using System Dynamics.* Chichester, UK: Wiley; 1996.

45. Anderson R. Populations, infectious disease and immunity: a very nonlinear world. *Philos Trans R Soc London B Biol Sci.* 1994;B346:457–505.

46. Kaplan EH, Craft DL, Wein LM. Emergency response to a smallpox attack: the case for mass vaccination. *Proc Natl Acad Sci.* 2002;99:10935–10940.

47. Gunning-Schepers LJ. The health benefits of prevention: a simulation approach. *Health Policy Special Issue.* 1989;12:1–255.

48. Naidoo B, Thorogood M, McPherson K, Gunning-Schepers LJ. Modelling the effects of increased physical activity on coronary heart disease in England and Wales. *J Epidemiol Commun Health.* 1997;51:144–150.

49. Honeycutt AA, Boyle JP, Broglio KR, et al. A dynamic Markov model for forecasting diabetes prevalence in the United States through 2050. *Health Care Manage Sci.* 2003;6:155–164.

50. Wolfson MC. POHEM: a framework for understanding and modelling the health of human populations. *World Health Stat Q.* 1994;47:157–176.

51. Halloran EM, Longini IM, Nizam A, Yang Y. Containing bioterrorist smallpox. *Science.* 2002;298:1428–1432.

52. Schlessinger L, Eddy DM. Archimedes: a new model for simulating health care systems—the mathematical formulation. *J Biomed Inf.* 2002;35:37–50.

53. Sterman JD. A skeptic's guide to computer models. In: Grant L, ed. *Foresight and National Decisions.* Lanham,

Md: University Press of America; 1988:133–169.

54. Eyre H, Kahn R, Robertson RM, et al. Preventing cancer, cardiovascular disease, and diabetes: a common agenda for the American Cancer Society, the American Diabetes Association, and the American Heart Association. *Diabetes Care.* 2004;27:1812–1824.

55. Brown R, Elixhauser A, Corea J, Luce B, Sheingod S. *National Expenditures for Health Promotion and Disease Prevention Activities in the United States.* Washington, DC: Battelle Medical Technology Assessment and Policy Research Center; 1991. Report BHARC-013/91-019.

56. Centers for Disease Control and Prevention. CDC's futures initiative. Webcast presentation slides and script. Available at: http://www.phppo.cdc.gov/phtn/webcast/futures. Accessed November 11, 2005.

57. Repenning N. Understanding fire fighting in new product development. *J Prod Innovation Manage.* 2001;18:285–300.

58. Repenning N, Sterman J. Nobody ever gets credit for fixing problems that never happened: creating and sustaining process improvement. *Calif Manage Rev.* 2001;43:64–88.

59. Kari NN, Boyte HC, Jennings B. Health as a civic question. American Civic Forum; November 28, 1994. Available at: http://www.cpn.org/topics/health/healthquestion.html. Accessed January 5, 2005.

60. Green LW, Kreuter MW. *Health Promotion Planning: An Educational and Ecological Approach.* 3rd ed. Mountain View, Calif: Mayfield; 1999.

61. Hirsch GB, Immediato CS. Design of simulators to enhance learning: examples from a health care microworld. In: *16th International System Dynamics Conference.* Quebec City, Que: System Dynamics Society; 1998.

62. Wallace R. A synergism of plagues. *Environ Res.* 1988;47:1–33.

63. Singer M. A dose of drugs, a touch of violence, a case of AIDS: conceptualizing the SAVA syndemic. *Free Inquiry.* 1996;24:99–110.

64. Singer M, Clair S. Syndemics and public health: Reconceptualizing disease in bio-social context. *Med Anthropol Q.* 2003;17:423–441.

65. Homer J, Milstein B. Communities with multiple afflictions: a system dynamics approach to the study and prevention of syndemics. In: Davidsen PI, Mollona E, Diker VG, Langer RS, Rowe JI. *20th International System Dynamics Conference.* Palermo, Italy: System Dynamics Society; 2002.

249

13

Chronic Illness in a Complex Health Economy: The Perils and Promises of Downstream and Upstream Reforms (2007)

Homer J, Hirsch G, Milstein B. *System Dynamics Review*, 23(2/3): 313-343

Related writing

Hirsch G, Homer J, McDonnell G, Milstein B. Achieving Health Care Reform in the United States: Toward a Whole-System Understanding. *Proceedings of the 23rd International System Dynamics Conference*, Boston, MA, 2005.

Chronic illness in a complex health economy: the perils and promises of downstream and upstream reforms

Jack Homer,[a]* Gary Hirsch[b] and Bobby Milstein[c]

Jack Homer has been an independent System Dynamics consultant to private and public organizations since 1989. In the public arena, his work focuses on public health and health care policy. He was formerly an assistant professor at the University of Southern California and received a PhD in management from MIT. He also holds BS and MS degrees in applied mathematics and statistics from Stanford University, and received the Jay W. Forrester Award in 1997.

Gary Hirsch is an independent consultant with 35 years of experience in applying System Dynamics to complex problems in health care, education, social services and industry. He holds SB and SM degrees from MIT's Sloan School of Management in System Dynamics. His work in health care has focused on improved management of chronic illness, strategic planning for

Abstract

Chronic illness is the largest cause of death and source of health care costs in developed countries and a growing problem in developing countries. Here we build on past work in system dynamics and present a generic model of chronic illness, its treatment and prevention, applied to the U.S. population. The model explains the rising prevalence of illness and responses to it, including the treatment of complications and management activities designed to reduce complications. We show how progress in treatment and disease management has slowed since 1980 in the U.S., largely due to competition between health care payers and providers, resulting in price inflation and an unstable climate for health care investments. We demonstrate the impact of moving "upstream" by managing known risk factors to prevent illness onset, and moving even further upstream by addressing behaviors and living conditions linked to the initial development of these risk factors. Copyright © 2007 John Wiley & Sons, Ltd.

Syst. Dyn. Rev. **23**, 313–343, (2007)

Introduction

Chronic illness is a major health challenge facing all countries. It is the largest cause of death and source of health care costs in developed countries and has become a significant and growing problem in developing countries as well (Mackenbach, 1994; Olshansky and Ault, 1986; Wild *et al.*, 2004; Mathers and Loncar, 2006). In the U.S., the Centers for Disease Control and Prevention (CDC) estimates that chronic illness is responsible for 70 percent of all deaths and 75 percent of all health care costs (CDC/NCCDPHP, 2007a). The aging of the U.S. population and increases in risk factors such as obesity suggest that chronic illnesses will be an even greater problem in future years. Already, according to the CDC, an estimated 32 percent of adults and 16% of children in the U.S. are obese (CDC/NCCDPHP, 2007b). The good news is that mortality rates from chronic illness have fallen significantly since 1970, dropping by about half for heart disease and stroke, for example (NIH/NHLBI, 2004). Even this good news must be tempered, however, since it means there are many more people living with chronic illness and its associated disabilities and health care costs.

Worldwide, the trends are even more stark. The World Health Organization (WHO) reports that 80 percent of deaths from chronic illness occur in

[a] Homer Consulting, 3618 Avalon Court, Voorhees, NJ 08043, U.S.A. E-mail: jhomer@comcast.net
[b] Wayland, MA 01778, U.S.A. E-mail: gbhirsch@comcast.net
[c] Centers for Disease Control and Prevention, Atlanta, GA, 30341, U.S.A. E-mail: bmilstein@cdc.gov
* Correspondence to: Jack Homer.
Received February 2007; Accepted April 2007

System Dynamics Review Vol. 23, No. 2/3, (Summer/Fall 2007): 313–343
Published online in Wiley InterScience
(www.interscience.wiley.com) DOI: 10.1002/sdr.379
Copyright © 2007 John Wiley & Sons, Ltd.

251

health care delivery systems, and development of interactive learning environments for helping managers and providers better understand those systems.

Bobby Milstein, PhD, MPH, works at the Centers for Disease Control and Prevention where he leads the Syndemics Prevention Network and also coordinates planning and evaluation activities for emerging investigations and policy initiatives in chronic disease, environmental health, emergency preparedness, and health system improvement. With an academic background that combines cultural anthropology, behavioral science, and systems science, he guided the development of CDC's framework for program evaluation and often consults on the role of democratic processes in protecting the public's health.

lower- and middle-income countries (WHO, 2005). In many developing countries, deaths from chronic illnesses now outstrip mortality from traditional health concerns such as injuries and communicable diseases (Yach *et al.*, 2004); and rates of chronic illness are rising in the developing countries, creating an additional burden of disease on top of high rates of acute illness. The worldwide prevalence of diabetes, for example, is projected to rise from 171 million (2.8 percent) in 2000 to 366 million (6.5 percent) in 2030, with over 80 percent of the projected cases in 2030 occurring in the developing world (Wild *et al.*, 2004). Future economic development is expected to bring with it increased risk of morbidity and mortality tied to chronic illness and driven by growth in obesity, tobacco use, and other risk factors.

In most nations, health care systems are organized in a way that makes them hard-pressed to respond to chronic illness. The shortcomings of health systems in dealing with chronic illness include a failure to empower patients and involve them in their own care, a lack of linkages between the health care system and other community agencies that should be involved, misaligned incentives for providers, and a failure to invest in prevention (WHO, 2002). In the U.S., the Institute of Medicine has detailed changes needed in the health care system to cope effectively with the increasing burden of chronic illness, including consistent provision of evidence-based care, reorganization of clinical office practices to provide for longer visits needed for patient education and follow-up, attention to the needs of patients when seeking lifestyle and other behavioral change, and implementation of supportive information systems (IOM, 2001).

This paper begins with a review of past work in system dynamics (SD) concerning populations with chronic illness. It then presents a generic model of illness in a population and its treatment and prevention, applied to the U.S. population. This model encompasses not only chronic illness but all illness and injury, primarily because of data limitations discussed below. The distinction between chronic and acute conditions is a somewhat arbitrary one in any event. Some chronic illnesses can nowadays be cured quickly once they are discovered and, conversely, some acute infections or injuries, if not treated quickly, can become chronic problems. Combining all afflictions into a single model requires only that the rates of death and cure reflect the entire continuous distribution of illness, from the very short-lived to the very long-lived, and from the easily cured to the incurable. Although our model covers all manner of illness and injury, our focus is on those chronic illnesses which are long-lived and incurable, which are responsible for the great majority of health impairment in the U.S.

Our model explains the rising prevalence of illness as well as responses to it, responses which include the treatment of complications as well as disease management activities designed to slow the progression of illness and reduce the occurrence of future complications. The model shows how progress in complications treatment and disease management has slowed since 1980 in the U.S., largely due to a behavioral tug-of-war between health care payers and

Published online in Wiley InterScience
(www.interscience.wiley.com) DOI: 10.1002/sdr

providers that has resulted in price inflation and an unstable climate for health care investments. The model is also used to demonstrate the impact of moving "upstream" by managing known risk factors to prevent illness onset, and moving even further upstream by addressing adverse behaviors and living conditions linked to the development of these risk factors in the first place.

Applications of system dynamics to chronic illness

A number of applications of SD to chronic illness extending over three decades provide a foundation for the concepts discussed in this paper. Dental care and oral health was a focus of early work. The most expansive of these studies (Hirsch *et al.*, 1975) explored feedback relationships among the supply of personnel and the availability of care, the distinction between preventive and symptomatic care, the oral health status of a population and prevalence of dental disease, and the workload of dental practices. This study also analyzed the impacts of various dental manpower policies on oral health outcome measures including prevalence of decayed, missing, and filled teeth.

Several SD modeling efforts have focused on cardiovascular disease. A model developed for the State of Indiana Health Planning Agency (Hirsch and Myers, 1975) projected the prevalence of heart disease and stroke in the state and evaluated the potential impact of different programs for reducing the costs and mortality due to these diseases. The model represented multiple stages through which people move as they develop cardiovascular disease from predisposing conditions such as hypertension to undetected and nonacute illness, acute incidents such as heart attacks and strokes, and rehabilitation and recovery after such attacks. Simulations with the model illustrated the value of comprehensive programs that combine preventive interventions such as hypertension screening and treatment with improved acute care.

Another model of cardiovascular disease (Luginbuhl *et al.*, 1981) used a similar structure to examine the impact of investing more resources in prevention and rehabilitation rather than more elaborate technologies for treating acute myocardial infarction. The model demonstrated how prevention and rehabilitation could lower the costs of heart disease in the U.S. more effectively than new technologies that only marginally extend the lives of people who are in the later stages of the disease.

Diabetes is another area in which SD modeling has been used to study chronic illness in populations. A model developed for a community coalition in Whatcom County in the state of Washington (Homer *et al.*, 2004) portrayed patients flowing through several stages as they moved from being at-risk for diabetes into diabetes and its complications and moved from having their blood sugar levels not under control to under control. The model demonstrated how the right combination of interventions for prevention and treatment could reduce the burden of diabetes in terms of both mortality and cost. A similar population

Published online in Wiley InterScience
(www.interscience.wiley.com) DOI: 10.1002/sdr

flow model of congestive heart failure—which, like diabetes, is another chronic illness producing high burden in the U.S.—was developed for Whatcom County and used for a similar analysis of interventions.

Another SD model of population flows in diabetes (Jones *et al.*, 2006) was conducted for the CDC and developed with experts at the federal, state, and local levels in the U.S. This model is similar to the Whatcom County diabetes model in many ways, but enables a closer look at primary prevention by delineating the condition of moderately elevated blood sugar known as pre-diabetes and by portraying the significant influence of obesity (the leading modifiable risk factor for diabetes) on the onset rates for pre-diabetes and diabetes.

Other prominent SD models exploring the epidemiology of particular chronic conditions have addressed obesity (Homer *et al.*, 2006) and smoking (Tengs *et al.*, 2001).

Some SD modeling has considered chronic illnesses more generally, rather than focusing on a specific disease. A Health Care Microworld developed by the New England Health Care Assembly and Innovation Associates (Hirsch and Immediato, 1999) portrays a population at different ages as they develop and move through increasingly severe stages of chronic illness. Users of the Microworld can employ a variety of medical and non-medical interventions to influence these population flows, including interventions that can mitigate social, behavioral, and environmental risk factors for chronic illness.

The common feature of many of these earlier efforts is the focus on a population developing a specific illness and then moving through one or more stages of increasing severity and complications. Movement between these stages occurs at rates that depend on behavioral and environmental factors as well as demographic characteristics. The models allow for multiple points of intervention, both downstream after the disease process has ensued and upstream at points when disease incidence can still be prevented or the well-being of people better protected. A common lesson is the value of balanced strategies that include preventive programs as well as care and treatment to produce the most net benefit in both the short term and the long term. In the next section, we present a model that, aggregating across all illnesses, demonstrates the potential impacts of attempting to improve downstream care or upstream prevention and describes the economic mechanisms for such interventions.

A national-level model of downstream care and upstream prevention

Model scope and historical evidence

The shortcomings of the U.S. and other health care systems in dealing more effectively with chronic illness are systemic and not confined to particular

Published online in Wiley InterScience
(www.interscience.wiley.com) DOI: 10.1002/sdr

localities or particular illnesses (IOM, 2001). They arise from the interactions of multiple stakeholders, including patients, providers, employers, third-party payers, makers of products, and regulatory and monitoring bodies as well as groups of ordinary citizens. Some of these actors are very influential, and their decisions can affect the health of an entire nation.

Accordingly, we have chosen to develop a model at the national level that aggregates across all illnesses to explore questions related to the evolution of downstream care and the potential benefits and costs of greater upstream effort. This model is based on data specific to the U.S., but its structure should be applicable to other countries. The past applications of SD to chronic illness have served as a useful background for this work, as have broader SD and non-SD studies that have considered the dynamics and economics of population health without a particular focus on chronic illness. The SD studies of this sort that have contributed to our thinking include (1) a simulation model of U.S. health care spending and finance (Ratanawijitrasin, 1993), (2) a simulation model of community health and the "syndemic" confluence of multiple interacting afflictions (Homer and Milstein, 2002, 2004), (3) a conceptual framework for thinking about U.S. health care reform (Hirsch et al., 2005), and (4) a conceptual framework for thinking about upstream–downstream dynamics (Figure 3 in Homer and Hirsch, 2006). The influential non-SD studies of health include books by Starr (1982) and Heirich (1999) and articles by Weisbrod (1991) and Cutler et al. (2006).

The SD works considering the health system broadly have contributed useful ideas and hypotheses. In our present work, we have looked more closely at historical data and sought to develop a model capable of reproducing key elements of that history so that we may better understand its underlying causes. While such empirical grounding does not guarantee that the model is adequate and useful for exploring the future, it is an important step toward that end (Homer, 1996).

As we have gained familiarity with the historical data, we have come to focus our modeling effort on a perplexing question: Why, with the tremendous growth in health spending since 1960, is the health of Americans not better than it is? More specifically, why has the U.S. health care system, for all its size and capability, not managed to subdue chronic illness more effectively?

A key source of historical data has been the National Health Expenditure Accounts (NHEA) (CMS, 2007), which measure annual health spending in the U.S. by category. From 1960 to 2004, total health spending grew (in per capita, constant dollar terms) by a factor of eight, and as a fraction of gross domestic product (GDP) tripled from 5.2 percent to 16.0 percent. Note that we are no longer speaking of chronic illness alone: the NHEA data cover all health spending and do not distinguish between expenditures for chronic illness and those for acute illness and injury. Although estimates of national spending exist for some individual chronic illnesses such as diabetes (ADA, 2003), these are generally only on a one-time snapshot basis, and no

Published online in Wiley InterScience
(www.interscience.wiley.com) DOI: 10.1002/sdr

comprehensive running audit of overall chronic illness spending is performed. Given this situation, and not wishing to abandon our desire to be empirically grounded, we have decided to expand the purview of the model to include all illness and injury, and not only chronic illness.

Total health spending grew rapidly from 1960 to 1990, slowed during the 1990s, then resumed more rapid growth in 2000. A consistent 82–85 percent of total health spending has been for what is known as personal health care (or what one might call health care consumption), which comprises hospital care (30 percent of health spending in 2004), nonhospital services (37 percent), drugs and health-related products and equipment purchased for individual use (13 percent), and miscellaneous personal health care (3 percent). Components of health spending in the NHEA other than personal health care include administration, public health, research, and capital investments. Rising costs for outpatient care have been responsible for much of the growth in health spending in the U.S. since 1980.

The recent historical record suggests the health of Americans has not improved as much as one might have expected from the dramatic growth in health care spending. We define illness or disease as a moderately or severely symptomatic biological or psychological condition—i.e., one associated with some reduction in perceived health-related quality of life. (A person with an asymptomatic or only mildly symptomatic condition is considered to be at risk for disease. Although not yet considered to have full-fledged disease, that person may be eligible for management or treatment of the risk condition.) Two of the CDC's large national annual health surveys—the National Health Interview Survey (NHIS) (CDC/NHIS, 2007) and the Behavioral Risk Factor Surveillance System (BRFSS) (CDC/BRFSS, 2007)—report the fractions of individuals describing their own health status as excellent, very good, good, fair, or poor. Research shows that these self-reported health metrics have desirable statistical properties and are predictive of adverse health events (Dominick *et al.*, 2002). The NHIS also publishes the self-reported prevalence of common chronic conditions. We have examined the reported results of other national surveys as well (Thorpe *et al.*, 2005; Hoffman *et al.*, 1996).

After considering all of these data sources, we have concluded that the NHIS sum of the poor, fair, and good health status categories (that is, people not reporting their health as excellent or very good) is the best indicator of the prevalence of illness as we have defined it, with a continuous span of reporting from 1982 to the present. Throughout this time period, this sum has remained within the relatively narrow range of 31–35 percent of the population, with some movement downward through 1990, upward through 1993, downward through 1998, then upward through 2004. Because the periods of downward movement are not consistent with some of the other measures described above, we are reluctant to emphasize the NHIS fluctuations before 1998. But the upward movement from 31 percent in 1998 to 33 percent in 2004 is clearly consistent with the other NHIS and BRFSS measures. We are thus confident in

Published online in Wiley InterScience
(www.interscience.wiley.com) DOI: 10.1002/sdr

saying two things about the prevalence of illness since 1982: (1) it has not varied by much and certainly has not declined significantly, if at all; and (2) it has increased somewhat since the late 1990s.

To address the question of why the U.S. has not been more successful in preventing and controlling chronic illness, we have constructed a simulation model that, although still a preliminary theory, can faithfully reproduce observed patterns of change in disease prevalence and mortality, and that can also reproduce the histories of the model's primary explanatory variables. The full model contains about 200 equations, including nine stocks and delay functions, 50 constants, and 11 exogenous time series. (The Vensim model is available in the online supplement at http://www.interscience.wiley.com/jpages/0883-7066/suppmat/sdr.379.html, or upon request from the authors.) Some of the exogenous time series ensure a closer model fit to history, while others represent potential policy levers. The exogenous time series do not affect the general findings discussed below; these findings are entirely determined by the model's feedback structure.

Conceptually, the model's hypothesized causal structure can be considered in three parts: (1) a population stock and flow structure; (2) feedback structure that explains the past and especially the growth of downstream care and spending; and (3) additional structure that can help explore the benefits and costs of upstream efforts to improve health.

Population stocks and flows

Figure 1 depicts all members of the population as being in one of three stocks: not at risk, at risk, or with disease. The population is increased by a net inflow rate, corresponding to births plus net immigration, and assumed for the sake of simplicity to flow entirely into the stock of population not at risk. The population is decreased by deaths, which are of two types: (1) deaths following

Fig. 1. Population stocks and flows as modeled

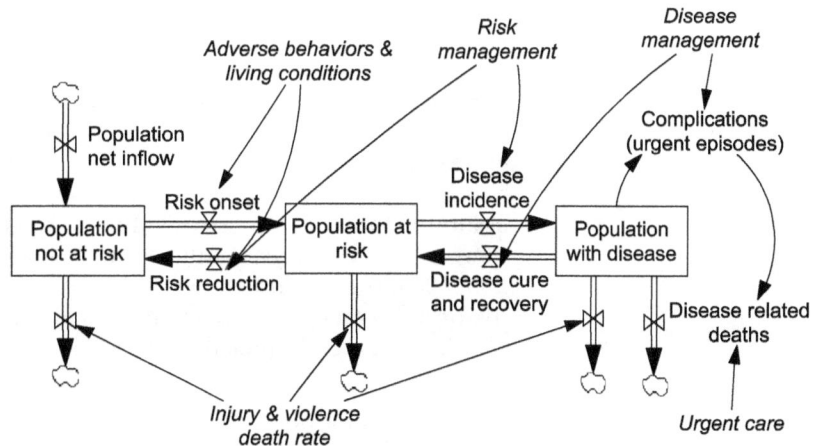

Published online in Wiley InterScience
(www.interscience.wiley.com) DOI: 10.1002/sdr

257

disease, affecting only the stock of population with disease; and (2) deaths from injury or violence, assumed to affect equally all three population stocks in the model. In 2003, the 1-year probability of death from injury or violence was 1 in 1743, and deaths from these causes accounted for only about 7 percent of all deaths (NSC, 2006; CDC/NVSR, 2006).

Disease prevalence is the fraction of the population with disease, while risk prevalence is the fraction of the population at risk. Disease is defined above. Risk refers to physical or psychological conditions or individual behaviors that may lead to disease. In particular, we have used as a proxy measure for risk prevalence the fraction of the adult population with one or more of the following cardiovascular risk factors: hypertension, high cholesterol, hyper-glycemia, obesity, and smoking. BRFSS data from self-reports indicate that this measure of risk prevalence grew continuously during the period 1991–1999, rising from 58 percent to 62 percent (Greenlund *et al.*, 2004).

Flows between the stocks, as well as disease-related deaths, may be affected by certain actions and factors that we will discuss in the remainder of this paper. The disease-related death rate is affected by the effectiveness of *urgent care* for disease complications, generally involving hospitalization. The frequency of complications, in turn, is affected by the effectiveness of *disease management*. In some cases, effective disease management may increase the likelihood of disease cure or recovery; this is certainly true for many acute infectious diseases and can also be true for chronic diseases, as in the case of organ transplantation or cancer chemotherapy. Effective *risk management* can reduce the flow of people from risk to disease, and may also in some cases allow people to return to a condition of being no longer at risk. Such management may include changes in nutrition or physical activity, stress management, or the use of medication.

Flows of risk onset and risk reduction are affected by *adverse behaviors and living conditions*. Adverse behaviors may include poor diet, lack of physical activity, or substance abuse. Adverse living conditions can encompass many factors, including crime, lack of access to healthy foods, inadequate regulation of smoking, weak social networks, substandard housing, poverty, or poor educational opportunities. In calibrating the model, we have found that the rise in risk prevalence for 1991–1999 described above can be explained by assuming that the onset of risk due to adverse behaviors and living conditions increased by 30 percent from 1980 to 1995, and by another 5 percent through 2005. The timing and shape of this increase correspond well to the apparent historical pattern of growth in net caloric intake that has driven the rise in obesity in the U.S. since the late 1970s (Homer *et al.*, 2006).

Downstream loops

Figure 2 presents a theory of the growth of downstream care and spending. This growth is affected by changes in disease prevalence, as well as by changes

Published online in Wiley InterScience
(www.interscience.wiley.com) DOI: 10.1002/sdr

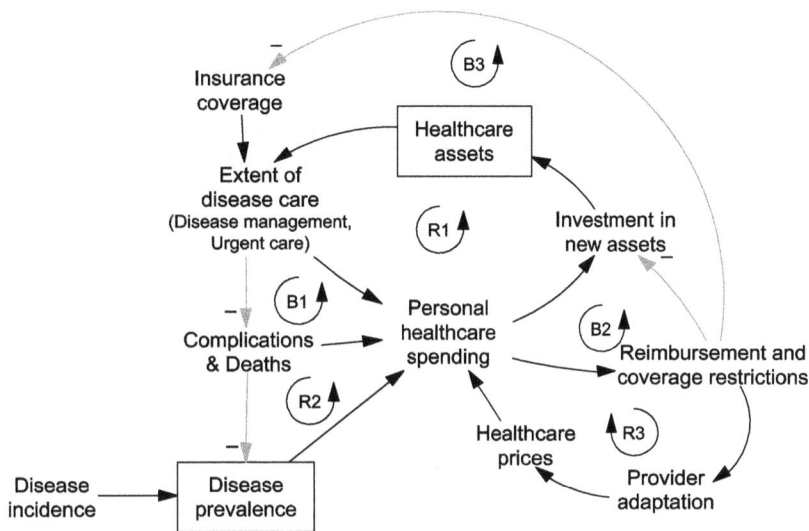

Fig. 2. Theory of downstream health care system expansion and adaptation

Key to downstream feedback loops:

R1: Health care revenues are reinvested for further growth
B1: Disease management reduces need for urgent care
R2: Disease care prolongs life and further increases need for care
B2: Reimbursement restriction limits spending growth
B3: Insurance denial limits spending growth
R3: Providers circumvent reimbursement restrictions

in the extent of care (disease management and urgent care) and in health care prices.

What are the drivers of extent of care? For the purposes of our model, we have reduced a large literature on health care quality (see IOM, 2001), of which extent of care is a part, down to just two factors: the abundance of health care assets, and insurance coverage. By health care assets we mean the structures and fixed equipment used directly for health care or for the production of health care products, as well as the human capital of personnel involved. A greater abundance of assets nationwide means that a larger number of people have access to a broad array of medical services, but beyond a certain point some of that greater abundance represents duplication, and as a result one reaches a point of diminishing returns to extent of care.

By insurance coverage we mean the fraction of the population with some form of health care insurance, either with a private insurer or through a government plan. (Government plans are available in the U.S. for those with lower income, the elderly, the disabled, and for military personnel and war veterans.) The uninsured are less likely than the insured to receive health care

Published online in Wiley InterScience
(www.interscience.wiley.com) DOI: 10.1002/sdr

services. The effect of insurance on extent of care is modeled as being relatively strong in the case of disease management services, for which the vast majority of providers require payment (something most of the uninsured cannot afford); while the effect is weaker in the case of urgent-care services, reflecting the fact that hospitals in the U.S. are required to provide emergency department access even to patients unable to pay.

HEALTH CARE ASSETS   The model includes two separate stocks of health care assets which differ in terms of their uses: a stock used for disease or risk management, and a stock used for urgent (complications) care. These two stocks have likely grown at different rates at different times over the years. Distinguishing these two stocks and their different growth rates in the model has helped us to explain the evolution of health care spending evident in NHEA data, from more rapid growth in urgent care in the 1960s and 1970s, focused on hospital-based life-saving interventions, to more rapid growth in disease and risk management since about 1983, focused on the development and use of diagnostic equipment and pharmaceuticals.

To calibrate the asset sector of the model, we have looked primarily to NHEA data on investments in structures and equipment (S&E). (We have found no dataset on human capital in health care that is complete and can be harmonized with the data on structures and equipment.) In particular, we have estimated (via spreadsheet calculations) the net value of health care assets by accumulating the S&E investments over time, decrementing for obsolescence or depreciation at an assumed rate of 5 percent per year, and initializing in 1960 at a level that permits smooth early growth of the estimate. The resulting estimate grows at an average rate of 4.1 percent per year during 1960–1980, and 3.2 percent per year during 1980–2004. This growth in assets is consistently less than that of personal health care spending (consumption), which grew 5.4 percent per year during 1960–1980 and 4.3 percent per year during 1980–2004. We hypothesize that this difference reflects a decline over time in the fraction of health care revenues reinvested in assets. In fact, we have found it is possible to reproduce the estimated time series for health care assets by assuming that the revenue reinvestment rate declined from 13–14 percent in the 1960s to 10 percent in 1980 and to 6 percent in 2004.

Why should the revenue reinvestment rate have declined in this way? We suggest that the cutback in investment has been the response by potential investors to various forms of cost control, including the restriction of insurance reimbursements, which affect the providers of health care goods and services.[1] With increasing controls and restrictions, these potential investors face greater risk and uncertainty about the future return on their investments, and the result is a greater reluctance to build a new hospital wing, or to purchase an expensive new piece of equipment, or even, at an individual level, to devote a decade or more of one's life to the hardship of medical education and training.

Published online in Wiley InterScience
(www.interscience.wiley.com) DOI: 10.1002/sdr

Taking one step back, why is it that cost controls started to take hold in the 1970s and not earlier? Several authors (e.g., Starr, 1982; Eckholm, 1993; Heirich, 1999) have described how economic power, starting in the 1970s, shifted from providers of medical care, who had been allowed to act freely for many decades, to employers and public agencies desiring to rein in costs. As Paul Starr (1982) puts it, "Until the 1970s . . . practitioners, hospitals, researchers, and medical schools enjoyed a broad grant of authority to run their own affairs. In the 1970s the mandate ran out." Max Heirich (1999) describes this shift as a reaction to the growth in health care costs relative to the rest of the economy beginning in the 1960s: "Where for decades [the costs of American health care] had consumed between 3.5 and 4.5 percent of GNP . . . by 1960 its share of the GNP had increased to 5.3 percent . . . and health care's share of GNP increased to 7.3 percent in 1970 . . . The American health-care system's non-equilibrium growth in costs now affected the rest of the economy."

HEALTH INSURANCE COVERAGE While some employers have reacted to high health care costs by selecting less generous, more cost-restrictive insurance plans for their employees, others have taken the more drastic action of not providing coverage at all to many of their workers. Surveys of insurance coverage taken annually since 1987 (U.S. Census Bureau, 2007) show that the fraction of the U.S. population covered to some degree by private (employer-provided or self-purchased) insurance fell from 75.5 percent in the late 1980s to 70 percent through most of the 1990s, rose briefly to 72 percent during 1999–2000, then declined again to 67.7 percent by 2005. This decline in private coverage is a serious matter affecting the ability of tens of millions of Americans to gain access to regular, good-quality health care. However, there is another dimension to the insurance story, and that is the growth of government-provided insurance. This growth started with the passage in 1965 of the federal Medicare and Medicaid programs to provide coverage for elderly and lower-income people, respectively. The Medicaid program in particular has grown over the years in terms of the fraction of the population it covers, from about 8.4 percent in 1987 to 12 percent in the early 1990s, declining to 10 percent in the late 1990s, and then rebounding to 13.0 percent by 2005. Thus, the Medicaid curve has for nearly 20 years moved consistently in a direction opposite to the curve for private insurance: a decline of 7.8 percentage points in private coverage has been countered by an increase of 4.6 percentage points in Medicaid coverage. As a result, the fraction of the population with any insurance coverage, private or public, has fallen by only 3 percentage points, from 87.1 percent in 1987 to 84.1 percent in 2005. Clearly, many of the people who have lost coverage from their employer or as a result of changing jobs, primarily wage-earners in lower paid positions, have been able to switch over to Medicaid as a fallback.[2]

Let us review the story of the health care system's evolution told thus far by walking through the hypothesized feedback loops in Figure 2:

- Loop R1 shows how the funds generated by health care lead to more investment in assets, and how the application of these new assets in the form of more extensive care can generate even more funds to support further growth. Even in today's more restrictive climate, this loop remains central to the story of progress in health care.
- Loops B1 and R2 show how more extensive care has effects on health and longevity that can moderate or reinforce Loop R1. Loop B1 indicates that increased disease management can prevent costly complications and thereby *reduce* spending and the need for investment in new assets for urgent care. Loop R2 indicates, however, that insofar as more extensive care prolongs life for people with disease, it tends to increase disease prevalence and thereby *increase* spending and investment in health care assets.
- Loop B2 shows how rising personal health care spending as a fraction of GDP triggers a backlash from employers and other payers, resulting in a more restrictive reimbursement climate that can suppress the rate of investment in new assets and thereby slow the growth in health care costs, although at the same time slowing further growth in the extent of care.
- Loop B3 shows how the denial of insurance coverage by some employers in reaction to high health care costs appears to be another route for slowing the growth in those costs, although, like Loop B2, it also slows growth in the extent of care.[3]

Taken together, one may view these loops as the story of a health care system that favors growth and investment until the resulting costs get to a point where further increases are perceived to be no longer worth the expected incremental improvements in health and productivity. That does not by itself sound like a story of dysfunction but rather one of progress followed by goal-seeking behavior. There is a potential for dysfunction in Loop B3, where a reduction in insurance coverage can drive up the unreimbursed costs of hospitals (resulting in a burden on the general public), and also create a situation of health inequity that separates the uninsured poor from the rest of society. But, although the insurance gap is certainly a matter of concern, that gap has been with us for decades, and its growth by 3 percentage points since 1987 is not by itself alarming. Because of this small magnitude of change, declining insurance coverage is unlikely to contribute much toward answering our question of how it is that health care spending can keep growing without doing much to improve health for the majority of the population.

HEALTH CARE PRICES   To find a more compelling causal mechanism behind this sort of system failure, we must go one step further and consider the dynamics of health care prices. Medical care is one of eight major groups in the Consumer Price Index (CPI) computed by the U.S. Bureau of Labor Statistics (BLS), measuring retail price changes over time "for a constant quality, constant quantity market basket of goods and services" (BLS, 2007). The medical care

Published online in Wiley InterScience
(www.interscience.wiley.com) DOI: 10.1002/sdr

CPI combines four major components, with approximate importance weights for 2005 as follows: professional services (2.8), hospital services (1.6), drugs and other personal use products (1.5), and health insurance (0.4). The medical care CPI has grown more rapidly than the general CPI for the overall economy, especially since 1980. For 1960–1980, inflation in medical care prices averaged 6.2 percent compared with general inflation of 5.3 percent, while for 1980–2004 inflation in medical care prices averaged 6.1 percent versus general inflation of 3.5 percent. Consequently, a fixed market basket of medical care goods and services costing $100 in 1960 had risen to $1391 in 2004, while a market basket for the general economy costing $100 in 1960 had risen to $638 in 2004.

Why has health care inflation exceeded that of the general economy? We have considered various possible explanations for why costs should have gone up so rapidly, particularly since 1980, for a given quality of care. These include increasing costs for drug development; more gadgetry in medical technology; the increased practice of "defensive medicine" by providers to avoid lawsuits alleging malpractice; the increase in medical malpractice insurance premiums; the shift of many procedures from inpatient settings to outpatient settings where prices may be less tightly regulated; and the use by providers of various methods to maintain their incomes in the face of greater restrictions on reimbursement. Although all of these phenomena have contributed to health care inflation, not all have contributed with sufficient magnitude or with the timing necessary to explain the historical pattern. One phenomenon that does appear to have such explanatory power, and which we have centered on for the purposes of this study, is the last one listed above, described in Figure 2 as "provider adaptation", or elsewhere as "the target income hypothesis" (Ratanawijitrasin, 1993, p. 77) or "the behavioral response" (Peter Passell in Eckholm, 1993, p. 285).

A variety of studies since the late 1970s provide strong support for the idea that, in response to cost containment efforts, providers may "increase fees, prescribe more services, prescribe more complex services (or simply bill for them), order more follow-up visits, or do a combination of these" (Ratanawijitrasin, 1993). Specific billing practices that can circumvent cost containment efforts include "upcoding" (billing with procedure codes that receive higher reimbursement rates) and "unbundling" (billing a single procedure in multiple parts to achieve a higher total) (Eckholm, 1993). Many tests and procedures are performed that contribute little or no diagnostic or therapeutic value, thereby inflating the cost per quality of care delivered. Writing in the *New York Times* in April of 1989, the former Secretary of Health, Education, and Welfare, Joseph Califano, Jr., claimed that "Americans would spend about $155 billion in 1989 for tests and treatments that would have little or no impact on the patients involved" (Heirich, 1999, p. 97). If correct, that unnecessary and inflationary expense would have represented 29 percent of all personal health care spending in that year.

Increased pressure on provider incomes comes not only from reduced reimbursements, but also from the administrative burden of dealing with many different insurance plans. With the era of cost containment also came greater competition between private insurers to offer employers acceptable benefits for their employees at the lowest price. One aspect of this competition is the creation of a broad and ever-changing menu of plans with different exclusions and different payment percentages for different health services. With this cacophony of payer fee schedules, the administrative overhead of providers in the U.S. has grown enormously, threatening to reduce provider incomes. (Woolhandler *et al.*, 2003, estimates administrative costs as 31 percent of provider revenue in the U.S. compared with 16 percent in Canada.) Providers have thus felt even more need to maintain their incomes through adaptation, and have consequently driven inflation in health care prices even further.

With the inclusion of provider adaptation in Figure 2 to explain health care inflation, a new loop is created: Loop R3. This loop describes the tug-of-war between payers restricting reimbursement in response to high health care costs, and providers adapting to these restrictions by effectively raising health care prices in an attempt to maintain their incomes. This loop has the effect of reducing the efficiency of health care spending and thus artificially raising the cost of health care to payers. The payers react to the magnified costs by seeking further restrictions on reimbursement, or by further denying insurance coverage. The net result is a reduction in health care assets and insurance coverage (through Loops B2 and B3, respectively), thus dampening growth in the extent of care. As shown below, this unintended chain of events might have been avoided or at least moderated had payers and providers not set Loop R3 in motion.

Baseline simulation and alternative tests of downstream behavior

In Figure 3 we present results from the baseline simulation for several of the model's key variables along with historical data. Results from the model are shown from 1960 through 2010.[4] We recognize that a couple of these data series are conceptually incomplete. In particular, the measure of health care investments does not include human capital, and the measure of the population at-risk fraction is based only on adults and on cardiovascular risk factors. Although more complete measures would likely show the same sorts of trends and have little or no effect on model findings, we would like to construct more complete data series, if possible, in future iterations of our model.

Having established the model's ability to do a good job of reproducing historical trends for a variety of key variables, let us examine how a few of the key feedback loops in Figure 2—in particular, those depicting the reactions of payers and providers—contribute to the overall simulated behavior. Shown in Figure 4 are results from the base run alongside results from alternative simulations for 1960–2010 in which one or more of these feedback loops has been cut. The assumptions and results for the simulations are as follows:

Published online in Wiley InterScience
(www.interscience.wiley.com) DOI: 10.1002/sdr

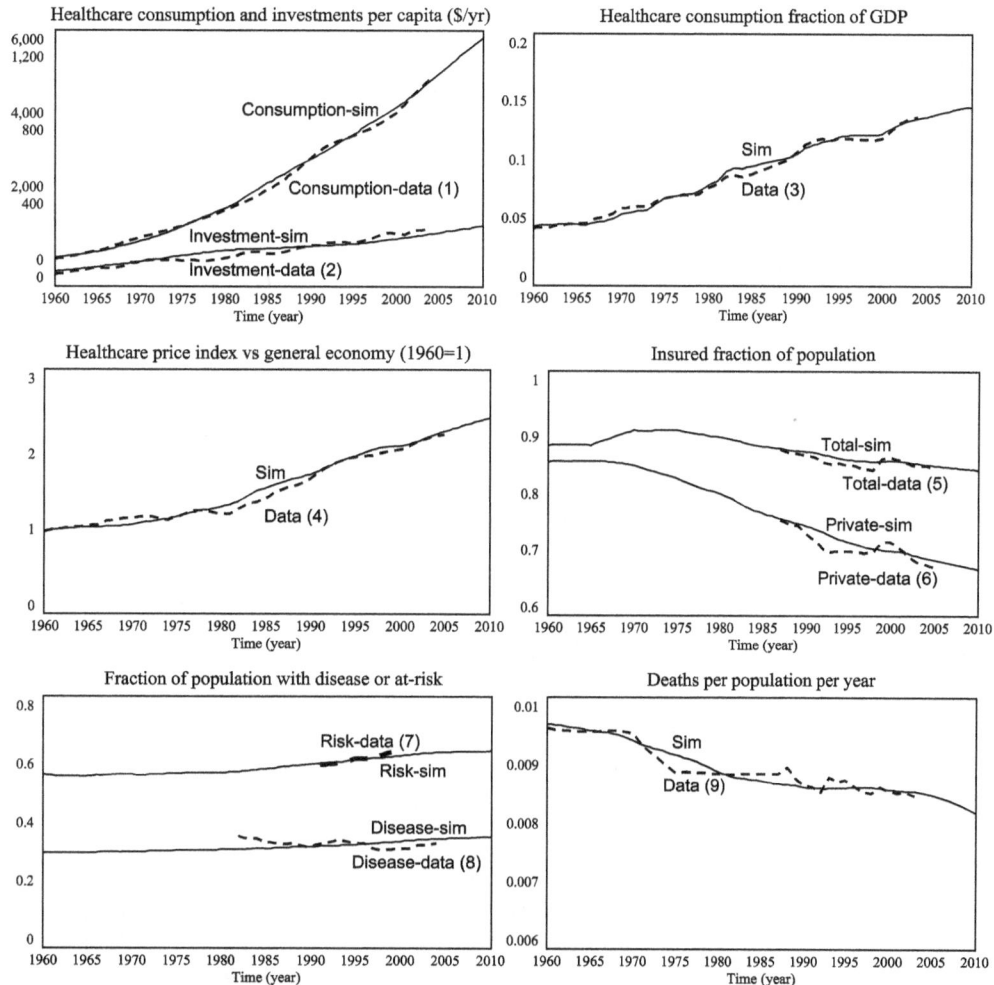

Fig. 3. Baseline simulation and historical data

Data sources are as follows:

(1) NHEA personal health care spending, 1960–2004 annual, divided by population and by GDP deflator (2000 = 1)
(2) NHEA investments in structures and equipment, 1960–2004 annual, divided by population and by GDP deflator
(3) NHEA personal health care spending divided by GDP, 1960–2004 annual
(4) BLS medical care CPI (1960 = 1) divided by general economy CPI (1960 = 1), 1960–2005 annual
(5) Census fraction of population of all ages covered by private or government health insurance, 1987–2005 annual
(6) Census fraction of population of all ages covered by private health insurance, 1987–2005 annual
(7) BRFSS fraction of adults who report having at least one of five specified cardiovascular risk factors, 1991–1999 odd years
(8) NHIS fraction of population of all ages who report their health as good, fair, or poor (i.e., not excellent or very good), 1982–2004 annual
(9) NVSR total deaths per year divided by population, 1960–1980 every 5 years, 1980–2003 annual

Published online in Wiley InterScience
(www.interscience.wiley.com) DOI: 10.1002/sdr

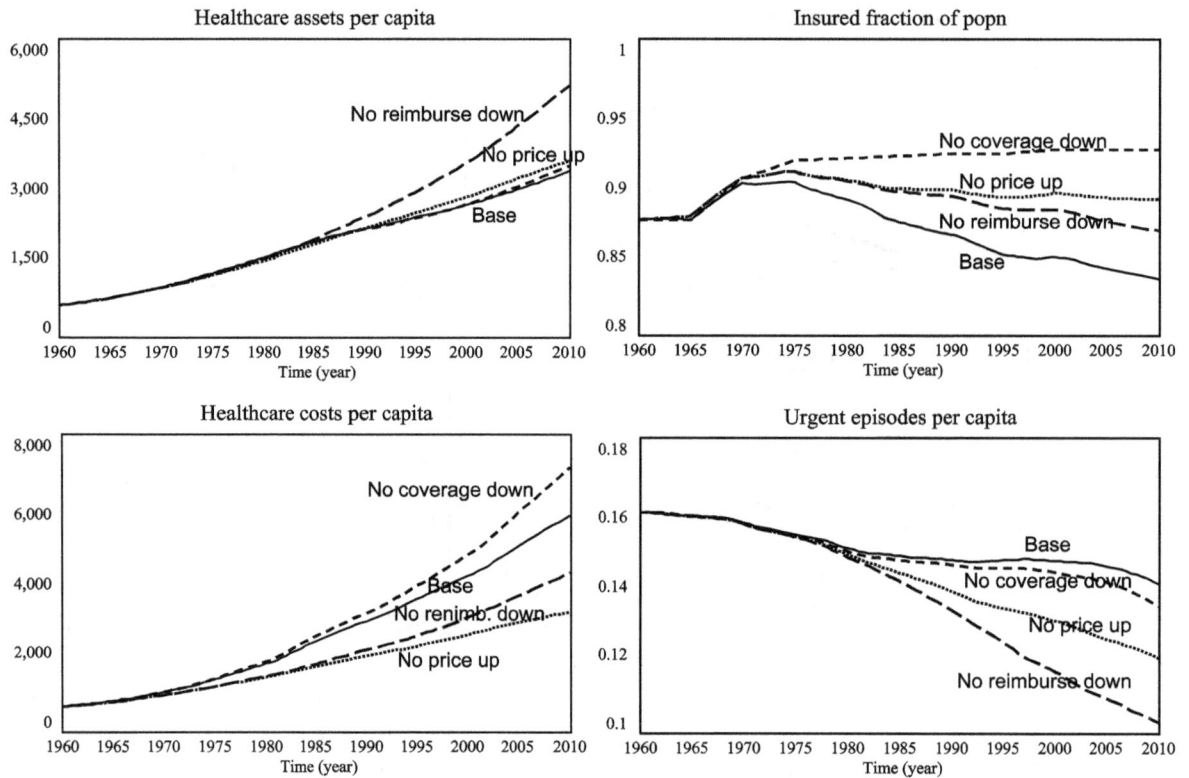

Fig. 4. Simulations exploring how reactions of payers, providers, and investors have determined health care system behavior

No coverage down. In this simulation, employers do not react to high health care costs by denying private insurance coverage; Loop B3 is cut. As a result, the insured fraction of the population does not decline, as it does in the base run; instead it continues to climb gradually to reflect the increasing availability of Medicaid coverage to those with lower incomes. With increased coverage, the extent of care—particularly disease management—is improved, and the rate of urgent episodes is therefore lower than in the base run. But this more extensive care costs more than it saves, and thus health care costs per capita increase relative to the base run. This outcome would seem to suggest that employers who have denied coverage to their employees have thereby saved money. Note, however, that the costs in the model do not include the sick days and losses of productivity that are much more likely to occur when disease is not well managed. This is why some of the nation's employers are taking another tack, providing free or low-cost primary health care in their own offices as a way of improving productivity and catching health problems before they get more serious and require expensive outside care (Freudenheim, 2007).

Published online in Wiley InterScience
(www.interscience.wiley.com) DOI: 10.1002/sdr

No price up. In this simulation, providers do not react to restrictions on reimbursement by raising their fees for a given quality of service; Loop R3 is cut. As a result, health care costs per capita grow much less than in the base run. Lower costs mean fewer revenues available for reinvestment but also less restriction of reimbursements and coverage. Because reimbursements are more stable, the investment rate does not decline as much as in the base run, and so, despite the decline in the revenue base, assets per capita increase no less than in the base run and even a bit more. With lower costs, there is also much less denial of insurance coverage. Because of the greater insurance coverage and the slightly greater assets, the extent of care is improved and urgent episodes per capita are reduced relative to the base run.

This simulation points to the importance of the "dysfunctional" Loop R3, but its results should not be taken too literally or as a prescription for policy. Legislators seeking to stabilize health care costs might be tempted to limit the autonomy of providers when it comes to billing and compensation, requiring that they be paid a fixed amount (as is done in some managed care organizations), perhaps through a single government payer (as is done in many countries). Some providers in the U.S. might welcome the predictability and reduced administrative burden such a simplified payment system would bring. Others, however, are likely to protest such loss of autonomy, especially the many who expect (and whose adaptive behavior to date has been based on the expectation of) high incomes in return for their long years of education and training. A national fixed-price policy might therefore be met by a decline in the supply of providers—an increased rate of retirement and decreased influx of medical students—leading perhaps to a severe shortage. In terms of Figure 2, if the adaptive responses of providers were no longer permitted, we might see a decline in the human capital component of health care assets; that is, a strengthening of Loop B2. Such a reaction could conceivably cause a fixed-price policy to do more harm than good, if the reaction were strong enough.

No reimburse down. In this simulation, employers and other payers do not react to high health care costs by restricting reimbursements; Loops B2 and R3 are cut. The stable reimbursement climate encourages more investment in assets as a fraction of revenues and also defuses the dysfunctional tug-of-war between payers and providers that leads to price inflation in Loop R3. The cutting of Loop R3 does keep health care costs down (as in the *No price up* simulation), but in so doing reduces health care revenues and therefore initially counteracts the effect of an increased investment fraction on asset formation, relative to the base run. By the 1980s, however, the stable investment fraction increasingly differentiates this scenario from the base run, and assets thus start to grow faster than in the base run. The rapid growth in assets per capita drives greater improvements in extent of care so that urgent episodes decline much more than in the base run. Also, the lower health care costs relative to the base run mean that there is less loss of insurance coverage, which improves the extent of care further.

The *No reimburse down* simulation underscores the importance of the dysfunctional payer–provider interaction in Loop R3 and also points to the importance of the impact of payers on investors in Loop B2. But, as above, the results should not necessarily be viewed as having direct policy implications. They seem to suggest—perhaps counterintuitively—that health insurance should be stable and nonrestrictive in its reimbursements, so as to avoid behavioral backlashes that can trigger health care inflation and underinvestment. However, few policymakers in the U.S. would at present be willing to mandate that private payers must provide plans of only a certain sort, as such a mandate would be seen as interference in a matter of private choice. Perhaps, then, the mandate could apply only to the government's own insurance programs. (Government reimbursement practices are often copied by private insurers, and so with such an approach one may end up with the desired effect on the private sector without having to interfere with it.) Even so, many policymakers might fear that such a mandate would open the door not only to beneficial investments, but also to indiscriminate and wasteful ones, such as occurred most prominently before the era of cost containment. Still, it is interesting to consider whether a more generous and stable approach to reimbursement could not only combat illness better than the current restrictive approach, but do it more efficiently and perhaps even at lower cost.

The above analysis suggests that there are no easy downstream fixes to the problem of an under-performing and expensive health care system. It is one thing to understand the dysfunctional tug-of-war between payers and providers, but quite another to defuse it. We have addressed the lagging extent of care in our model by looking at the influences of health care assets and insurance coverage, but we have not explored improvements in the efficacy and safety of that care. Such improvements can include better information and decision-support systems, better payment incentives, and better clinical training. Local implementations of such improvements indicate their promise for reducing the burden of disease and providing more effective care for the health care dollar (IOM, 2001). One wonders, though, just how much we can hope to gain from such downstream measures, when they may appear to payers or providers as an even greater expense to bear (at least initially) and could therefore end up feeding into the system's divisiveness and dysfunction.

Potential upstream loops and tests of their behavior

Let us turn, then, to the upstream prevention of disease incidence, to see what promise it may hold for lessening our dependence on a costly and inefficient

Published online in Wiley InterScience
(www.interscience.wiley.com) DOI: 10.1002/sdr

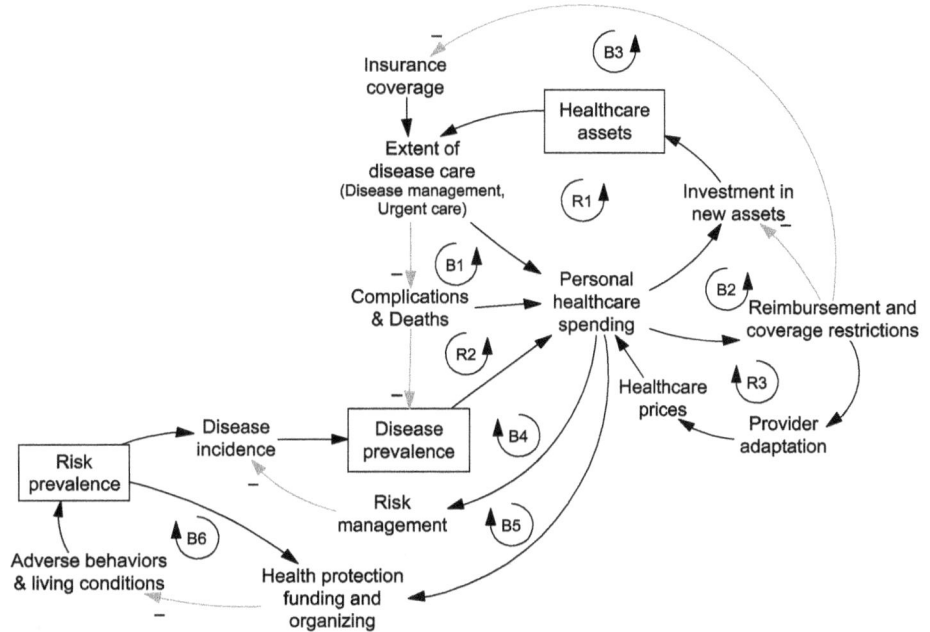

Fig. 5. Extending the theory to allow for upstream responses

Key to upstream feedback loops:
B4: Risk management proportional to downstream spending can help limit it
B5: Health protection proportional to downstream spending can help limit it
B6: Health protection (via sin taxes) proportional to risk prevalence can help limit it

system of downstream care (Fries *et al.*, 1998; McKinlay, 1979; McKinlay and Marceau, 2000). Illustrated in Figure 5 are two broad categories of such efforts: risk management for people already at risk, and health protection for the population at large. The literature identifies significant opportunities for medically oriented risk management for a variety of diseases, through improved nutrition and exercise, smoking cessation, and the appropriate use of drugs (Eyre *et al.*, 2004; Hajjar *et al.*, 2006; Leonhardt, 2007). For example, the fraction of people with hypertension whose condition is considered under control stood at 29 percent for 1999–2002, up only a few percentage points from 25 percent for 1988–1991 (Hajjar *et al.*, 2006).

The literature also describes opportunities for socially oriented health protection, which may include efforts to change adverse behaviors and mitigate unhealthy conditions in homes, schools, workplaces, and neighborhoods and to alter macroeconomic forces and the media so that they are more health promoting (Northridge *et al.*, 2003; Gerberding, 2005; Yach *et al.*, 2005; Simon, 2006; CDC, 1999, 2006; IOM, 2002; Smedley and Syme, 2000; Wilkinson and Marmot, 2003; Evans *et al.*, 1994; Hanna and Coussens, 2001). Note that,

Published online in Wiley InterScience
(www.interscience.wiley.com) DOI: 10.1002/sdr

unlike downstream interventions, health protection efforts rely on the actions of individuals and organizations most of whom are not health care professionals.

What can the data tell us about the history of upstream spending? Much upstream work involves population-based public health efforts emphasizing health promotion and disease prevention. Public health spending has grown as a fraction of total NHEA spending from 1.5 percent in the early 1960s to a fairly constant 3 percent since the early 1990s. Another contributor to upstream spending is risk management. We have data from various reports, both public (e.g., NIHCM, 2002) and proprietary, that have allowed us to assemble a partial time series on spending on drugs for treating hypertension and high cholesterol. These data suggest that the use of these drugs grew from a negligible amount before 1980 to at least 1.5 percent of NHEA spending by 2004. Risk management in total would also include prescribed treatments for weight loss and smoking cessation, for which we have not yet assembled historical data. Thus, we estimate that upstream spending has grown to more than 4.5 percent (=3 percent population-based public health + more than 1.5 percent risk management) of total health spending. This amount is larger than the 3 percent upstream spending that was estimated in a 1991 report (Brown *et al.*, 1991). The data thus show that upstream spending has grown as a fraction of total health spending since 1960, even if it is still a relatively small fraction. This conclusion is significant because it stands counter to an impression we had before this study, that upstream spending had in recent decades been "squeezed out" by downstream spending (Homer and Hirsch, 2006).

Three balancing feedback loops have been included in Figure 5 and in our model to indicate how, in general terms, efforts in risk management and health protection might be funded or resourced more systematically and in proportion to indicators of capability or relative need. Funding is not the only prerequisite for such efforts, which also depend upon the enthusiasm and organization of the people involved (providers and patients in the case of risk management, and the general public in the case of health protection), but it is the leading requirement for most initiatives. Loop B4 suggests that funding for programs promoting risk management could be made proportional to spending on downstream care, so that when downstream care grows, funding for risk management would grow as well. Loop B5 suggests something similar for health protection, supposing that government budgets and philanthropic investments for health protection could be set in proportion to recent health care spending. Loop B6 takes a different approach to the funding of health protection, linking it not to health care spending but to risk prevalence (the stock which health protection most directly seeks to reduce). The linkage to risk prevalence can be made fiscally through "sin taxes" on unhealthy items, such as cigarettes (already taxed throughout the U.S. to varying extents; see Lindblom, 2006) and fatty foods (Marshall, 2000). In theory, the optimal magnitude of such taxes may be rather large in some cases, as the taxes can be used both to discourage unhealthy activities and promote healthier ones (O'Donoghue and Rabin, 2006).

Published online in Wiley InterScience
(www.interscience.wiley.com) DOI: 10.1002/sdr

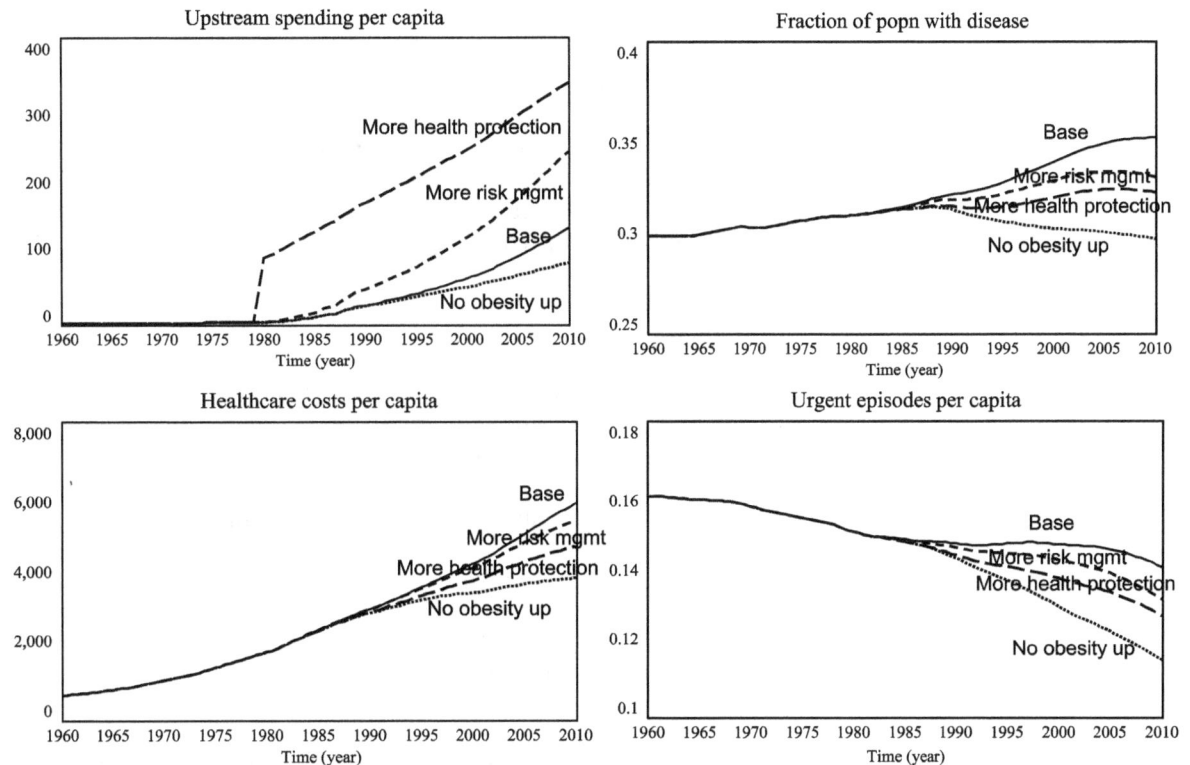

Fig. 6. Simulations exploring how upstream responses could alter health care system behavior

Presented in Figure 6 are results from simulations in which we ask how much the prevalence and burden of disease might have been diminished relative to the base run (through the year 2010) if greater upstream efforts at risk management or health protection had been made starting in 1980. These results may also be compared to a scenario, *No obesity up*, in which we assume that the base run's exogenous increase in the onset of risk by 35 percent from 1980 to 2005—representing a host of socioeconomic factors that have led to greater net caloric intake and obesity—had never occurred.

In these simulations we make assumptions about the degree to which upstream spending can affect rates of the onset of risk, reduction of risk, and onset of disease. In particular, we assume that maximum risk management could reduce the onset of disease by 40 percent and enhance reduction of risk by 40 percent, and that maximum health protection efforts could reduce onset of risk onset by 50 percent and enhance reduction of risk by 50 percent. For risk management, the assumptions, although uncertain, have been informed by studies focusing on the cost-effectiveness of risk management for patients

Published online in Wiley InterScience
(www.interscience.wiley.com) DOI: 10.1002/sdr

with diabetes (CDC/DCEG, 2002; Hayashino *et al.*, 2004). For health protection, our assumptions are more uncertain, because relatively little is known about the required cost and potential impact of measures that could prevent the onset of various risk factors for disease; somewhat more is known in this regard about preventing smoking than about preventing obesity (Tengs *et al.*, 2001; Homer *et al.*, 2006). The Tengs analysis, focusing on a school-based anti-smoking program for young teens, estimated a cost of about $50 per student per year and projected long-term benefits in terms of reduced medical costs and increased quality-adjusted life years.

Because our assumptions about upstream efforts are associated with significant uncertainties, we do not purport here to provide accurate cost-effectiveness estimates, but only to illustrate how such estimates may be generated by our model. It is interesting to ask not only to what degree upstream efforts can improve health but also whether, and over what time frame, increases in upstream spending can be justified in terms of subsequent reductions in downstream spending. The model calculates upstream spending as the sum of risk management and health protection spending, and calculates downstream spending as the sum of all personal health care spending less spending on risk management. These measures of spending are accumulated over time, starting in 1980, as a way of quantifying overall costs and benefits; in the current model, no discount rate is applied to these cumulative measures.

No obesity up. This simulation is presented as a "best case" alternative history to the base run. Relative to the base run, the fraction of the population with disease grows more slowly during the 1980s, and this fraction declines from the 1990s onward rather than continuing to grow as in the base run. The result is much more progress starting in the late 1980s in reducing the number of urgent episodes, as well as a significant slowing in the growth of health care costs. This simulation indicates the extent to which increasing risk prevalence has undermined progress on health and has pushed health care costs upward since the late 1980s.

More risk management. In this simulation, the strength of the assumed linkage between personal health care spending (specifically, the non-urgent portion of that spending) and risk management is doubled relative to the base run; thus, the strength of Loop B4 in Figure 5 is doubled. By 2010, upstream spending per capita is increased by $108 relative to the base run (see Figure 6), and the effectiveness of risk management (in terms of reducing disease incidence and enhancing risk reduction) is increased to 51 percent of its assumed potential, versus 27 percent in the base run. The increase in risk management leads to slower growth of disease prevalence starting in the late 1980s. But, with the onset of risk left unaddressed in this scenario, disease prevalence does grow rather than decline. Urgent episodes and health care costs are somewhat improved relative to the base run, but not dramatically so. By 2010, a cumulative

Published online in Wiley InterScience
(www.interscience.wiley.com) DOI: 10.1002/sdr

additional $359 billion in upstream spending since 1980 has led to a reduction in downstream spending of $1140 billion. The increased spending in risk management is not paid back immediately, however. Not until 1995, 15 years after the policy is initiated, does the cumulative reduction in downstream spending exceed the cumulative increase in upstream spending.

More health protection. In this simulation, health protection is much enhanced through a proportional funding program, starting in 1980, that devotes $5 to health protection for every $100 of personal health care spending; thus, Loop B5 in Figure 5 is activated. The result is an immediate $90 per capita increase in upstream spending in 1980, which grows to a $203 per capita increase by 2010 relative to the base run. By 2010, the effectiveness of health protection (in terms of reducing risk incidence and enhancing risk reduction) is increased from its baseline value of 19 percent to 48 percent of its assumed potential. The increase in health protection goes a long way but does not quite offset the adverse socioeconomic influences (such as changes in food and activity environments) that increase the onset of risk in the base run. As shown in Figure 6, this simulation produces improvements about halfway between the base run and *No obesity up* with regard to disease prevalence, urgent episodes, and personal health care costs. By 2010, a cumulative additional $1288 billion in upstream spending since 1980 has led to a reduction in downstream spending of $2750 billion. The breakeven year does not occur until 2002, however—22 years after the policy is initiated.

The 22-year payback period under *More health protection* is notably greater than the 15-year payback period in the *More risk management* simulation. Much of this additional payback time comes from the fact that health protection acts further upstream than risk management does, as seen in Figure 5. Some of the additional payback time for health protection is also likely a reflection of the fact that there is much earlier (1980s) upstream spending under the health protection scenario than there is in the risk management scenario. The 1980s spending is arguably less cost-effective than spending is during the 1990s, the period of most rapid growth in disease prevalence. As a partial test of this idea, we have performed another simulation in which the health protection program is implemented not in 1980 but in 1985. By 2010, a cumulative additional $1209 billion in upstream spending has led to a reduction in downstream spending of $2250 billion, with breakeven occurring in 2004, 19 years after the policy is initiated. Thus, the payback period is a few years less (19 versus 22) in this simulation, but the breakeven year has actually been pushed back further (2004 versus 2002).

In any event, whether the approach to upstream action is risk management or health protection, the model suggests that the payback time, purely in terms of health care costs, may be a relatively long one. It should be noted, however, that our model does not include losses in productivity to employers and society at large. Another SD model suggests that when these losses are taken into account, the payback on upstream action may shrink to a much shorter time period (Homer *et al.*, 2004), a length of time that may be acceptable to the public as well as to those employers in a position to put upstream efforts into effect.

A couple of broad conclusions may be drawn from the model simulations presented here in Figures 4 and 6. First, we see that cost-containment measures in the U.S. have thus far been futile, and they have done more to limit growth in the extent of care than they have to limit costs. In this sense, the existing market for health care services has been dysfunctional, and it would appear that societal measures to stabilize and simplify this market might be considered. Second, we see that progress is possible even within the current system to reduce costs and improve health through increased investments in upstream risk management and health protection measures. The financial payback on such investments may take some years but could ultimately be very large.

Conclusion

We have sought to explain why chronic illness is such a difficult problem to deal with and why the U.S. in particular has stumbled both in producing better health outcomes and in controlling the cost of illness.[5] Part of the problem comes from the growth in health risk that leads to greater incidence of disease, as exemplified by the rise in obesity. But another aspect of the problem is that progress in improving the treatment of existing illness seems to have stalled in recent years. Growth in health care assets has historically been a key driver of improved extent of care, but with rising costs reimbursement has become constrained, thereby creating uncertainty in the minds of potential investors regarding future revenues, and slowing investment in assets. Another driver of extent of care is insurance coverage, and private coverage, like reimbursement, has declined in response to rising costs. However, the impact of such decline has been mitigated by the availability of the government's Medicaid program as fallback coverage for many lower-income workers.

If rising costs are a great stumbling block to progress, why do they keep rising? Some of the increase is simply a reflection of past progress; namely, growth in health care assets, and increased longevity for those with chronic illness due to those health- and life-saving improvements. If this were the whole story, one could view the recent slowing in asset growth as part of an orderly process by which the health care system moves toward an acceptable maximum spending level relative to GDP. However, the data show that health care costs have continued to rise rapidly even as asset growth has slowed. The explanation for this continued rapid growth in costs may lie in income-maintaining adaptations by providers, who have been able to raise prices and

Published online in Wiley InterScience
(www.interscience.wiley.com) DOI: 10.1002/sdr

service volumes for a given quality of care, especially in the less well-regulated outpatient sector. These adaptations have come in direct response to the attempts by payers to control costs through restrictions of reimbursement.

This tug-of-war between payers and providers, permitted by the current system of payment in the U.S., has had damaging effects in terms of limiting the supply of care, and it has also increased the administrative overhead of providers. In the absence of effective controls, health care costs as a fraction of GDP in the U.S. accelerated ahead of those in other industrialized countries starting in the 1980s, without delivering better care (Docteur and Oxley, 2003). In this sense, the entrepreneurial U.S. health care system which made such great progress in the past has now become bloated and inefficient. As many have come to realize, the time is overdue for a fundamental change in this dysfunctional system.

The difficulty of controlling costs and improving outcomes in the U.S. suggests the need for an innovative approach to health reform—one that emphasizes upstream efforts to reduce the health risks that may lead to chronic illness. While spending on population-based health protection and risk management programs has grown somewhat, it still represents a small fraction of total U.S. health care spending. Our model suggests that policies that shift the balance toward more upstream programs can have beneficial impacts on both health care costs and the population's health status. Although such upstream investments may take several years or even decades to come to fruition, it is important to recognize that improved health is a chief aspiration of all people and therefore deserves a commitment to strategies that will benefit both current and future generations.

Disclaimer

The findings and conclusions in this report are those of the authors and do not necessarily represent the views of the Centers for Disease Control and Prevention.

Notes

1. Despite the clear importance of cost controls in the evolution of U.S. health care since the 1970s, we have no hard data on how they have changed over time. Consequently, although reimbursement and coverage restrictions play a central role in Figure 2, they are not included explicitly in the simulation model. Instead, in the model we focus on *personal health care costs as a fraction of GDP* as the key factor to which other variables in Figure 2 react, including the reinvestment rate in new assets, private insurance coverage, and inflation in health care prices. Each one of these responses to increased

costs is modeled with a delay of 3 or 4 years, reflecting the adjustment times of the relevant stakeholders, who (depending upon the variable in question) may be employers, prospective investors, or health care providers.

2. We have modeled the fraction covered by any insurance ("total") as the sum of (1) the fraction covered by private insurance and (2) the fraction covered by government insurance but not by private insurance. (A large fraction of the elderly have both Medicare and private insurance to supplement Medicare's copay requirements and gaps in coverage. Because this population has both public and private insurance, one cannot simply add up the different categories of insurance to get the total covered population.) In line with the "fallback" argument in the main text, the government-only coverage is modeled as a fraction of those not covered by private insurance, a fraction specified by an exogenous time series. Based on Census data from 1987–2005, we estimate that this fallback fraction has risen only slightly in recent years, from 48 percent in the late 1980s to 50 percent by 2000. For the years preceding 1987, we have examined the NHEA, which provides a breakdown of health care spending by major payer category, to see how the balance of private and government-paid spending has changed over time. Based on these data in conjunction with the Census data for the late 1980s, we estimate that the fallback fraction sat at about 18 percent during the years preceding the establishment of Medicare and Medicaid in 1965 but rose rapidly thereafter to 37 percent in 1970 and 45 percent in 1975.

3. Although Loops B2 and B3 appear to act similarly, their impacts on disease management or urgent care, as subsets of disease care, are rather different. With regard to Loop B3, we have noted previously that the uninsured poor have greater access to urgent care, through hospital emergency departments, than they do to disease management. Thus, when a person loses insurance coverage, this will tend to lead to more of a reduction in disease management than in urgent care. Lacking disease management, this person becomes more prone to complications of disease leading to expensive hospitalization, the costs of which are shifted from the employer to hospitals and the general public. With regard to Loop B2, the trend in reduced investment appears to have been more benign, more genuinely cost saving due to limitations on urgent care. Our model-based analysis of NHEA data suggests that the fraction of investments directed to disease management rather than urgent care was roughly 30 percent through 1980 but by 2004 had increased to 35 percent. That is, investors appear to have moved gradually more in the direction of disease management and away from urgent care.

4. Simulation beyond 2005 requires assumptions for several different input time series. For future deaths and total population, we have used U.S. Census projections. For future growth in real GDP per capita, we have assumed a flat 1.9 percent per year, which was the average growth rate for 1985–2005. For future changes in risk onset and disease onset due to social

Published online in Wiley InterScience (www.interscience.wiley.com) DOI: 10.1002/sdr and economic influences beyond the scope of the model, we have assumed no further increase after 2005. For the government coverage fraction of people not covered by private insurance plans, we have assumed no further change beyond 2000, at which time we estimate the fraction at 50 percent. And for risk management spending as a fraction of total disease and risk management spending, we assume linear growth extrapolating from the past: from 1.5 percent in 1989 to 3.1 percent in 2004 to 3.8 percent in 2010.

5. The world's developing countries lack the resources to duplicate the expensive patterns of care that emerged in the U.S. and will need to find their own path. One study indicates that developing countries will require very different prevention strategies for cardiovascular diseases than those of higher income countries (Reddy and Yusuf, 1998). The authors suggest that currently low levels of cardiovascular risk factors in the large rural segments of the developing countries offer a window of opportunity for early and effective control of the epidemic. They state: "At the present levels of these risk factors in the developing countries, the approach would be predominantly non-pharmacological, population based, and lifestyle linked. This would largely avoid the biologic and economic costs of a pharmacological approach warranted by high levels of these risk factors in the developed countries."

References

ADA (American Diabetes Association). 2003. Economic costs of diabetes in the U.S. in 2002. *Diabetes Care* **26**(3): 917–932.

Brown R, Elixhauser A, Corea J, Luce B, Sheingod S. 1991. National expenditures for health promotion and disease prevention activities in the United States. *Report BHARC-013/91-019*. Battelle Medical Technology Assessment and Policy Research Center: Washington, DC.

BLS (Bureau of Labor Statistics). 2007. Consumer Price Index: all urban consumers (Series IDs CUUR0000SA0 and CUUR0000SAM). http://data.bls.gov/cgi-bin/surveymost?cu [24 January 2007].

CDC (Centers for Disease Control and Prevention). 1999. Ten great public health achievements: United States, 1900–1999. *Morbidity and Mortality Weekly Report* **48**(12): 241–243.

——. 2006. Health protection goals. http://www.cdc.gov/about/goals/goals.htm [5 October 2006].

CDC/BRFSS (Centers for Disease Control and Prevention). 2007. Behavioral Risk Factor Surveillance System. Prevalence data: general health. http://apps.nccd.cdc.gov/brfss/list.asp?cat=HS&yr=2005&qkey=1100&state=All [24 January 2007].

CDC/DCEG (CDC Diabetes Cost-Effectiveness Group). 2002. Cost-effectiveness of intensive glycemic control, intensified hypertension control, and serum cholesterol level reduction for Type 2 diabetes. *Journal of the American Medical Association* **287**(19): 2542–2551.

CDC/NCCDPHP (Centers for Disease Control and Prevention, National Center for

Chronic Disease Prevention and Health Promotion). 2007a. Chronic Disease Overview. http://www.cdc.gov/nccdphp overview.htm [24 July 2007].

———. 2007b. Overweight and Obesity: Obesity Trends. http://www.cdc.gov/nccdphp/dnpa/obesity/trend/index.htm [24 July 2007].

CDC/NHIS (Centers for Disease Control and Prevention). 2007. National Health Interview Survey, Series 10 (annual). Respondent-Assessed Health Status and Reported Chronic Conditions. http://www.cdc.gov/nchs/products/pubs/pubd/series/sr10/ser10.htm [24 January 2007].

CDC/NVSR (Centers for Disease Control and Prevention). 2006. *National Vital Statistics Reports*, Vol. 54(13). http://www.cdc.gov/nchs/data/nvsr/nvsr54/nvsr54_13.pdf. [24 January 2007].

CMS (Center for Medicare and Medicaid Services). 2007. National Health Expenditure Data. http://www.cms.hhs.gov/NationalHealthExpendData/02_ [24 January 2007].

Cutler DM, Rosen AB, Vijan S. 2006. The value of medical spending the United States, 1960–2000. *New England Journal of Medicine* **355**(9): 920–927.

Docteur E, Oxley H. 2003. Health-care systems: lessons from the reform experience. Organization for Economic Cooperation and Development (OECD), Economics Department Working Papers No. 374. http://www.oecd.org/eco [24 July 2007].

Dominick KL, Ahern FM, Gold CH, Heller DA. 2002. Relationship of health-related quality of life to health care utilization and mortality among older adults. *Aging Clinical and Experimental Research* **14**: 499–508.

Eckholm E and *New York Times* staff. 1993. *Solving America's Health-Care Crisis*. Times Books: New York.

Evans RG, Barer ML, Marmor TR. 1994. *Why are Some People Healthy and Others Not? The Determinants of Health of Populations*. A de Gruyter: New York.

Eyre H, Kahn R, Robertson RM *et al.* 2004. Preventing cancer, cardiovascular disease, and diabetes: a common agenda for the American Cancer Society, the American Diabetes Association, and the American Heart Association. *Circulation* **109**: 3244–3255.

Freudenheim M. 2007. Company clinics cut health costs. *New York Times* 14 January.

Fries JF, Koop CE, Sokolov J, Beadle CE, Daniel W. 1998. Beyond health promotion: reducing need and demand for medical care. *Health Affairs* **17**(2): 70–84.

Gerberding JL. 2005. Protecting health: the new research imperative. *Journal of the American Medical Association* **294**(11): 1403–1406.

Greenlund KJ, Zheng ZJ, Keenan NL *et al.* 2004. Trends in self-reported multiple cardiovascular disease risk factors among adults in the United States, 1991–1999. *Archives of Internal Medicine* **164**: 181–188.

Hajjar I, Kotchen JM, Kotchen TA. 2006. Hypertension: trends in prevalence, incidence, and control. *Annual Review of Public Health* **27**: 465–490.

Hanna KE, Coussens C. 2001. *Rebuilding the Unity of Health and the Environment: A New Vision of Environmental Health for the 21st Century*. Institute of Medicine, National Academy Press: Washington, DC.

Hayashino Y, Nagata-Kobayashi S, Morimoto T *et al.* 2004. Cost-effectiveness of screening for coronary artery disease in asymptomatic patients with Type 2 diabetes and additional atherogenic risk factors. *Journal of General Internal Medicine* **19**(12): 1181–1191.

Heirich M. 1999. *Rethinking Health Care*. Westview Press: Boulder, CO.

Hirsch G, Homer J, McDonnell G, Milstein B. 2005. Achieving health care reform in the United States: toward a whole-system understanding. In *23rd International Conference of the System Dynamics Society*, Boston, MA.

Hirsch GB, Bergan TA, Goodman MR. 1975. Examining Alternatives for Improving the Nation's Oral Health: A System Dynamics Model of the Dental Care Delivery System. A report to the Division of Dentistry, Bureau of Health Manpower, DHEW: Washington, DC.

Hirsch GB, Immediato CS. 1999. Microworlds and generic structures as resources for integrating care and improving health. *System Dynamics Review* **15**(3): 315–330.

Hirsch GB, Myers R. 1975. Designing Strategies for Particular Health Problems: The Indiana Cardiovascular Disease Model. Indiana Health Planning and Development Agency.

Hoffman C, Rice D, Sung H. 1996. Persons with chronic conditions: their prevalence and costs. *Journal of the American Medical Association* **276**(18): 1473–1479.

Homer J, Hirsch G, Minniti M, Pierson M. 2004. Models for collaboration: how system dynamics helped a community organize cost-effective care for chronic illness. *System Dynamics Review* **20**(3): 199–222.

Homer JB. 1996. Why we iterate: scientific modeling in theory and practice. *System Dynamics Review* **12**(1): 1–19.

Homer JB, Hirsch GB. 2006. System dynamics modeling for public health: background and opportunities. *American Journal of Public Health* **95**: 452–458.

Homer JB, Milstein B. 2002. Communities with multiple afflictions: a system dynamics approach to the study and prevention of syndemics. In *Proceedings of the 20th International Conference of the System Dynamics Society*, Palermo, Italy. System Dynamics Society. http://www.systemdynamics.org/conferences/2002/papers/Homer1.pdf [6 August 2007].

Homer J, Milstein B. 2004. Optimal decision making in a dynamic model of community health. In *37th Hawaii International Conference on System Sciences*. IEEE: Waikoloa, HI. http://csdl.computer.org/comp/proceedings/hicss/2004/2056/03/205630085a.pdf [24 July 2007].

Homer J, Milstein B, Dietz W, Buchner D, Majestic E. 2006. Obesity population dynamics: exploring historical growth and plausible futures in the U.S. In *24th International Conference on SD Society*, Nijmegen, Netherlands.

IOM (Institute of Medicine). 2001. *Crossing the Quality Chasm: A New Health System for the 21st Century*. National Academy Press: Washington, DC.

——. 2002. *The Future of the Public's Health in the 21st Century*. National Academy Press: Washington, DC.

Jones AP, Homer JB, Murphy DL, Essien JDK, Milstein B, Seville DA. 2006. Understanding diabetes population dynamics through simulation modeling and experimentation. *American Journal of Public Health* **96**(3): 488–494.

Leonhardt D. 2007. What's a pound of prevention really worth? *New York Times* 24 January.

Lindblom E. 2006. State cigarette excise tax rates and rankings. Campaign for Tobacco-Free Kids: Washington, DC. http://www.tobaccofreekids.org/research/factsheets/pdf/0097.pdf [5 February 2007].

Luginbuhl WH, Forsyth BR, Hirsch GB, Goodman MR. 1981. Prevention and rehabilitation as a means of cost containment: the example of myocardial infarction. *Journal of Public Health Policy* **2**(2): 103–115.

Published online in Wiley InterScience
(www.interscience.wiley.com) DOI: 10.1002/sdr

Mackenbach JP. 1994. The epidemiologic transition theory. *Journal of Epidemiology and Community Health* **48**(4): 329–331.

Marshall T. 2000. Exploring a fiscal food policy: the case of diet and ischaemic heart disease. *British Medical Journal* **320**: 301–305.

Mathers CD, Loncar D. 2006. Projections of global mortality and burden of disease from 2002 to 2030. *PLoS Medicine* **3**(11): 2011–2030.

McKinlay JB. 1979. A case for refocusing upstream: the political economy of illness. In *Patients, Physicians, and Illness: A Sourcebook in Behavioral Science and Health* (3rd edn.), Jaco EG (ed.). Free Press: New York; 9–25.

McKinlay JB, Marceau LD. 2000. Upstream healthy public policy: lessons from the battle of tobacco. *International Journal of Health Services* **30**(1): 49–69.

NIHCM (National Institute for Health Care Management Foundation). 2002. Prescription Drug Expenditures in 2001: Another Year of Escalating Costs. Washington, DC. http://www.nihcm.org/spending2001.pdf [24 January 2007].

NIH/NHLBI (National Institutes of Health, National Heart, Lung and Blood Institute). 2004. Morbidity and Mortality: 2004 Chart Book on Cardiovascular, Lung, and Blood Diseases. Washington, DC. http://www.nhlbi.nih.gov/resources/docs/cht-book.htm [24 July 2007].

Northridge ME, Sclar ED, Biswas P. 2003. Sorting out the connections between the built environment and health: a conceptual framework for navigating pathways and planning healthy cities. *Journal of Urban Health* **80**(4): 556–568.

NSC (National Safety Council). 2006. What are the odds of dying? (Odds of death due to injury). http://www.nsc.org/lrs/statinfo/odds.htm [24 January 2007].

O'Donoghue T, Rabin M. 2006. Optimal sin taxes. *Journal of Public Economics* **90**: 1825–1849.

Olshansky SJ, Ault AB. 1986. The fourth stage of the epidemiologic transition: the age of delayed degenerative diseases. *Milbank Quarterly* **64**(3): 355–391.

Ratanawijitrasin S. 1993. The dynamics of health care finance: a feedback view of system behavior. PhD dissertation, State University of New York at Albany, NY.

Reddy KS, Yusuf S. 1998. Emerging epidemic of cardiovascular disease in developing countries. *Circulation* **97**: 596–601.

Simon E. 2006. Bosses push staff to eat right, exercise. *Associated Press (New York)* 4 December.

Smedley BD, Syme SL. 2000. *Promoting Health: Intervention Strategies from Social and Behavioral Research*. Institute of Medicine. National Academy Press: Washington, DC.

Starr P. 1982. *The Social Transformation of American Medicine*. Basic Books: New York.

Tengs TO, Osgood ND, Chen LL. 2001. The cost-effectiveness of intensive national school-based anti-tobacco education: results from the tobacco policy model. *Preventive Medicine* **33**: 558–570.

Thorpe KE, Florence CS, Howard DH, Joski P. 2005. The rising prevalence of treated disease: effects on private health insurance spending. *Health Affairs (Web Exclusive)* **W5**: 317–325.

U.S. Census Bureau. 2007. Health Insurance Coverage Status and Type of Coverage by Age: 1987 to 2005 (Table HI-7). http://www.census.gov/hhes/www/hlthins/historic/hihistt7.html [24 January 2007].

Weisbrod BA. 1991. The health care quadrilemma: an essay on technological change, insurance, quality of care, and cost containment. *Journal of Economic Literature* **29**: 523–552.

WHO (World Health Organization). 2002. Innovative Care for Chronic Conditions: Building Blocks for Action. http://www.who.int/chp/knowledge/publications/icccglobalreport.pdf [24 July 2007].

———. 2005. Preventing Chronic Diseases: A Vital Investment. http://www.who.int/chp/chronic_disease_report/en [24 July 2007].

Wild S, Roglic G, Green A, Sicree R, King H. 2004. Global prevalence of diabetes: estimates for the year 2000 and projections for 2030. *Diabetes Care* **27**: 1047–1053.

Wilkinson RG, Marmot MG (eds). 2003. *The Solid Facts: Social Determinants of Health* (2nd edn.). Centre for Urban Health. World Health Organization: Copenhagen.

Woolhandler S, Campbell T, Himmelstein DU. 2003. Costs of health care administration in the United States and Canada. *New England Journal of Medicine* **349**: 768–775.

Yach D, Hawkes C, Gould CL, Hoffman KJ. 2004. The global burden of chronic diseases: overcoming impediments to prevention and control. *Journal of the American Medical Association* **291**(21): 2616–2622.

Yach D, McKee M, Lopez AD, Novotny T. 2005. Improving diet and physical activity: 12 lessons from controlling tobacco smoking. *British Medical Journal* **330**: 898–900.

14

Reply to Jay Forrester's "System Dynamics—the Next Fifty Years" (2007)

Homer J. *System Dynamics Review*, 23(4): 465-467

Reply to Jay Forrester's "System dynamics— the next fifty years"

Jack Homer*

Jack Homer has been an independent System Dynamics consultant to private and public organizations since 1989. In the public arena, his work focuses on public health and health care policy. He was formerly an assistant professor at the University of Southern California and received a PhD in management from MIT. He also holds BS and MS degrees in applied mathematics and statistics from Stanford University, and received the Jay W. Forrester Award in 1997.

Jay Forrester (2007) has assayed the current state of our field and issued a sobering diagnosis leading to a radical prescription for reform. He feels that we are on an "aimless plateau", doing work that is overly cautious and compromised by superficiality and mediocre quality. Academics are hemmed in by a conservative institution that punishes bold work and does not allow system dynamics to be taught beyond an introductory level. Practitioners are constrained by clients who appear to have power but who ask only small questions and fear rocking the boat. Jay's solution to this apparent state of stagnation is to develop a long-term plan for (a) raising large sums of money from private donors, (b) creating new schools of system dynamics, and (c) writing brave iconoclastic books to convert key opinion leaders to our cause.

These ideas are provocative and should cause us to reflect upon our own experiences and expectations for the field. Jay, as founder of the field, has the longest purview of us all, and so when he says we are in trouble, we must take notice. Yet, it is possible that some of us, his followers, may be involved in developments in the field that give us a different but equally valid perspective.

My own perspective is that of a consultant and researcher who has worked extensively with government agencies in the area of health, one of the areas that Jay names as needing a bolder push into the big issues. Much of my work since 2002 has been with the U.S. Centers for Disease Control and Prevention (CDC), where my clients and I have not shied away from big issues, but in fact have addressed such difficult issues as the growth of diabetes and obesity, the high burden of chronic illness in poor neighborhoods, and the resistance of the health care system to investing in primary prevention (see Homer and Hirsch, 2006; Jones *et al.*, 2006). This work has come to the attention of the CDC's director, Dr Julie Gerberding, who for a few years now has used it to explain to Congress and to her own staff her strategic vision for the agency (Gerberding 2004, 2005, 2007). Senior managers at the CDC regularly express keen interest in system dynamics, and even works not done under their direct auspices (such as the paper on health care economics published in the 50th Anniversary special issue of SDR; Homer *et al.*, 2007) are read and recommended.

How have we managed to make such inroads at the CDC these past several years? One answer to this question, as Jay would appreciate, is doing quality work that is both rigorous and policy relevant, a combination that gives system

* Correspondence to: Jack Homer, Homer Consulting, 3618 Avalon Court, Voorhees, NJ 08043, U.S.A.
E-mail: jhomer@comcast.net
Received October 2007; Accepted November 2007

System Dynamics Review Vol. 23, No. 4, (Winter 2007): 465–467
Published online in Wiley InterScience
(www.interscience.wiley.com) DOI: 10.1002/sdr.388
Copyright © 2007 John Wiley & Sons, Ltd.

dynamics an advantage over other methodologies. Rigor in this setting has meant taking the time to become expert in the subject matter, including all available historical evidence. Yes, we fit curves to historical data, but only when we can also establish that the model's structure is realistic, insightful, robust, and parsimonious. Under these conditions, demonstrating historical fit to multiple time series is a powerful way to establish model adequacy and is not a trivial matter. Moreover, if we find that the fit to historical evidence is forcing model parameters outside of their plausible range, we do not take this as a sign that the evidence should be ignored, but we rather take it as an indication that perhaps the model should be revised. That, to my clients at the CDC, is a solid scientific approach (see Homer, 1996).

If one secret to our success at the CDC is respecting and understanding the evidence, another is respecting, understanding and, when appropriate, using alternative modeling methodologies. I do not hesitate to use estimates from statistical regressions or from the forecasts of a highly disaggregated Markov model when they have been properly done. At the same time, I am careful to point out the limitations of those other methodologies and models, so that my clients can understand the contribution of a system dynamics model which may have less fine-grained detail but has a broader boundary capable of delivering more robust long-term projections and greater policy insights.

In the early days of system dynamics, perhaps a more isolated and icono-clastic approach was justified for making rapid progress, capturing the public's imagination, and stirring up greater interest in the field. Now, our field has grown up, and I think that we mostly need a less splashy but more sustainable approach. To have a more lasting influence among key opinion leaders, we need to be viewed not merely as having bold new ideas, but as credible researchers who know our material inside and out and who, if we say surprising things, do so based on a careful analysis of the evidence.

In sum, I think we can work within the existing system without compromising our integrity and our desire to do great things. I suspect this is as true in academia as I have found it to be in work with clients. Although junior faculty may, as Jay suggests, be unlikely to gain tenure by straying far from convention in their research and writing, there are now numerous examples of senior faculty who have written important books and brought great benefits to our field through their leadership and teaching. Their books typically have not attempted to attack sacred cows so much as they have sought to present system dynamics as a rigorous and well-developed field.

This is the hopeful future I see for system dynamics. Not attempting to strike it rich with a big private donor, but rather establishing credibility through good, diligent work so that our funding sources become many and varied. Not creating new schools of system dynamics, but rather expanding our opportunities within existing collegial networks. Not expecting junior faculty to write politically incorrect books, but encouraging them to learn their craft as researchers and teachers respected by their colleagues in related fields.

Published online in Wiley InterScience
(www.interscience.wiley.com) DOI: 10.1002/sdr

The superficiality and mediocre quality of much work in system dynamics is, of course, a real concern. This is an issue that must be faced by any field still coming into its maturity. But it is an issue we can face without a radical change in our approach. The path up the mountain requires patience and tenacity, and may involve joining forces with other climbers who use different tools. I don't think the right next step is jumping off and hoping for a sudden gust of wind to carry us above the fray.

References

Forrester JW. 2007. System dynamics—the next fifty years. *System Dynamics Review* **23**(2–3): 359–370.

Gerberding JL. 2004. *CDC's Futures Initiative.* Public Health Training Network: Atlanta, GA, 12 April 2004.

——. 2005. Protecting health: the new research imperative. *JAMA* **294**(11): 1403–1406.

——. 2007. FY 2008 CDC Congressional Budget Hearing. Testimony before the Committee on Appropriations, Subcommittee on Labor, Health and Human Services, Education and Related Agencies, United States House of Representatives: Washington, DC, 9 March 2007.

Homer J. 1996. Why we iterate: scientific modeling in theory and practice. *System Dynamics Review* **12**(1): 1–19.

Homer J, Hirsch G. 2006. System dynamics modeling for public health: background and opportunities. *American Journal of Public Health* **96**(3): 452–458.

Homer J, Hirsch G, Milstein B. 2007. Chronic illness in a complex health economy: the perils and promises of downstream and upstream reforms. *System Dynamics Review* **23**(2–3): 313–343.

Jones A, Homer J, Murphy D, Essien J, Milstein B, Seville D. 2006. Understanding diabetes population dynamics through simulation modeling and experimentation. *American Journal of Public Health* **96**(3): 488–494.

Published online in Wiley InterScience
(www.interscience.wiley.com) DOI: 10.1002/sdr

15

Simulating and Evaluating Local Interventions to Improve Cardiovascular Health (2010)

Homer J, Milstein B, Wile K, Trogdon J, Huang P, Labarthe D, Orenstein D. *Preventing Chronic Disease,* 7(1): A18. Available at: http://www.cdc.gov/pcd/issues/2010/jan/08_0231.htm

---

Related writings

Hirsch G, Homer J, Evans E, Zielinski A. A System Dynamics Model for Planning Cardiovascular Disease Interventions. *American Journal of Public Health*, 100(4): 616-622, 2010.

Homer J, Milstein B, Wile K, Pratibhu P, Farris R, Orenstein D. Modeling the Local Dynamics of Cardiovascular Health: Risk Factors, Context, and Capacity. *Preventing Chronic Disease*, 5(2), April 2008. Available at: http://www.cdc.gov/pcd/issues/2008/apr/07_230.htm.

PREVENTING CHRONIC DISEASE

PUBLIC HEALTH RESEARCH, PRACTICE, AND POLICY

VOLUME 7: NO. 1   JANUARY 2010

SPECIAL TOPIC

Simulating and Evaluating Local Interventions to Improve Cardiovascular Health

Jack Homer, PhD; Bobby Milstein, PhD, MPH; Kristina Wile, MS; Justin Trogdon, PhD; Philip Huang, MD, MPH; Darwin Labarthe, MD, MPH, PhD; Diane Orenstein, PhD

Suggested citation for this article: Homer J, Milstein B, Wile K, Trogdon J, Huang P, Labarthe D, et al. Simulating and evaluating local interventions to improve cardiovascular health. Prev Chronic Dis 2010;7(1). http://www.cdc.gov/pcd/issues/2010/jan/08_0231.htm. Accessed [*date*].

PEER REVIEWED

Abstract

Numerous local interventions for cardiovascular disease are available, but resources to deliver them are limited. Identifying the most effective interventions is challenging because cardiovascular risks develop through causal pathways and gradual accumulations that defy simple calculation. We created a simulation model for evaluating multiple approaches to preventing and managing cardiovascular risks. The model incorporates data from many sources to represent all US adults who have never had a cardiovascular event. It simulates trajectories for the leading direct and indirect risk factors from 1990 to 2040 and evaluates 19 interventions. The main outcomes are first-time cardiovascular events and consequent deaths, as well as total consequence costs, which combine medical expenditures and productivity costs associated with cardiovascular events and risk factors. We used sensitivity analyses to examine the significance of uncertain parameters. A base case scenario shows that population turnover and aging strongly influence the future trajectories of several risk factors. At least 15 of 19 interventions are potentially cost saving and could reduce deaths from first cardiovascular events by approximately 20% and total consequence costs by 26%. Some interventions act quickly

to reduce deaths, while others more gradually reduce costs related to risk factors. Although the model is still evolving, the simulated experiments reported here can inform policy and spending decisions.

Introduction

Conditions in particular neighborhoods or cities can profoundly enhance or impede people's prospects for a healthy life (1). This dependence on local context is especially evident in cardiovascular health, for which behavioral, social, and environmental factors combine to affect the likelihood of developing disease or dying prematurely (2). Heart disease and stroke are largely preventable, but they remain the first and third leading causes of death in the United States, partly because we have yet to establish living conditions that minimize such modifiable risks as smoking, obesity, stress, air pollution, poor diet, and physical inactivity. The importance of intervening to limit these risks is highlighted in *A Public Health Action Plan to Prevent Heart Disease and Stroke* (3).

The notion that cardiovascular disease (CVD) can be prevented through local actions raises practical questions that can be examined through systems modeling and simulation. Working closely with colleagues in Austin/Travis County, Texas, and subject matter experts at the Centers for Disease Control and Prevention and the National Institutes of Health, we developed a system dynamics simulation model to answer the following questions:

- How does local context affect the major risk factors for CVD and, in turn, population health status and costs?

- How might local interventions affect CVD risk, health status, and costs over time?
- How might local health leaders better balance their policy efforts given limited resources?

Methods

System dynamics models improve our ability to anticipate the likely effects of interventions in dynamically complex situations, where the pathways from interventions to outcomes may be indirect, delayed, and possibly affected by nonlinearities or feedback loops (4). System dynamics has been used effectively since the 1970s to model many areas of public health and social policy, including CVD (5).

Model structure

We previously described a framework for understanding cardiovascular health in a local context (6). That framework has been refined and quantified by using additional literature and input from veteran health planners and analysts. The resulting simulation model (Figure 1) focuses on primary prevention; it does not address people who have experienced a CVD event. Causal influences move down and to the right, ending with 2 outcomes: 1) first-time cardiovascular events and consequent deaths and 2)

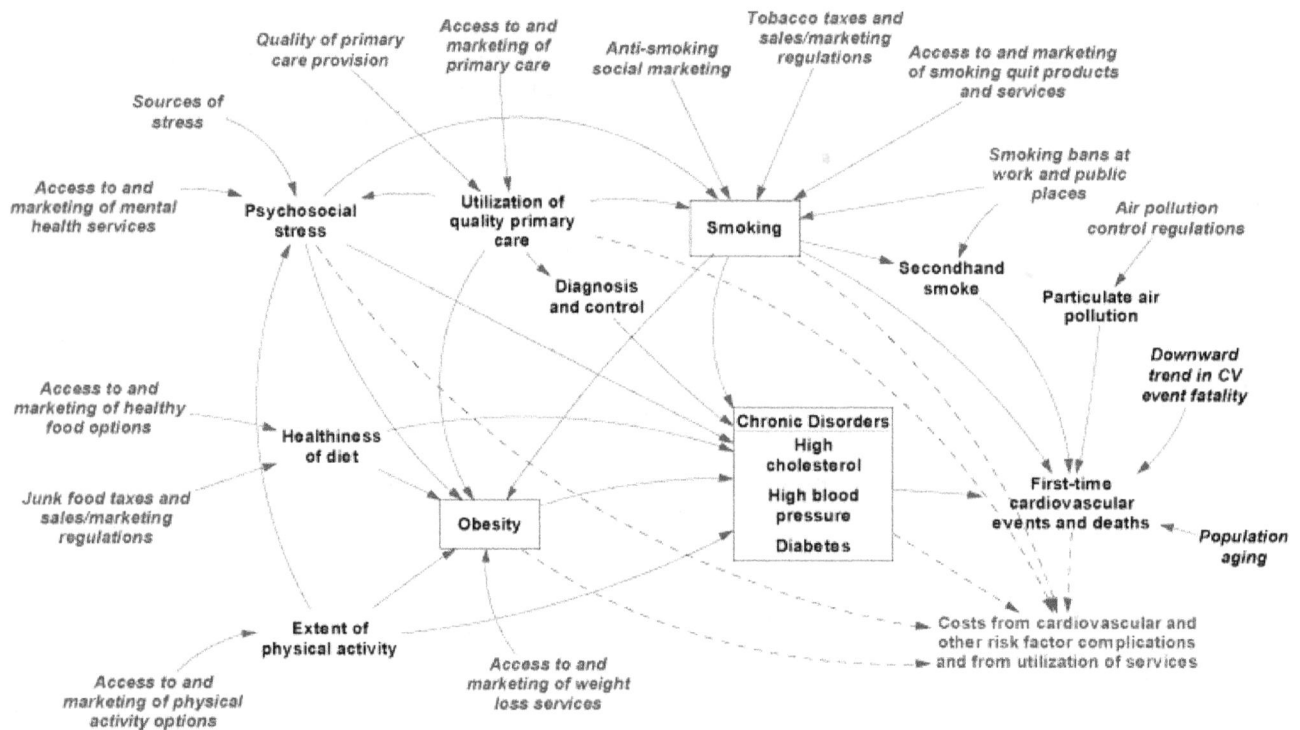

Figure 1. Simulation model for cardiovascular disease (CVD) outcomes. This diagram depicts major health conditions related to CVD and their causes. Boxes identify risk factor prevalence rates modeled as dynamic stocks. The population flows associated with these stocks — including people entering the adult population, entering the next age category, immigration, risk factor incidence, recovery, cardiovascular event survival, and death — are not shown.

Key:
Blue solid arrows: causal linkages affecting risk factors and cardiovascular events and deaths.
Brown dashed arrows: influences on costs.
Purple italics: factors amenable to direct intervention.
Black italics (population aging, cardiovascular event fatality): other specified trends.
Black nonitalics: all other variables, affected by italicized variables and by each other.

costs associated with these events and with the identified risk factors.

The model starts with conditions in the United States in 1990 and simulates them continuously through 2040. The population without CVD and risk factor prevalence rates are represented as dynamic stock or state variables, subdivided by sex and age group (18-29 y, 30-64 y, and ≥65 y). Smoking and obesity are viewed as reversible conditions, whereas diabetes, high blood pressure, and high cholesterol are viewed as chronic conditions that are not reversible but that can be controlled, with the help of good-quality primary health care, to reduce CVD risk.

The incidence of first-time CVD events in the model is driven by the effect of several direct risk factors, based on a widely used risk calculator from the Framingham Heart Study (7). We modified that calculator in several ways for this study, most fundamentally by estimating annual risks at a population level on the basis of risk factor prevalence rates, rather than at the level of an individual. (A detailed description of the modified calculator is available at http://sustainer.org/cvd/documents/SELI_App1.pdf). We also recognized the direct effect on CVD events, especially myocardial infarctions, from secondhand smoke and small particulate matter (PM 2.5) air pollution (8-11). Furthermore, because deaths from CVD have declined partly because of improvements in emergency and acute care, we incorporated a downward trend in the CVD case-fatality rate for 1990 through 2003 (12).

Obesity contributes to CVD, largely through diabetes, high blood pressure, and high cholesterol levels (13). Other indirect influences in the model are physical inactivity, poor diet, psychosocial stress, and smoking as it affects diabetes and obesity (14-21).

Both the direct and indirect influences in the model may be modified by 19 interventions (Table 1). These 19 interventions could be implemented at a city, county, or state level rather than requiring changes nationwide. In functional terms, the interventions are of a few basic types: those that provide broader access to health-promoting services, those that promote desirable behaviors, and those that tax or regulate to deter undesirable behaviors.

Cost calculation

We used a common metric — constant 2005 dollars — to track medical and productivity costs (for morbidity and mortality) that might be affected by the 19 interventions. We measured the societal value of morbidity (sick days) and premature mortality (years of life lost) using a human capital approach, which estimates the market value of lost productivity at work and at home (22). (A detailed description of cost calculations is available at http://sustainer.org/cvd/documents/SELI_App2.pdf.) This summary of medical and productivity costs can determine whether any intervention, or package of interventions, is justified by its likely aggregate consequences, or "total consequence costs." We did not estimate the costs of interventions. However, the total consequence costs can inform spending decisions. For example, suppose that for a given intervention the model calculates a total consequence cost savings of $50 per capita. Planners may then conclude that up to $50 per capita could be justifiably spent to implement that intervention and still create positive net benefits to society.

The model tracks 3 types of intervention consequences:

- Medical and productivity costs attributable to fatal and nonfatal CVD events.
- Medical and productivity costs attributable to noncardiovascular complications of smoking (eg, lung cancer), diabetes (eg, blindness), high blood pressure (eg, kidney failure), and obesity (eg, colorectal cancer). We have thus far been able to quantify these costs, but not yet the costs related to noncardiovascular complications of stress (eg, depression), physical inactivity (eg, back pain), or poor diet (eg, colorectal cancer).
- Costs of services and products to manage risk factors. These include medications and visits for managing chronic disorders, mental health services, weight-loss services, and smoking cessation services and products.

Model calibration

Although the model is meant to investigate interventions in localities such as Austin/Travis County, we began by calibrating it to represent the entire United States. This approach enabled more precise estimation, given that certain data were either unreliable or unavailable at the local level. The results are generally reported as per capita estimates to facilitate interpretation at a local level. Table 2 lists the major information sources on which the model is based (23-32).

The model specifies initial (1990) incidence rates for

smoking, obesity, diabetes, high blood pressure, and high cholesterol, as well as cessation rates for smoking and obesity. These parameters have been set so that the model accurately simulates the observed changes in prevalence rates in the National Health and Nutrition Examination Survey from 1988-1994 to 1999-2004.

The model also contains 56 causal links requiring the estimation of relative risks, effect sizes, or initial values. Many of these parameters were estimated through the use of published studies, meta-analyses, and in some instances, ad hoc surveys of veteran practitioners (33). Because most of these parameter estimates have some level of uncertainty, we also identified lower and upper bounds to be used for sensitivity analysis.

Model testing

Having calibrated the model to accurately reproduce observed trends in risk factor prevalence rates as well as CVD events and deaths, we then explored plausible futures. A base case scenario assumed no changes after 2004 in many of the local determinants of risk in the adult population, including healthiness of diet, extent of physical activity, stress, use of quality primary care, air pollution, and the prevalence rates of smoking and obesity among incoming 18-year-olds. This base case should not be taken as a statement about what is most likely to happen in the absence of intervention, but rather serves as a straightforward and easily understood benchmark against which to compare intervention scenarios.

We tested interventions singly and in groups of similar interventions. For all interventions, we assumed a 1-year ramp-up during 2009, followed by full implementation from 2010 through 2040. The significance of full implementation depends on the intervention, but in all cases is based on effect sizes that the research literature or veteran practitioners indicate should be possible:

- For the 7 marketing interventions and for taxes on tobacco or junk food: doing the maximum that has been demonstrated or seriously proposed somewhere in the United States.
- For the 6 access interventions: raising access to 100%.
- For smoking restriction: reducing secondhand exposures in workplaces and public places to zero.
- For air pollution: reducing small particulate matter by 50% from its 2001-2003 value.

- For sources of chronic stress: a 50% reduction.
- For the quality of primary care (ie, adherence to guidelines): improvement from a national average of 54% (27) to 75%.

For each intervention scenario, we conducted separate simulations using the midpoint, lower-end, and upper-end values for all uncertain parameters. This method yielded a range of plausible outcomes for each intervention scenario.

Discussion

Base case results

The base case projects that even after 2004, when we assume no further changes to the model's inputs, historical trends in the model's risk factor prevalence rates will continue through 2040, although with diminishing slopes. In particular, the model projects further declines in smoking (and, thus, secondhand smoke exposure) and high cholesterol, and at the same time further growth in high blood pressure and diabetes. The projected continuation of past trends reflects the eventual death of older cohorts and their replacement by younger cohorts with different habits and characteristics. For instance, the continued decline of smoking prevalence reflects the lower rate of smoking among teens and young adults today than in previous decades. Such demographic turnover also helps explain the continued growth of high blood pressure and diabetes, which occurs in the model as a legacy of the increase in obesity — a leading risk factor for both disorders — from 1980 to 2004. The projected continuation of trends also reflects the future aging of the population; the over-65 population will increase from 2010 through 2030. This aging effect contributes to the projected decline in smoking because smoking is much less common among the elderly. It also contributes to the projected increase in high blood pressure and diabetes because the prevalence of these disorders is higher with increasing age.

Deaths from first-time CVD events, which declined by 35% from 1990 to 2004, are projected in the base case to rebound by 33% from 2004 to 2040. Much of the past decline is attributable to a 28% reduction in the event fatality rate, from improvements in emergency and acute care. But it also reflects an 11% decline in the rate of CVD events that occurred, despite increases in high blood pressure and diabetes, because of decreases in smoking, secondhand smoke, PM 2.5 air pollution, and uncontrolled high cholesterol.

The potential future rebound in deaths anticipated by our model reflects a 17% increase in fatality from CVD events per capita and a 15% increase in the rate of CVD events because of the aging of the population. Although the base case projects no future increase in CVD events or deaths within each age group, the aging of the population will lead to an increase in the overall rate of CVD events and deaths.

Per capita total consequence costs, which the model calculates to have declined by 25% from 1990 to 2004, are projected in the base case to decline by another 5% from 2004 to 2040. Total consequence costs encompass not only CVD events (which account for 44% of the total costs in 2004) but also noncardiovascular complications of risk factors (also 44%) and management of risk factors (12%). Although per capita CVD event costs are projected to increase by 12% from 2004 to 2040 (reflecting the increase in the frequency of the events themselves) and per capita risk management costs are projected to increase by 8% (reflecting the growing demand for blood pressure and diabetes treatment), these increases are more than offset by a 25% decrease in noncardiovascular complications. This decrease is due to the projected decline of smoking, which in 2004 was responsible for more than 400,000 noncardiovascular deaths, primarily from lung cancer and chronic obstructive pulmonary disease. These premature non-CVD deaths from smoking account for a large fraction (about 28% in 2004) of the total consequence costs calculated in the model.

Intervention scenario results

Individual tests of the 19 interventions suggest that each can reduce deaths from first-time CVD events, and most can reduce total consequence costs. Four of the interventions, however, raise total consequence costs, meaning that they increase risk factor management costs more than they decrease the costs of medical events and complications. These 4 interventions include the 2 that encourage use of mental health services and the 2 that encourage use of weight-loss services. However, because of limitations in the model, planners should not dismiss these interventions in the real world. In the case of mental health services, we have not yet estimated the noncardiovascular costs of depression. In the case of weight-loss

services for obese people, our estimates of cost and benefit are based on conventional dieting and exercise programs, rather than on bariatric surgery, which, although more costly, also appears to be more effective (34).

We present simulation results for only the 15 interventions that in the model do not increase total consequence costs (Figure 2). Such a multipronged approach may be challenging to implement, given resource limitations, but it is useful to look at what could be achieved.

The model suggests that if all risk factors in the model were eliminated, the death rate could be reduced by approximately 60% below the base case, which falls between the 50% to 75% rate that other authors have suggested (35). This model dichotomizes blood pressure, cholesterol, and diabetes as "high" or "not high" and does not further subdivide the "not high" into normal and borderline. Reducing borderline conditions (prehypertension, borderline cholesterol, prediabetes) to normal could further reduce CVD, but we cannot explore this possibility with this model. (A static analysis of the potential benefits of reducing both high and borderline conditions is available at http://sustainer.org/cvd/documents/SELI_App3.pdf.)

The model projects that a 15-component intervention could reduce the first-time CVD event death rate relative to the base case by 20% (range based on sensitivity analysis, 15%-26%) in 2015 and by 19% (range, 14%-25%) in 2040. Thus, the interventions that could reduce CVD deaths have a relatively rapid effect.

The effect of the interventions is more gradual with regard to total consequence costs than it is with regard to CVD deaths; nearly 40% of the eventual effect on costs occurs after 2015 (Figure 2). If all risk factors in the model were eliminated, consequence costs could be reduced by approximately 80% below the base case. Relative to the base run, the 15-component intervention reduces consequence costs by 16% (range, 12%-23%) in 2015, eventually reaching 26% (range, 19%-33%) in 2040. The reduction in consequence costs is $348 per capita (range, $254-$514) in 2015 and $565 per capita (range, $416-$722) in 2040.

The 15-component intervention may be better understood by examining the incremental contributions of its components grouped by topical cluster (Figure 3). We used the same base case graph as in Figure 2 and then incre-

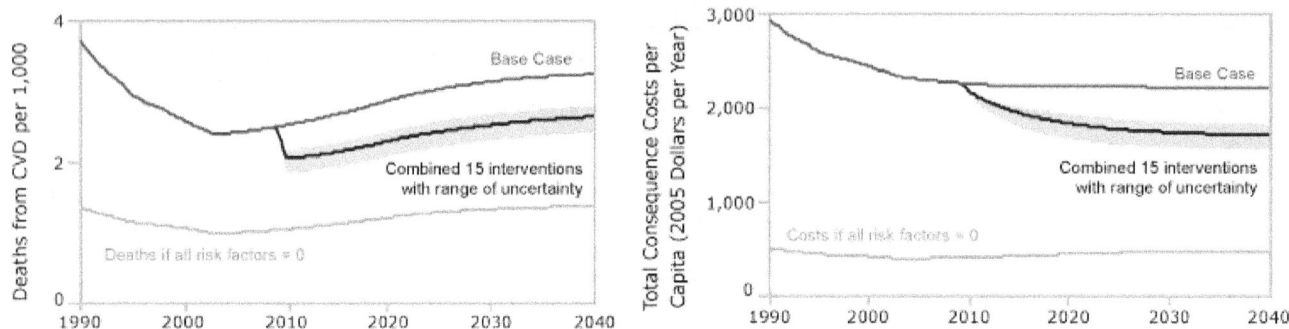

Figure 2. Estimated impacts of a 15-component intervention, with ranges based on sensitivity testing, simulation model for cardiovascular disease (CVD) outcomes. The 15 interventions are listed in Table 1 under the topical clusters of Care, Air, and Lifestyle.

Key:
Blue line = base case results.
Black line = expected reduction in death rate or costs from the 15-component intervention when the uncertain parameters are all set to their baseline values.
Orange shaded area around the black line = envelope of plausible outcomes in the 15-component intervention outcomes based on sensitivity testing. Upper edge (least impact) results when all uncertain impact parameters are set to their lowest values, while lower edge (most impact) results when all uncertain impact parameters are set to their greatest values.
Gray line = the model's calculation of what the death rate or costs would be if all of the risk factors in the model — smoking, small particulate matter (PM 2.5) air pollution, high blood pressure, high cholesterol, diabetes, obesity, poor nutrition, inactivity, and stress — were reduced to zero.

mentally added the following topical clusters:

1) The 3 interventions that improve the use and quality of primary care (Care).
2) The 6 interventions related to air quality and smoking (Air).
3) The 5 interventions related to improved nutrition and physical activity and the 1 intervention that would reduce sources of stress (Lifestyle).

The relative effects of the clusters are different for CVD deaths than they are for total consequence costs. Of the 3 topical clusters, the largest contributor to projected reductions in CVD deaths, both in 2015 and 2040, is Care, followed by Air and then a smaller (though growing) contribution from Lifestyle. In contrast, the largest contributor to projected cost reductions, both in 2015 and 2040, is Air, followed by Lifestyle and then a smaller (and ultimately negligible) contribution from Care. The contributions to per capita cost reduction in 2015 are $235 from Air, $71 from Lifestyle, and $42 from Care. The contributions in 2040 are $393 from Air, $165 from Lifestyle, and $7 from Care.

Conclusions

The major factors that affect cardiovascular health at a population level interact through causal pathways and develop through gradual accumulations that defy simple calculation. This dynamic complexity — and not just gaps in data — is a challenge for local leaders who want to intervene most effectively given limited resources. Our simulation model helps meet this challenge by integrating what is known about the various risk factors in a single testable framework for prospective policy analysis.

The simulations reported here point to several conclusions that local leaders and national allies may find valuable.

1) The CVD death rate has declined in recent years, not only because of improvements in emergency and acute care but also because of reductions in the CVD event rate itself, due to reductions in smoking, secondhand smoke, particulate air pollution, and uncontrolled high cholesterol. If this progress does not continue at a similar pace in the future, however, the CVD death rate will likely rebound strongly as the population ages.

2) Medical and productivity costs associated with CVD risk factors have declined because of declines in first-time CVD events and consequent deaths, and because of reductions in non-CVD deaths (especially lung cancer and chronic obstructive pulmonary disease) associated with

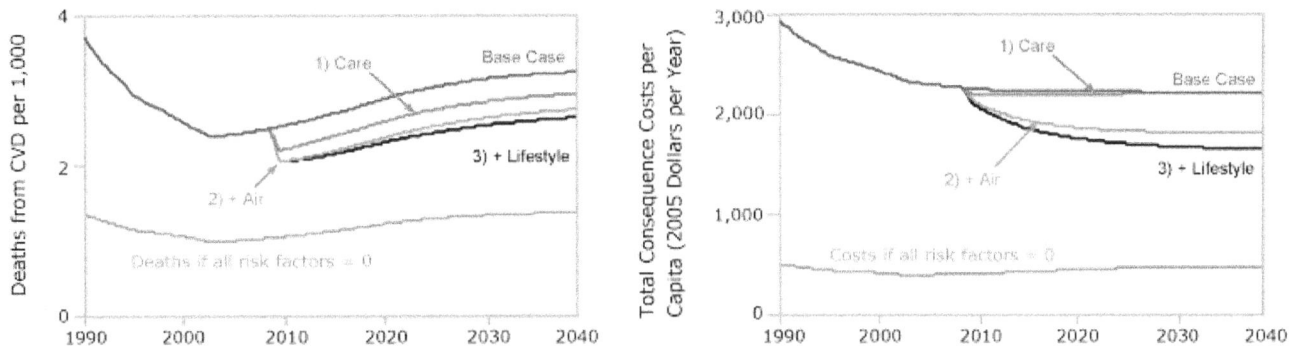

Figure 3. Projected changes in the death rate from first-time cardiovascular disease (CVD) events and in total consequence costs per capita when 15 interventions are combined, expressed in terms of clusters of interventions, simulation model for cardiovascular health outcomes.

Key:
Blue line = base case results.
Gray line = outcomes if all risk factors were reduced to zero.
Red line = implement the 3 interventions that improve the use and quality of primary care (Care).
Green line = add the 6 interventions related to air quality and smoking (Air).
Black line = add the 5 interventions related to improved nutrition and physical activity and the 1 intervention that would reduce sources of stress (Lifestyle). This scenario includes all 15 interventions and is identical to the black line in Figure 2.

smoking. Population aging will likely keep smoking prevalence on a path of decline into the future, so that even if CVD deaths rebound, the total consequence costs need not rebound.

3) Of 19 interventions that local planners may consider for lowering CVD risk, at least 15 could reduce CVD deaths without increasing total consequence costs.

4) Interventions aimed at reducing smoking and improving indoor and outdoor air quality can save lives relatively quickly and can justify intervention spending equivalent to as much as $300 per capita per year for 30 years (in 2005 constant dollars, without time discounting) to achieve the full implementation targets. Most local health leaders are already aware of the need for tobacco control and smoking bans, but many may not be aware of the contribution of particulate air pollution to CVD risk, even in areas like Austin/Travis County without heavy pollution.

5) Interventions aimed at improving the use and quality of primary care to diagnose and control high blood pressure, high cholesterol, and diabetes can save lives quickly but should not be expected to save much on total costs, primarily because of the high cost of medications. Consequently, the intervention spending to achieve and maintain such improvement should not exceed the

equivalent of $25 per capita per year for 30 years. Other researchers have similarly found that good preventive care for chronic conditions may be cost-effective but is not necessarily cost-saving (36,37).

6) Interventions to improve nutrition and physical activity and to reduce sources of stress take more time to affect CVD deaths, as they gradually reduce obesity and other chronic disorders. Nonetheless, their contribution grows over time and may justify intervention spending equivalent to as much as $100 per capita per year for 30 years.

The ability of particular localities to achieve full implementation within these cost limits may vary depending on context and implementation factors. Potential extensions and improvements to the model include the following:

- Modeling medical and personal costs for the post-CVD event population and targeted interventions for secondary prevention to reduce the rate of recurrent CVD events.
- Modeling the prevalence rates of borderline conditions (prehypertension, borderline cholesterol, prediabetes) and incorporating them in the CVD risk calculations.
- Modeling the prevalence of former smokers and incorporating their differential risks in the CVD event and cost calculations.

293

- Incorporating the non-CVD consequences of stress, physical inactivity, and poor diet.
- Estimating intervention implementation costs to better inform intervention priorities.
- Incorporating additional independent risk factors for CVD (eg, excess sodium intake, excess trans fat intake, vitamin D deficiency, periodontal disease).

The model described here was created through a close collaboration with health planners in Austin/Travis County, who are now using a locally calibrated version of the model to support local strategy design and leadership development. We plan to pursue similar engagements with colleagues elsewhere. With more widespread use, this tool may help health planners across the country transform local contexts to most effectively improve cardiovascular health.

Acknowledgments

We acknowledge the contributions of Terry Pechacek, Dave Buchner, Roseanne Farris, Parakash Pratibhu, Deb Galuska, Adolfo Valadez, Karina Loyo, Rick Schwertfeger, Cindy Batcher, Ella Pugo, Jessie Patton-Levine, Josh Vest, Patty Mabry, John Robitscher, Alyssa Easton, Nancy Williams, and Larry Fine.

Author Information

Corresponding Author: Jack Homer, PhD, Homer Consulting, 4016 Hermitage Dr, Voorhees, NJ 08043. Telephone: 856-810-7673. E-mail: jhomer@comcast.net.

Author Affiliations: Bobby Milstein, Darwin Labarthe, Diane Orenstein, Centers for Disease Control and Prevention, Atlanta, Georgia; Kristina Wile, Sustainability Institute, Stow, Massachusetts; Justin Trogdon, RTI International, Research Triangle Park, North Carolina; Philip Huang, Austin/Travis County Health and Human Services Department, Austin, Texas.

References

1. Bell J, Rubin V. Why place matters: building a movement for healthy communities. Oakland (CA): PolicyLink; 2007.

2. Labarthe D. Epidemiology and prevention of cardiovascular diseases: a global challenge. Gaithersburg (MD): Aspen Publishers; 1998.

3. A public health action plan to prevent heart disease and stroke. Centers for Disease Control and Prevention; 2003. http://www.cdc.gov/dhdsp/library/action_plan/. Accessed August 14, 2009.

4. Sterman JD. Business dynamics: systems thinking and modeling for a complex world. Boston (MA): Irwin McGraw-Hill; 2000.

5. Homer JB, Hirsch GB. System dynamics modeling for public health: background and opportunities. Am J Public Health 2006;96(3):452-8.

6. Homer J, Milstein B, Wile K, Pratibhu P, Farris R, Orenstein D. Modeling the local dynamics of cardiovascular health: risk factors, context, and capacity. Prev Chronic Dis 2008;5(2). http://www.cdc.gov/pcd/issues/2008/apr/07_0230.htm. Accessed August 14, 2009.

7. Anderson KM, Odell PM, Wilson PW, Kannel WB. Cardiovascular disease risk profiles. Am Heart J 1991;121(1 Pt 2):293-8.

8. Pope CA 3rd, Burnett RT, Thurston GD, Thun MJ, Calle EE, Krewski D, et al. Cardiovascular mortality and long-term exposure to particulate air pollution: epidemiological evidence of general pathophysiological pathways of disease. Circulation 2004;109(1):71-7.

9. Dominici F, Peng RD, Zeger SL, White RH, Samet JM. Particulate air pollution and mortality in the United States: did the risks change from 1987 to 2000? Am J Epidemiol 2007;166(8):880-8.

10. Mowery PD, Babb S, Bishop EE, Pechacek TF. Comparison of two methods for estimating prevalence of exposure to secondhand smoke in the US population. Paper presented at Towards a Smokefree Society Conference; September 9-12, 2007; Edinburgh, Scotland.

11. The health consequences of involuntary exposure to tobacco smoke: a report of the Surgeon General. US Department of Health and Human Services, Public Health Service; 2006. http://www.surgeongeneral.gov/library/secondhandsmoke/. Accessed August 14, 2009.

12. Morbidity and mortality: chart book on cardiovascular, lung, and blood diseases. National Heart, Lung, and Blood Institute; 2007. http://www.nhlbi.nih.gov/resources/docs/cht-book.htm. Accessed August 14, 2009.

13. Thompson D, Edelsberg J, Colditz GA, Bird AP, Oster G. Lifetime health and economic consequences of obe-

sity. Arch Intern Med 1999;159(18):2177-83.

14. Katzmarzyk PT, Gledhill N, Shephard RJ. The economic burden of physical inactivity in Canada. CMAJ 2000;163(11):1435-40.

15. Fleshner M. Physical activity and stress resistance: sympathetic nervous system adaptations prevent stress-induced immunosuppression. Exerc Sport Sci Rev 2005;33(3):120-6.

16. Elmer PJ, Obarzanek E, Vollmer WM, Simons-Morton D, Stevens VJ, Young DR, et al. Effects of comprehensive lifestyle modification on diet, weight, physical fitness, and blood pressure control: 18-month results of a randomized trial. Ann Intern Med 2006;144(7):485-95.

17. Rozanski A, Blumenthal JA, Kaplan J. Impact of psychological factors on the pathogenesis of cardiovascular disease and implications for therapy. Circulation 1999;99(16):2192-217.

18. Björntorp P. Do stress reactions cause abdominal obesity and comorbidities? Obes Rev 2001;2(2):73-86.

19. Kouvonen A, Kivimaki M, Virtanen M, Pentti J, Vahtera J. Work stress, smoking status, and smoking intensity: an observational study of 46,190 employees. J Epidemiol Community Health 2005;59(1):63-9.

20. Willi C, Bodenmann P, Ghali WA, Faris PD, Cornuz J. Active smoking and the risk of type 2 diabetes: a systematic review and meta-analysis. JAMA 2007;298(22):2654-64.

21. Flegal KM. The effects of changes in smoking prevalence on obesity prevalence in the United States. Am J Public Health 2007;97(8):1510-4.

22. Haddix AC, Teutsch SM, Corso PS. Prevention effectiveness: a guide to decision analysis and economic evaluation. 2nd edition. New York (NY): Oxford University Press; 2003.

23. Thom T, Haase N, Rosamond W, Howard VJ, Rumsfeld J, Manolio T, et al. Heart disease and stroke statistics — 2006 update: a report from the American Heart Association Statistics Committee and Stroke Statistics Subcommittee. Circulation 2006;113(6):e85-151.

24. Mendez D, Warner KE, Courant PN. Has smoking cessation ceased? Expected trends in the prevalence of smoking in the United States. Am J Epidemiol 1998;148:249-58.

25. Sloan FA, Ostermann J, Conover C, Taylor Jr DH, Picone G. The price of smoking. Cambridge (MA): MIT Press; 2004.

26. Homer J, Milstein B, Dietz W, Buchner D, Majestic E. Obesity population dynamics: exploring historical growth and plausible futures in the US. Paper presented at 24th International System Dynamics Conference; July 23-27, 2006; Nijmegen, The Netherlands.

27. Asch SM, Kerr EA, Keesey J, Adams JL, Setodji CM, Malik S, et al. Who is at greatest risk for receiving poor-quality health care? New Engl J Med 2006;354(11):1147-56.

28. Russell MW, Huse DM, Drowns S, Hamel EC, Hartz SC. Direct medical costs of coronary artery disease in the United States. Am J Cardiol 1998;81(9):1110-5.

29. Sasser AC, Rousculp MD, Birnbaum HG, Oster EF, Lufkin E, Mallet D. Economic burden of osteoporosis, breast cancer, and cardiovascular disease among postmenopausal women in an employed population. Womens Health Issues 2005;15(3):97-108.

30. Flegal KM, Graubard BI, Williamson DF, Gail MH. Cause-specific excess deaths associated with underweight, overweight, and obesity. JAMA 2007;298(17):2028-37.

31. American Diabetes Association. Economic costs of diabetes in the US in 2002. Diabetes Care 2003;26:917-32.

32. Clausen J, Jensen G. Blood pressure and mortality: an epidemiological survey with 10 years follow-up. J Hum Hypertens 1992;6:53-9.

33. Homer J. A dynamic model of cardiovascular risk: reference guide for model version 8m. Atlanta (GA): Centers for Disease Control and Prevention; 2008.

34. Sjostrom L, Lindroos AK, Peltonen M, Torgerson J, Bouchard C, Carlsson B, et al. Lifestyle, diabetes, and cardiovascular risk factors 10 years after bariatric surgery. N Engl J Med 2004;351(26):2683-93.

35. Magnus P, Beaglehole R. The real contribution of the major risk factors to the coronary epidemics: time to end the "only-50%" myth. Arch Intern Med 2001;161(22):2657-60.

36. Russell LB. Preventing chronic disease: an important investment, but don't count on cost savings. Health Aff (Millwood) 2009;18(1):42-5.

37. Kahn R, Robertson RM, Smith R, Eddy D. The impact of prevention on reducing the burden of cardiovascular disease. Circulation 2008;118(5):576-85.

Tables

Table 1. Interventions Used in Simulation Model for Cardiovascular Health Outcomes, Organized by Topical Cluster

Topical Cluster	Intervention
Care	Access to affordable primary care Promotion of primary care use Good-quality primary care
Air	Tobacco taxes and sales restrictions Social marketing against smoking Access to affordable smoking cessation services and products Promotion of smoking cessation Bans on smoking in public places Regulations and incentives that reduce air pollution
Lifestyle	Access to affordable healthy foods Promotion of healthy diet Junk food taxes and sales restrictions Access to safe and affordable physical activity Promotion of physical activity Reduced sources of chronic stress through improved living conditions and social supports
Weight-loss and mental health services	Access to affordable weight-loss services for the obese Promotion of weight-loss services Access to affordable mental health services for the chronically stressed Promotion of mental health services

Table 2. Information Sources Used in Simulation Model for Cardiovascular Health Outcomes

Topic	Source
Population size, growth, and aging, and health care coverage	US Census
Rates of cardiovascular events and deaths	Reports from American Heart Association (23) and National Institutes of Health (12)
Prevalence rates of smoking, obesity, and chronic disorders, and rates of diagnosis and control of chronic disorders	National Health and Nutrition Examination Survey (NHANES), 1988-1994 and 1999-2004
Fraction of 18-year-olds who smoke, are obese, or have chronic disorders	NHANES, Youth Risk Behavior Surveillance System
Prevalence of psychosocial stress	Behavioral Risk Factor Surveillance System (BRFSS)
Access to and use of good nutrition, physical activity, and primary care	BRFSS
Rates of smoking cessation	Mendez et al (24), Sloan et al (25)
Rates of people moving from obese to nonobese	Homer et al (26)
Trend in fraction of workplaces allowing smoking	Surgeon General's report (11)
Trend in small particulate matter (PM 2.5) air pollution	Dominici et al (9)
Average quality of primary care	Asch et al (27)
Medical costs, sick days, and years of life lost due to CVD events and deaths	Social Security actuarial life tables, Haddix et al (22), Russell et al (28), Sasser et al (29)
Noncardiovascular medical costs and sick days due to smoking, obesity, diabetes, and high blood pressure	Linked files of Medical Examination Panel Survey, National Health Interview Survey
Noncardiovascular mortality and years of life lost due to smoking, obesity, diabetes, and high blood pressure	Centers for Disease Control and Prevention Smoking-Attributable Mortality, Morbidity, and Economic Costs (SAMMEC), World Health Organization Statistical Information System, Flegal et al (30), American Diabetes Association (31), Clausen et al (32)

16

Analyzing National Health Reform Strategies with a Dynamic Simulation Model (2010)

Milstein B, Homer J, Hirsch G. *American Journal of Public Health*, 100(5): 811-819

Related writings

Milstein B, Homer J, Briss P, Burton D, Pechacek T. Why Behavioral and Environmental Interventions are Needed to Improve Health at Lower Cost. *Health Affairs*, 30(5), May 2011. DOI: 10.1377/hlthaff.2010.1116.

Milstein B, Homer J, Hirsch G. The "HealthBound" Policy Simulation Game: An Adventure in U.S. Health Reform. *Proceedings of the 27th International System Dynamics Conference*, Albuquerque, NM, 2009.

Analyzing National Health Reform Strategies With a Dynamic Simulation Model

| Bobby Milstein, PhD, MPH, Jack Homer, PhD, and Gary Hirsch, SM

Proposals to improve the US health system are commonly supported by models that have only a few variables and overlook certain processes that may delay, dilute, or defeat intervention effects. We use an evidence-based dynamic simulation model with a broad national scope to analyze 5 policy proposals. Our results suggest that expanding insurance coverage and improving health care quality would likely improve health status but would also raise costs and worsen health inequity, whereas a strategy that also strengthens primary care capacity and emphasizes health protection would improve health status, reduce inequities, and lower costs. A software interface allows diverse stakeholders to interact with the model through a policy simulation game called HealthBound. (*Am J Public Health.* 2010;100:811–819. doi:10.2105/AJPH.2009.174490)

The multiple shortcomings of the US health system are well known. US health care spending per capita is the highest in the world, but Americans have comparatively high rates of morbidity and premature mortality,[1] along with persistent disparities among subgroups. People with lower socioeconomic status, for example, are much more likely to develop disease and injury and to become disabled or die prematurely as a result, in part because they face greater health threats and are also less likely to have access to high-quality health care.[2,3]

Various theories have been offered to explain why the US health system performs so poorly and is so costly.[4,5] Many point to the lack of health insurance for millions as the system's chief problem.[6] Some criticize the medical industry and the public at large for overemphasizing disease detection and treatment while missing opportunities to reduce preventable risk and protect people's health.[7,8] Others blame perverse incentives and community norms that encourage physician entrepreneurship and profit making over collaboration, coordination, or conservative practice.[9,10] Still others say that there are too few primary care providers or that the providers we do have are underpaid and unable to offer the highest-quality care.[11,12] And some fault private insurers, who pass along high overhead costs to consumers, are unwilling to reimburse adequately for preventive care, and offer a confusing array of coverage plans, creating a substantial administrative burden for providers.[13]

Likewise, reform proposals vary widely in their goals and policy levers. One leading proposal calls for wider health insurance coverage and better quality of care through computerization and payment incentives for providers. Supporters claim that these 2 changes will improve people's health and reduce health care costs, effectively remaking the health delivery system.[14] Other proposals run the gamut, from outlining 15 ways to cut costs[15] to focusing on relatively inexpensive population-based programs to increase physical activity, improve nutrition, and prevent tobacco use.[16]

Such proposals are usually justified by calculating their costs and benefits over time. However, the mathematical models commonly used to generate these projections typically examine only a small number of variables over a relatively short time, rather than portraying the health system as the diverse and dynamic enterprise that it is.[17] In particular, these models usually ignore the effects of accumulations, time delays, resource constraints, and behavioral feedback. System scientists have shown how such processes must be considered to reach correct conclusions about the net impacts of interventions in systems with many interacting actors, multiple goals, and conflicting interests.[18–20] They have also challenged the methodological convention of using artificially short time frames for prospective evaluation, such as when studying investments in youth or interventions to prevent slow-moving chronic diseases.[21,22]

We present the results of simulations of several intervention scenarios using a computer model that accounts for many factors that make the US health system hard to understand and difficult to change. The model we developed has a relatively long time horizon (25 years) and a causal structure that is broad in scope and rich in realistic detail. Still, the model is sufficiently streamlined to allow extensive testing and analysis. Interventions are represented at a general level without detailing how they might be implemented. Such simplification allowed us to evaluate various intervention strategies and study their fundamentally different impacts on the health system before considering tactical details.

Strategic modeling of this sort is not intended to predict the future but rather to help us learn how the complex US health system tends to respond over time to different reform strategies. It supports critical thinking and more robust conclusions about how selected intervention options are likely—in directional terms and relative to one another—to affect performance of the health system in the short and long term. Sterman presented a rationale for using simulation models in this way, i.e., as tools to improve understanding of complex systems.[20]

Planners may use our model to simulate many types of interventions, but just 5 scenarios are explored here, starting with a strategy to expand insurance coverage and improve the quality of health care. Four additional strategies are then tested to determine whether alternative approaches might yield better results.

METHODS

We used the well-established methodology of system dynamics simulation modeling, combining established evidence and insights from previous research, to create a realistic but simplified model of the US health system.[23] Figure 1 displays the major elements of this simulated system, and Table 1 lists the interventions that may be tested, individually or in combination.

Our intent was to trace the consequences of selected interventions in sufficient detail to calculate their likely combined impact over time on morbidity, mortality, health equity, and cost. Several hundred elements were needed to adequately represent the concepts and interactions depicted in Figure 1. Early in the process of developing the model, the Centers for Disease Control and Prevention (under whose auspices the model was developed) convened a stakeholder review involving about 25 policy leaders—including the senior staff of national health organizations, university faculty, and health policy analysts—to help ensure that the model's boundaries, relationships, and level of aggregation were appropriate for representing the most important interventions and outcomes.

National-level data from the late 1990s and early 2000s were used to calibrate the model's input parameters and to confirm that its output faithfully reproduced key historical metrics. These data included measures of death rates (*National Vital Statistics Reports*) and their disparity by income level,[24] unhealthy days and access to primary care providers by income level (Behavioral Risk Factor Surveillance System), rates of health care utilization (National Ambulatory Medical Care Survey, National Hospital Ambulatory Medical Care Survey, National Nursing Home Survey, National Home Health Care Survey), prevalence of disease and asymptomatic disorders (National Health Interview Survey, National Health and Nutrition Examination Survey) and their disparity by income level,[2] and health care costs by category (National Health Expenditure Accounts).

Model Logic and Assumptions

The model simulates changes over a 25-year period for the entire US population, which is divided into 2 socioeconomic groups—the advantaged and the disadvantaged—and further subdivided among 3 states of health: those with (1) symptomatic disease or injury, (2) asymptomatic disorders that may lead to symptomatic disease, and (3) no significant health problems. In the model's causal structure, disadvantage erodes health partly by making life more stressful.[25] Disadvantage also makes it harder for people to choose healthier behaviors and exposes them to more hazardous environments, leaving them more vulnerable to disease and injury.[26,27] The disadvantaged also have less access to health care than do the advantaged, because they have less insurance coverage and less sufficiency of primary care providers (PCPs) to meet current patient demand.[3] The adverse conditions of the disadvantaged combined with poor access to health care means that the disadvantaged experience higher morbidity and mortality than do the advantaged, resulting in substantial health inequity.[24,28,29]

Another factor in the model that affects health outcomes is the quality or thoroughness of care delivered, reflecting the extent to which providers follow guidelines for best practices in detecting and preventing illness, managing chronic diseases, and handling urgent events.[30,31]

The model captures all reported health care costs associated with personal health services and supplies, plus administrative costs of health insurance. Together these costs account for more than 90% of all health-related spending in the United States.[32] The model differentiates among payments to hospitals, office-based physicians, nursing homes and home health care, and dental and other health services, as well as spending on prescription drugs and other personal medical products.

Better access to and quality of care can improve health outcomes, but the effects of better access and quality on costs are less certain. For example, better preventive and chronic care can reduce the frequency of more costly acute complications and urgent hospital visits,[33,34] but better preventive and chronic care require additional visits and medications. As a result, although good preventive and chronic care is typically cost-effective (improving health at reasonable cost and thus arguably worth doing), it does not typically reduce total health care costs.[35–38]

For learning purposes, we have initialized the model in a "status quo" equilibrium, with all outcome variables unchanging and closely matching historical values circa 2003, the latest year for which some key data were available. The equilibrium setting lets planners easily understand the effects of their simulated interventions by comparing output values with the corresponding starting values. Interventions begin immediately at time zero and continue to the end of the 25 years of simulation, with specified direct effects and time delays.

Evaluating Interventions

Summary measures of population health simulated by the model include the number of unhealthy days per month and the annual death rate per 1000 persons. A less familiar but

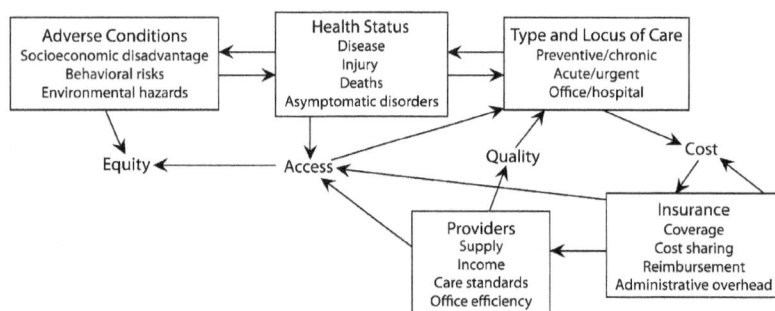

Note. The following elements are stratified by socioeconomic status: behavioral risks, environmental hazards, health status, type and locus of care received, primary care providers, access, insurance coverage, and cost sharing. For detailed explanation of the relationships among the elements, see the technical appendix (available as a supplement to the online version of this article at http://www.ajph.org).

FIGURE 1—Connections between the major elements of a simulated US health system model.

TABLE 1—Intervention Options and Effects in Model Simulating the US Health System

Intervention	Effects
Insurance related	
Expand insurance coverage[a]	Would improve access to quality care, meaning better health but also more spending on visits, procedures, and medications. Also would put more demand on limited supply of PCPs and would increase total insurance administration expenses.
Change self-pay fraction for the insured[a] (cost sharing)	Increased cost sharing would reduce utilization of high-quality office care, meaning worse health but less spending. Also would alleviate some demand on limited supply of PCPs.
Change reimbursement rate for office visits	A lower rate would reduce spending on office visits but also would reduce PCP net income, which may lead to a decline in primary care supply.
Change reimbursement rate for hospital visits	A lower rate would reduce spending on hospital visits but also would hurt the quality of urgent care. A lower rate also would lead to reduced elective hospital capacity, thereby impairing the effectiveness of disease and injury management in some cases.
Simplify insurance (reduce administrative overhead)	Insurance plan standardization would reduce PCP administrative costs and thereby improve PCP income. Single-payer approach would do the same and also would eliminate the marketing, eligibility review, and negotiation costs of private insurance administration.
Provider related	
Increase preventive and chronic care quality	Would slow progression of asymptomatic disorders into disease and reduce frequency of acute and urgent episodes, but also would increase spending on office visits and medications while increasing demand on limited supply of PCPs.
Improve urgent care quality	Would reduce mortality and need for inpatient stays and extended care.
Expand primary care supply[a]	Would alleviate shortages of PCPs, but if a surplus results, would reduce PCP average net income.
Improve primary care efficiency	Would alleviate shortage of PCPs and increase PCP net income.
Coordinate health care	Would reduce volume of office visits as well as elective hospital procedures and inpatient stays without adversely affecting quality of care. Could help alleviate PCP shortages, but if a surplus results, would reduce PCP net income.
Adverse conditions related	
Enable healthier behaviors[a]	Would reduce the fraction of the population at elevated risk for asymptomatic disorders or disease and injury resulting from unhealthy behavior.
Build safer environments[a]	Would reduce the fraction of the population at elevated risk for disease and injury, and for unhealthy behavior, resulting from an unsafe environment.
Create pathways to advantage	Would increase the flow of people from disadvantaged to advantaged.

Note. PCPs = primary care providers.
[a]For advantaged or disadvantaged subgroups.

equally important metric is health inequity, which we calculate as the fraction of unhealthy days in the overall population attributable to the disparity between the advantaged and disadvantaged subgroups.

Another key simulation output is the net benefit of interventions, which is our summary measure of cost-effectiveness. Net benefit is calculated by subtracting the simulated change in total costs (health care costs plus direct costs of interventions) from the monetized change in quality-adjusted life-years (QALYs). We used a value of $75 000 per QALY, which falls in the middle of the range recommended by health economists.[39,40] The change in QALYs resulting from intervention is calculated on the basis of changes in the simulated number of

unhealthy days and in the simulated number of deaths.

Proposed interventions are evaluated according to 5 outcome measures: unhealthy days, deaths, health inequity, health care costs, and net benefit. Multiple metrics are needed to reflect distinct values at stake in the health system, as well as the potential trade-offs among them. Even if an intervention produces positive net benefit (improving health at reasonable cost), it may still be unacceptable to some stakeholders if it raises total health care spending or worsens inequity.

Many of the model's input parameters are subject to some imprecision, often because diverse concepts have been aggregated. In particular, several interventions listed in Table

1 may encompass an array of activities having different costs and direct effects. For example, the cost of enabling healthier behaviors may range from relatively inexpensive population-wide measures (e.g., social marketing) to more costly individual interventions (e.g., smoking cessation therapy). Accordingly, sensitivity tests were performed around the ranges of possible parameter values to determine what effect, if any, the uncertainties may have on the rank ordering of scenarios according to various performance measures. For each scenario, the model was run under 3 settings: baseline, pessimistic (i.e., assuming weaker intervention effects with higher costs), and optimistic (i.e., assuming stronger intervention effects with lower costs). Specific values for these

configurations and additional details on the model's information sources, logic, and assumptions are included in the technical appendix (available as a supplement to the online version of this article at http://www.ajph.org).

RESULTS

The results of 5 simulated intervention scenarios are presented below, starting with a strategy to expand insurance coverage and improve the quality of health care services.[41] Table 2 shows the percentage changes in selected outcomes for the 5 scenarios at year 5 and year 25 relative to starting values, with all uncertain parameters set to baseline values.

Figure 2 shows the impacts of the intervention scenarios over time in terms of per capita costs, QALYs, and net benefit. Cumulative 25-year impacts on costs, QALYs, and net benefit are presented in Table 3, which shows results corresponding to the baseline parameter settings as well as to the optimistic and pessimistic settings, to assess the effects of numerical uncertainties.

Coverage Plus Quality

In the scenario of coverage plus quality, insurance coverage was expanded for the entire population, advantaged and disadvantaged, causing the uninsured fraction to drop from 15.5% to less than 2% (data not shown).

At the same time, initiatives were implemented to improve the quality of preventive, chronic, and urgent care. The fraction of office-based physicians following guidelines rose from 80% to 88%, leading to better diagnosis and care of asymptomatic disorders, disease, and injury, which in turn reduced the volume of visits for acute and urgent care. Also, the fraction of hospitals following guidelines rose from 80% to 88%, which reduced fatalities and ensured fewer inpatient stays and less need for extended care.

This strategy produced mixed results (Table 2). On the positive side, morbidity (unhealthy days) declined by about 6% by year 25, and mortality declined by about 12%. This

TABLE 2—Percentage Changes From Initial Value in Selected Outcomes in Model Simulating the US Health System at Year 5 and Year 25, by Scenario

	Coverage + Quality, % Change	Coverage + Quality + Capacity, % Change	Coverage + Quality + Capacity + Reimbursement Cut, % Change	Coverage + Quality + Capacity + Protection, % Change	Capacity + Protection, % Change
Unhealthy days per capita per mo (initial = 5.26)					
Year 5	−4.7	−6.4	−3.4[a]	−9.5[b]	−4.5
Year 25	−6.1	−11.1	−5.3[a]	−20.7[b]	−13.0
Deaths per 1000 persons per y (initial = 7.46)					
Year 5	−9.7	−11.3	−5.4	−14.2[b]	−4.5[a]
Year 25	−11.8	−16.4	−8.2[a]	−25.5[b]	−13.1
Health inequity[c] (initial = 0.143)					
Year 5	+2.3[a]	−7.1	+0.5	−14.2	−15.2[b]
Year 25	+6.9[a]	−20.6	+3.3[a]	−32.9[b]	−26.8
Health care costs per capita per y (initial = $5434)					
Year 5	+4.9[a]	+3.5[a]	−8.5[b]	+1.3[a]	−3.4
Year 25	+5.8[a]	+2.6[a]	−8.3[b]	−4.6	−8.8
Disease and injury prevalence (initial = 0.378)					
Year 5	−0.2[a]	−0.3	−0.4	−3.3[b]	−3.1
Year 25	−1.0[a]	−2.0	−1.7	−12.6[b]	−11.1
Effectively managed fraction of disease and injury (initial = 0.583)					
Year 5	+14.8	+20.3	+10.0	+20.8[b]	+4.6[a]
Year 25	+17.2	+30.3[b]	+12.0	+30.2	+6.9[a]
Fraction of acute nonurgent visits to hospital versus physician office (initial = 0.143)					
Year 5	+16.6[a]	−15.3	+0.6	−18.5	−31.5[b]
Year 25	+26.7[a]	−40.1	+6.2[a]	−41.0	−41.7[b]
Sufficiency of PCPs (initial = 0.905)					
Year 5	−6.8[a]	+7.0	−0.1	+8.1	+10.5[b]
Year 25	−7.8[a]	+10.5[b]	−1.3[a]	+10.5[b]	+10.5[b]

Note. PCP = primary care provider. All scenarios computed according to baseline parameter settings.
[a]Worst result among the 5 scenarios or an undesirable result.
[b]Best result among the 5 scenarios.
[c]Health inequity is the fraction of unhealthy days in the overall population that is attributable to the disparity between the advantaged and disadvantaged subgroups.

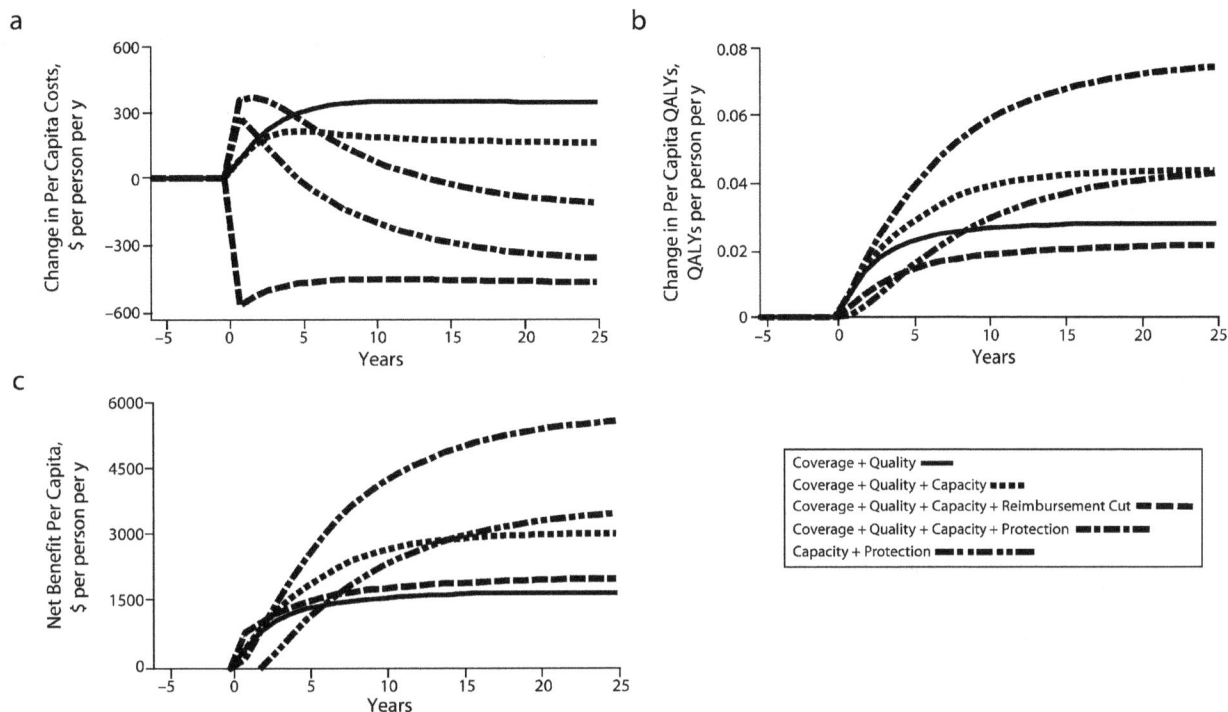

Note. The change in per capita costs equals the change in health care costs plus spending on intervention programs. The change in QALYs results from fewer deaths plus fewer unhealthy days. Net benefit is calculated as the product of the change in QALYs × $75 000 minus the change in costs. All 3 measures are presented here on a per capita basis, using total population as the denominator. All outcomes shown are computed according to the baseline parameter setting.

FIGURE 2—Outcomes in model simulating the US health system, by scenario for (a) per capita costs, (b) quality-adjusted life-years (QALYs), and (c) net benefit.

improvement reflected greater utilization of higher-quality care, including a 17% increase in the effectively managed fraction of disease and injury. On the negative side, health care costs rose nearly 6% (5% because of increased coverage alone; data not shown). This result is in line with Hadley and Holahan's statistical analysis indicating that extended coverage would increase health care spending by 3% to 6%.[42] Health inequity in this scenario also worsened by 7%.

If preventive and chronic care prevented costly urgent events, why did costs rise? First, as noted above, good-quality preventive and chronic care, though cost-effective, do not in general reduce costs. Second, improved chronic and urgent care extends the lives of persons who often have costly conditions.[43] Third, higher care utilization resulting from increased coverage and quality of care reduced the sufficiency of PCPs in the simulation, especially

for disadvantaged patients. As a result, minor acute events that would otherwise have resulted in a visit to a PCP increasingly shifted to more costly care in hospital emergency departments.

If expanded coverage helped the disadvantaged more than the advantaged, why did health inequity become worse? Overloaded PCPs are at the root of this problem as well. Increased coverage and quality of care put some strain on PCPs for the advantaged population, but the increased demand for care for this subgroup was gradually met by an increased supply of PCPs, so the strain was minimized. By contrast, the supply of PCPs serving the disadvantaged was far from adequate to meet demand, even before our interventions. Because PCPs lacked the capacity to improve preventive and chronic care for all of the additional patients who obtained coverage, the health of the disadvantaged population

did not improve nearly as much as that of the advantaged population. Recent experience in Massachusetts confirms that limited PCP capacity undercuts the effect of expanded insurance coverage.[44,45] There, expanded coverage enrolled more than 300 000 people, but a lack of PCP availability meant that many of the newly insured could not find a regular source of primary care.

The coverage plus quality scenario resulted in an increase in QALYs, but it also resulted in an increase in costs. Both QALYs and costs grew for the first 5 or so years and then remained at their increased levels (Figure 2). The net benefit for this scenario was positive throughout the simulation, amounting cumulatively to trillions of dollars over 25 years (baseline: $10 trillion; optimistic: $20 trillion; pessimistic: $5 trillion; Table 3). The positive net benefit tells us that this approach was cost-effective, though it appears to have serious drawbacks. Other

TABLE 3—Cumulative 25-Year Outcomes for Costs, Quality-Adjusted Life-Years (QALYs), and Net Benefits in Model Simulating the US Health System, by Scenario

	Coverage + Quality	Coverage + Quality + Capacity	Coverage + Quality + Capacity + Reimbursement Cut	Coverage + Quality + Capacity + Protection	Capacity + Protection
Baseline parameter settings					
Change in costs[a] ($ billions)	+2111[d]	+1175	−3240[e]	+508	−1178
Change in QALYs[b] (millions)	+164	+240	+115[d]	+376[e]	+191
Net benefit[c] ($ billions)	+10 175[d]	+16 842	+11 836	+27 689[e]	+15 488
Optimistic parameter settings					
Change in costs[a] ($ billions)	+1750[d]	+515	−3533[e]	−1147	−2368
Change in QALYs[b] (millions)	+285	+394	+285	+528[e]	+214[d]
Net benefit[c] ($ billions)	+19 640	+29 003	+24 934	+40 717[e]	+18 414[d]
Pessimistic parameter settings					
Change in costs[a] ($ billions)	+2196	+1563	−3096[e]	+3330[d]	+1432
Change in QALYs[b] (millions)	+92	+136	+8[d]	+266[e]	+163
Net benefit[c] ($ billions)	+4704	+8651	+3716[d]	+16 630[e]	+10 785

[a]Change in costs equals the change in health care costs plus spending on intervention programs.
[b]Change in QALYs equals the change resulting from fewer deaths plus the change resulting from fewer unhealthy days.
[c]Net benefit equals (change in QALYs × $75 000) minus change in costs.
[d]Worst result among the 5 scenarios.
[e]Best result among the 5 scenarios.

interventions would be needed to reverse the adverse cost and equity impacts.

Coverage Plus Quality Plus Capacity

The next scenario combined the coverage plus quality scenario with 2 interventions that increase primary care capacity. One intervention subsidizes the training and placement of new PCPs for the disadvantaged population, and the other makes existing primary care offices more efficient. A more efficient office has lower operating expenses and allows PCPs to see more patients in a day without skimping on quality. Reduced expenses, in turn, raise PCP incomes and encourage more medical students to become PCPs. As a result of these interventions, the shortages of PCPs seen in the coverage plus quality scenario (Table 2) were eliminated in this second scenario by year 8 (data not shown).

Eliminating PCP shortages improved access for the disadvantaged and eventually reduced health inequity by 21%, leading to more reduction of unhealthy days (11% decrease) and deaths (16% decrease). The effective management of disease and injury expanded by 30%, and improved preventive care among the disadvantaged allowed disease and injury prevalence to be reduced by 2%.

The extension of preventive and chronic care to disadvantaged persons caused no net increase in health care costs and in fact reduced the 6% increase seen in coverage plus quality alone to only a 3% increase. Why did the cost situation improve? When the shortage of PCPs was eliminated, many of the disadvantaged shifted their locus of care from hospital emergency departments to less expensive PCPs. This shift from emergency care to PCPs was the opposite of what occurred in the coverage plus quality scenario.

Adding capacity interventions to coverage plus quality increased net benefit by both increasing QALYs and reducing total costs (Figure 2). The cumulative net benefit increased by 48% to 84%, depending on the uncertainty settings (Table 3). Costs were still higher than in the status quo with no intervention at all, but with the elimination of PCP shortages, this approach improved health outcomes more cost-effectively while also reducing health inequity.

Coverage Plus Quality Plus Capacity Plus Reimbursement Cut

A third scenario combined the previous interventions with 20% cuts in reimbursement rates. These cuts applied to both office and

hospital care and for all insurance types (both private and public), with no change in the fee-for-service payment scheme. The previous interventions improved health outcomes and equity but did not lower overall health care costs. We wanted to determine whether including reimbursement cuts would reduce health care costs while maintaining the previously achieved benefits.

These cuts did lower health care costs, by about 8% relative to the baseline. However, they also impeded the ability of hospitals to provide first-class urgent and elective care, thus undermining the quality interventions. The entire package with lower reimbursement reduced morbidity and mortality relative to the baseline, but by less than in the previous scenarios. The cuts also reduced PCP incomes, thereby inhibiting PCP supply and ultimately undermining the capacity interventions as well. The sufficiency of PCPs for the disadvantaged increased, but only negligibly and only in the short term, erasing the equity gains seen in the previous simulation. In addition, the cost-saving movement of disadvantaged patients from hospital emergency departments to PCPs seen in the coverage plus quality plus capacity scenario no longer occurred. In this way, the reimbursement cuts somewhat undermined their own intended purpose (to reduce health care costs).

Although the reimbursement cuts reduced total costs below the starting level, they undercut QALYs to such an extent that they would do more harm than good in terms of net benefit (Figure 2). Under all uncertainty assumptions, the cumulative net benefit over 25 years (Table 3) was lower than it was in the coverage plus quality plus capacity scenario by 14%–57%.

Coverage Plus Quality Plus Capacity Plus Protection

A fourth scenario combined the coverage, quality, and capacity interventions with 2 broad-based interventions for health protection, a concept encompassing both promotion of health and the ensuring of safer, healthier living conditions.[46–48] One of the interventions enables healthier behaviors, reducing the prevalence of unhealthy behavior by 30% by year 5 and 43% by year 25. The other intervention builds safer environments, reducing the fraction of the population living in an unsafe environment by 36% by year 5 and 52% by year 25.

Adding health protection to the mix of interventions reduced the onset of disease and injury, causing the prevalence of these conditions to decline over time, by about 3% relative to the baseline by year 5 and 13% by year 25. By reducing the prevalence of disease and injury, the health protection interventions improved most of the metrics in Table 2, including unhealthy days (21% decrease relative to baseline by year 25), deaths (26% decrease), health inequity (33% decrease), and health care costs (5% decrease). Inequity was reduced because the disadvantaged, who have a larger share of behavioral and environmental risks than the advantaged do to begin with, benefited disproportionately when these problems were addressed. Costs went down for 2 reasons. First, lower disease and injury prevalence translated directly into lower costs.[49] Second, lower prevalence also meant less demand on PCPs, which further alleviated PCP shortages (eliminated by year 7 of the simulation) and further reduced the fraction of visits for acute care going to more expensive hospital emergency departments rather than PCPs.

Effective health protection interventions may be costly to implement, and we have estimated conservatively that they would cost hundreds of dollars per beneficiary, rising to thousands in the pessimistic setting. Greater spending was required in the earlier years, when there were more potential recipients, than in later years, after the interventions had time to reduce the number in need. This initial investment explains why, as Figure 2 shows, simulated total costs rose for the first several years above those of the previous scenarios. However, total costs dropped rapidly as the beneficial impacts of the combined investment began lowering health care costs, and as the magnitude of the investment itself declined. By year 14, total costs crossed the zero line into the region of net savings.

This scenario produced the greatest net benefit, under all uncertainty settings, and the greatest increase in QALYs (Figure 2; Table 3). The fact that this conclusion held true even under the pessimistic setting, with its considerably larger assumed intervention costs, underscores how powerfully cost-effective these population-based interventions could be.

Capacity Plus Protection

Although the greatest overall improvement was delivered by combining interventions from 4 different intervention categories (coverage, quality, capacity, and protection), this combination did not reduce health care costs in the early years of the simulation; they were still up 1% as of year 5. During those early years, the reductions in health care costs resulting from the capacity and protection interventions did not reach their full potential to overcome the increases in costs caused by expanding coverage and improving quality.

A final scenario considered how much capacity plus protection by themselves could accomplish. Improvements in morbidity and mortality were not as substantial as for the more comprehensive 4-component scenario, of course, but over time these 2 measures did reduce morbidity and mortality more than did the coverage plus quality approach. Also, absent the additional demands on PCPs from expanded coverage and quality, the capacity and protection interventions caused PCPs for the disadvantaged to be sufficient by year 4—earlier than it was in any other scenario. As a consequence, health inequity was impressively reduced (27% by year 25), despite the lack of expansion in insurance coverage.

Health care costs declined in this scenario, even early on (down more than 3% in year 5), and increasingly over time (9% by year 25). Initially, the cost reduction came primarily from patient visits shifting from hospital emergency departments to PCPs as a result of the capacity interventions; but later, additional cost reductions came from the declining prevalence of disease and injury as a result of the protection interventions.

The capacity plus protection approach produced less net benefit than the full 4-component combination (Figure 2; Table 3), as expected, but it ultimately produced more net benefit than did the coverage plus quality approach. The capacity plus protection approach quickly (after the first couple of years of initial investment) became lower in total costs, and increasingly so over time (Figure 2). And though it took capacity plus protection a longer time to generate QALY improvements, by year 9 it surpassed coverage plus quality in this respect as well. Consequently, the net benefit from capacity plus protection started below that of coverage plus quality but surpassed it by year 7. Our simulations suggest that, over 25 years, capacity plus protection was likely to contribute significantly more cumulative net benefit than coverage plus quality alone, but with some uncertainty (Table 3).

DISCUSSION

These simulations suggest that a strategy of expanding insurance coverage and improving health care quality is clearly cost-effective (providing reasonable value for the money), but if implemented without other interventions would likely yield only modest improvements in health while increasing costs and worsening health inequities. One roadblock to the success of this approach is an insufficiency of PCPs to handle the additional workload, especially for the disadvantaged portion of the population. Expanding primary care capacity for the disadvantaged could dramatically improve access and equity and would help to lower costs by encouraging a shift to PCPs and away from expensive hospital emergency departments for the care of minor ailments.

This analysis also suggests that population-based health protection efforts enabling healthier behaviors and ensuring safer environments are indispensable for reducing the prevalence of disease and injury and thereby for improving health and reducing inequity while lowering costs. Improvements in health care quality would do little by themselves to reduce disease prevalence, because these improvements would do nearly as much to extend life as they would to reduce the onset of new disease and injury.

Broad cuts in insurance reimbursement rates to providers would likely be counterproductive, reducing costs at the expense of health and health equity. A previous dynamic analysis[50] suggested that reimbursement cuts may not even accomplish their primary aim of reducing health care costs and instead may incite an inflationary tug of war between insurers and providers, who have a variety of ways to circumvent cost-containment efforts. Such cuts would undermine the benefits of the other interventions by impairing hospital services and making primary care an even more undesirable choice for medical students than it already is.

In the midst of an active national debate over health reform, it is useful to consider how the elements of coverage, quality, capacity, and protection could be addressed to maximize the opportunity for improvement and avoid the mistake of doing too little, as coverage plus quality alone would likely be. Indeed, with cost a primary concern, the cost-saving interventions of capacity and protection take on additional importance and are needed now, not later, to offset the likely cost increases from coverage and quality.

Limitations and Extensions

Our dynamic model of the US health system is, like all models, a simplified version of reality, with many limitations and uncertainties that make it impossible to use the model to predict the future with precision.[51] Although this model is simplified, it integrates a wider spectrum of what is known about health system dynamics than do narrower analyses and thus more fully captures the potential impacts of possible interventions. Moreover, sensitivity testing demonstrated that the conclusions drawn from the

model are robust to many numerical uncertainties. However, the model does have limitations and is a work in progress that will continue to evolve with new information, policies, and stakeholder perspectives.

One limitation is the use of an equilibrium base case, which required that we exclude several known trends affecting the health system. These include, for example, the spread of new medical technologies, population growth and aging, the practice of defensive medicine, the medicalization of common ailments through direct-to-consumer advertising, changing lifestyles, and changes in economic prosperity. To gauge the significance of these exclusions, we developed an extended model that includes 2 key trends: health care price inflation and aging of the population. Analyses using the extended model confirmed that the general conclusions described in this article are not affected by either of these exacerbating trends (data not shown).

Another possible limitation is that the model represents interventions in broad strokes, without detailing precisely how they might be implemented. We contend that there is value in first understanding the likely consequences of major intervention options before delving into specific tactics. Nevertheless, some policy leaders may want greater specificity.

An Opportunity for Game-Based Learning

Intervention planners and change agents can use this simulation model to improve their understanding of how the US health system may respond to different interventions and why. An Internet-based user interface allows nonmodelers to interact with the tool in the form of a policy simulation game called HealthBound.[52] "Serious games" such as this one are gaining popularity, in part because they support both cognitive and experiential learning without relying on expert analysts as go-betweens. As players explore scenarios for themselves, they learn how to act more effectively in the real world.[53,54]

One of the biggest impediments to health reform initiatives is that different stakeholders tend to approach the challenge differently, each focused only on a few parts of the overall system. By contrast, HealthBound offers a more comprehensive and neutral framework for intervention planning that can be especially

useful for unfreezing attitudes and expanding perceptions. When used in multistakeholder groups led by a trained facilitator, HealthBound allows stakeholders to learn not only from the game but also from each other, as players test and refine their ideas in this simplified but realistic version of the US health system. ∎

About the Authors

Bobby Milstein is with the Syndemics Prevention Network, Centers for Disease Control and Prevention, Atlanta, GA. Jack Homer is with Homer Consulting, Voorhees, NJ. Gary Hirsch is an independent consultant in Wayland, MA.

Correspondence should be sent to Jack Homer, Homer Consulting, 4016 Hermitage Drive, Voorhees, NJ 08043 (e-mail: jhomer@comcast.net). Reprints can be ordered at http://www.ajph.org by clicking the "Reprints/Eprints" link.

This article was accepted October 20, 2009.

Note. *The findings and conclusions in this article are those of the authors and do not necessarily represent the official position of the Centers for Disease Control and Prevention.*

Contributors

All authors collaborated equally in conceptualizing and conducting this study, and in writing and revising the article. B. Milstein coordinated the project and helped with scenario development. J. Homer led simulation modeling and interpretation. G. Hirsch gathered empirical information and analyzed policy implications.

Acknowledgments

The authors thank Brad Perkins, Julie Gerberding, Casey Chosewood, Stephanie Bailey, Donna Garland, Kathleen Toomey, Darwin Labarthe, Paul Halverson, Mary Pittman, Jeff McKenna, Cathleen Walsh, Richard Klein, John Anderton, Julia Smith-Easley, Linda Carnes, Alvin Hall, Diane Orenstein, and the many colleagues who have championed our health policy modeling at the Centers for Disease Control and Prevention. The name HealthBound is used courtesy of Associates and Wilson, Inc.

Human Participant Protection

No human research participants were involved in this study.

References

1. Nolte E, McKee CM. Measuring the health of nations: updating an earlier analysis. *Health Aff (Millwood).* 2008;27(1):58–71.

2. Banks J, Marmot M, Oldfield Z, Smith JP. Disease and disadvantage in the United States and in England. *JAMA.* 2006;295(17):2037–2045.

3. National Association of Community Health Centers, Robert Graham Center. Access denied: a look at America's medically disenfranchised. Washington, DC: American Academy of Family Physicians; 2007. Available at: http://www.graham-center.org/online/graham/home/publications/monographs-books/2007/rgcmo-access-denied.html. Accessed November 1, 2009.

4. The Commonwealth Fund Commission on a High Performance Health System. Why not the best? Results

from the National Scorecard on US Health System Performance, 2008. New York, NY: Commonwealth Fund; 2008. Available at: http://www.commonwealthfund.org/publications/publications_show.htm?doc_id=692682. Accessed November 1, 2009.

5. Farrell D, Jensen E, Kocher B, et al. Accounting for the cost of US health care: a new look at why Americans spend more. New York, NY: McKinsey Global Institute; 2008. Available at: http://www.mckinsey.com/mgi/publications/US_healthcare. Accessed November 1, 2009.

6. Baicker K, Chandra A. Myths and misconceptions about US health insurance. *Health Aff (Millwood)*. 2008; 27(6):w533–w543.

7. Schroeder SA. We can do better: improving the health of the American people. *N Engl J Med.* 2007; 357(12):1221–1228.

8. Lantz PM, Lichtenstein RL, Pollack HA. Health policy approaches to population health: the limits of medicalization. *Health Aff (Millwood)*. 2007;26(5):1253–1257.

9. Fisher ES, Bynum JP, Skinner JS. Slowing the growth of health care costs: lessons from regional variation. *N Engl J Med.* 2009;360(9):849–853.

10. Gawande A. The cost conundrum: what a Texas town can teach us about health care. *New Yorker.* June 1, 2009. Available at: http://www.newyorker.com/reporting/2009/06/01/090601fa_fact_gawande. Accessed November 1, 2009.

11. Showstack JA, Rothman AA, Hassmiller S, eds. *The Future of Primary Care.* San Francisco, CA: Jossey-Bass; 2004:xxiii, 328.

12. Bodenheimer T, Berenson RA, Rudolf P. The primary care–specialty income gap: why it matters. *Ann Intern Med.* 2007;146(4):301–306.

13. Woolhandler S, Campbell T, Himmelstein DU. Costs of health care administration in the United States and Canada. *N Engl J Med.* 2003;349(8):768–775.

14. Sack K. Necessary medicine. *New York Times.* December 13, 2008. Available at: http://www.nytimes.com/2008/12/14/weekinreview/14sack.html?_r=1. Accessed November 1, 2009

15. Schoen C, Guterman S, Shih A, et al. Bending the curve: options for achieving savings and improving value in US health spending. New York, NY: Commonwealth Fund; 2007. Available at: http://www.commonwealthfund.org/publications/publications_show.htm?doc_id=620087. Accessed November 1, 2007.

16. Levi J, Segal LM, Juliano C. *Prevention for a Healthier America: Investments in Disease Prevention Yield Significant Savings, Stronger Communities.* Washington, DC: Trust for America's Health; 2008.

17. Heirich M. *Rethinking Health Care: Innovation and Change in America.* Boulder, CO: Westview Press; 1999:xi, 452.

18. Richardson GP. *Feedback Thought in Social Science and Systems Theory.* Philadelphia, PA: University of Pennsylvania Press; 1991:x, 374.

19. Forrester JW. Counterintuitive behavior of social systems. In: Forrester JW, ed. *Collected Papers of Jay W. Forrester.* Waltham, MA: Pegasus Communications; 1975:211–244.

20. Sterman JD. Learning from evidence in a complex world. *Am J Public Health.* 2006;96(3):505–514.

21. Huang ES, Basu A, O'Grady MJ, Capretta JC. Using clinical information to project federal health care spending. *Health Aff (Millwood)*. 2009;28(5):w978–w990.

22. Marks JS. What if Benjamin Franklin ran the Congressional Budget Office? *Huffington Post.* Published August 5, 2009. Available at: http://www.huffingtonpost.com/james-s-marks/what-if-benjamin-franklin_b_251721.html. Accessed November 1, 2009.

23. Homer JB, Hirsch GB. System dynamics modeling for public health: background and opportunities. *Am J Public Health.* 2006;96(3):452–458.

24. Lochner K, Pamuk E, Makuc D, Kennedy BP, Kawachi I. State-level income inequality and individual mortality risk: a prospective, multilevel study. *Am J Public Health.* 2001;91(3):385–391.

25. Szanton SL, Gill JM, Allen JK. Allostatic load: a mechanism of socioeconomic health disparities? *Biol Res Nurs.* 2005;7(1):7–15.

26. Mechanic D, Tanner J. Vulnerable people, groups, and populations: societal view. *Health Aff (Millwood)*. 2007;26(5):1220–1230.

27. Centers for Disease Control and Prevention. Racial/ethnic and socioeconomic disparities in multiple risk factors for heart disease and stroke—United States, 2003. *MMWR Morb Mortal Wkly Rep.* 2005;54(5):113–117.

28. Lantz PM, House JS, Lepkowski JM, Williams DR, Mero RP, Chen J. Socioeconomic factors, health behaviors, and mortality: results from a nationally representative prospective study of US adults. *JAMA.* 1998; 279(21):1703–1708.

29. Mechanic D. Disadvantage, inequality, and social policy. *Health Aff (Millwood)*. 2002;21(2):48–59.

30. Institute of Medicine Committee on Quality Health Care in America. *Crossing the Quality Chasm: A New Health System for the 21st Century.* Washington, DC: National Academies Press; 2001:xx, 337.

31. Asch SM, Kerr EA, Keesey J, et al. Who is at greatest risk for receiving poor-quality health care? *N Engl J Med.* 2006;354(11):1147–1156.

32. Centers for Medicaid and Medicare Services. National health expenditures by type of service and source of funds, CY 1960–2008. Available at: http://www.cms.hhs.gov/NationalHealthExpendData/downloads/nhe2008.zip. Accessed November 1, 2009.

33. Cunningham PJ. What accounts for differences in the use of hospital emergency departments across US communities? *Health Aff (Millwood)*. 2006;25(5):w324–w336.

34. Weber EJ, Showstack JA, Hunt KA, et al. Are the uninsured responsible for the increase in emergency department visits in the United States? *Ann Emerg Med.* 2008;52(2):108–115.

35. Russell LB. Preventing chronic disease: an important investment, but don't count on cost savings. *Health Aff (Millwood)*. 2009;28(1):42–45.

36. Kahn R, Robertson RM, Smith R, Eddy DM. The impact of prevention on reducing the burden of cardiovascular disease. *Circulation.* 2008;118(5):576–585.

37. van Baal PH, Polder JJ, de Wit GA, et al. Lifetime medical costs of obesity: prevention no cure for increasing health expenditure. *PLoS Med.* 2008;5(2):e29.

38. Cohen JT, Neumann PJ, Weinstein MC. Does preventive care save money? Health economics and the presidential candidates. *N Engl J Med.* 2008;358(7): 661–663.

39. Ubel PA, Hirth RA, Chernew ME, Fendrick AM. What is the price of life and why doesn't it increase at the rate of inflation? *Arch Intern Med.* 2003;163(14):1637–1641.

40. McGregor M. Cost-utility analysis: use QALYs only with great caution. *CMAJ.* 2003;168(4):433–434.

41. Lambrew JM, Podesta JD, Shaw TL. Change in challenging times: a plan for extending and improving health coverage. *Health Aff (Millwood)*. 2005;(suppl web exclusives):W5-119–W5-132.

42. Hadley J, Holahan J. Covering the uninsured: how much would it cost? *Health Aff (Millwood)*. 2003;(suppl web exclusives):W3-250–W3-265.

43. Kaiser Family Foundation. US health care costs. Available at: http://www.kaiseredu.org/topics_im.asp?imID=1&parentID=61&id=358. Updated July 2009. Accessed November 1, 2009.

44. Steinbrook R. Health care reform in Massachusetts: expanding coverage, escalating costs. *N Engl J Med.* 2008;358(26):2757–2760.

45. Kowalczyk L. ER visits, costs in Mass. climb. *Boston Globe.* April 24, 2009. Available at: http://www.boston.com/news/local/massachusetts/articles/2009/04/24/er_visits_costs_in_mass_climb. Accessed November 1, 2009.

46. McKinlay JB, Marceau LD. Upstream healthy public policy: lessons from the battle of tobacco. *Int J Health Serv.* 2000;30(1):49–69.

47. Smedley BD, Syme SL, eds; Institute of Medicine Committee on Capitalizing on Social Science and Behavioral Research to Improve the Public's Health. Promoting health: intervention strategies from social and behavioral research. Washington, DC: National Academies Press; 2000. Available at: http://bob.nap.edu/books/0309071755/html. Accessed November 1, 2009.

48. Gerberding JL. Protecting health: the new research imperative. *JAMA.* 2005;294(11):1403–1406.

49. Thorpe KE. The rise in health care spending and what to do about it. *Health Aff (Millwood)*. 2005;24(6): 1436–1445.

50. Homer J, Hirsch G, Milstein B. Chronic illness in a complex health economy: the perils and promises of downstream and upstream reforms. *Syst Dynamics Rev.* 2007;23(2–3):313–343.

51. Sterman JD. All models are wrong: reflections on becoming a systems scientist. *Syst Dynamics Rev.* 2002; 18(4):501–531.

52. Milstein B, Homer J, Hirsch G. HealthBound Web site. Available at: http://www.cdc.gov/healthbound. Accessed March 31, 2010.

53. Woodrow Wilson International Center for Scholars, Science and Technology Innovation Program. Serious games: improving public policy through game-based learning and simulation. Available at: http://www.wilsoncenter.org/index.cfm?topic_id=1414&fuseaction=topics.documents&group_id=10264. Accessed November 1, 2009.

54. Serious Games Initiative Web site. Available at: http://www.seriousgames.org/index2.html. Accessed November 1, 2009.

17

Writings on other topics

Homer J. *Home Insurance in a Changing Residential Community: A System Dynamics Approach and Case Study.* Sloan School Working Paper WP 1107-80, MIT, Cambridge, MA, December 1979.

Homer J. *The Role of Consumer Demand in Business Cycle Entrainment.* System Dynamics Group Paper D-3227-1, MIT, Cambridge, MA, October 1980.

Homer J. Theories of the Industrial Revolution: A Feedback Perspective. *DYNAMICA*, 8(1): 30-35, 1982.

Roberts E, Homer J, Kasabian A, Varrel M. A Systems View of the Smoking Problem. *International Journal of Biomedical Computing*, 13: 69-86, 1982.

Homer J. Partial-Model Testing as a Validation Tool for System Dynamics. *Proceedings of the 1st International System Dynamics Conference,* Chestnut Hill, MA, 1983.

Homer J, John R, Cotreau W. A Dynamic Model for Understanding Eating Disorders. *Proceedings of the 4th International System Dynamics Conference*, Seville, Spain, 1986.

Homer, J, Sterman J, Greenwood B, Perkola M. Delivery Time Reduction in Pulp and Paper Mill Construction Projects: A Dynamic Analysis of Alternatives. *Proceedings of the 11th International System Dynamics Conference*, Cancún, Mexico, 1993.

Homer J, Keane T, Lukianseva N, Bell D. Evaluating Strategies to Improve Railroad Performance: A System Dynamics Approach. *Proceedings of the Winter Simulation Conference*, Phoenix, AZ, 1999.

Homer J, Milstein B. Optimal Decision Making in a Dynamic Model of Community Health. *Proceedings of the Hawaii International Conference on System Sciences*, Waikoloa-Kona, Hawaii, 2004.

Manley W, Homer J, Hoard M, Roy S, Furbee PM, Summers D, Blake R, Kimble M. A Dynamic Model to Support Surge Capacity Planning in a Rural Hospital. *Proceedings of the 23rd International System Dynamics Conference*, Boston, MA, 2005.

Homer J, Milstein B, Dietz W, Buchner D, Majestic E. Obesity Population Dynamics: Exploring Historical Growth and Plausible Futures in the U.S. *Proceedings of the 24th International System Dynamics Conference*, Nijmegen, The Netherlands, July 2006.

www.ingramcontent.com/pod-product-compliance
Lightning Source LLC
Chambersburg PA
CBHW080513220326
41599CB00032B/6067

9780615679280